T0122425

Lecture Notes in Statistics 104

Edited by P. Diggle, S. Fienberg, K. Krickeberg, I. Olkin, N. Wermuth

Lecture Notes in Statistics

104

Edited by P. Diggle, S. Fienberg, K. Krickeberg,
I. Olkin, N. Wermuth

G.U.H. Seeber
B.J. Francis
R. Hatzinger
G. Steckel-Berger (Editors)

Statistical Modelling

Proceedings of the

10th International Workshop on

Statistical Modelling

Innsbruck, Austria, 10-14 July, 1995

Springer-Verlag
New York Berlin Heidelberg London Paris
Tokyo Hong Kong Barcelona Budapest

Gilg U.H. Seeber
Institut für Statistik
Universität Innsbruck
Christoph-Probst-Platz
A-6020 Innsbruck
Austria

Brian J. Francis
Centre for Applied Statistics
Lancaster University
Fylde College
Lancaster, LA1 4YF
England

Reinhold Hatzinger
Institut für Statistik
Wirtschaftsuniversität
Augasse 2-6
A-1090 Wien
Austria

Gabriele Steckel-Berger
Institut für Statistik
Universität Innsbruck
Christoph-Probst-Platz
A-6020 Innsbruck
Austria

Library of Congress Cataloging-in-Publication Data Available
Printed on acid-free paper.

© 1995 Springer-Verlag New York, Inc.
All rights reserved. This work may not be translated or copied in whole or in part without the written permission of the publisher (Springer-Verlag New York, Inc., 175 Fifth Avenue, New York, NY 10010, USA), except for brief excerpts in connection with reviews or scholarly analysis. Use in connection with any form of information storage and retrieval, electronic adaptation, computer software, or by similar or dissimilar methodology now known or hereafter developed is forbidden.
The use of general descriptive names, trade names, trademarks, etc., in this publication, even if the former are not especially identified, is not to be taken as a sign that such names, as understood by the Trade Marks and Merchandise Marks Act, may accordingly be used freely by anyone.

Camera ready copy provided by the author.
Printed and bound by Braun-Brumfield, Ann Arbor, MI.
Printed in the United States of America.

9 8 7 6 5 4 3 2 1

ISBN 0-387-94565-2 Springer-Verlag New York Berlin Heidelberg

Preface

This volume presents the published proceedings of the 10th International Workshop on Statistical Modelling, to be held in Innsbruck, Austria from 10 to 14 July, 1995.

This workshop marks an important anniversary. The inaugural workshop in this series also took place in Innsbruck in 1986, and brought together a small but enthusiastic group of thirty European statisticians interested in statistical modelling. The workshop arose out of two GLIM conferences in the U.K. in London (1982) and Lancaster (1985), and from a number of short courses organised by Murray Aitkin and held at Lancaster in the early 1980s, which attracted many European statisticians interested in Generalised Linear Modelling. The inaugural workshop in Innsbruck concentrated on GLMs and was characterised by a number of features – a friendly and supportive academic atmosphere, tutorial sessions and invited speakers presenting new developments in statistical modelling, and a very well organised social programme. The academic programme allowed plenty of time for presentation and for discussion, and made available copies of all papers beforehand.

Over the intervening years, the workshop has grown substantially, and now regularly attracts over 150 participants. The scope of the workshop is now much broader, reflecting the growth in the subject of statistical modelling over ten years. The elements of the first workshop, however, are still present, and participants always find the meetings relevant and stimulating. The number of submitted papers has grown with the number of participants, but successful contributed papers still receive a relatively generous 30 minutes of presentation time, and invited speakers one hour. Parallel sessions have been avoided, allowing everyone both to learn and to contribute. Poster sessions are now held, and software demonstrations and displays are organised. One change is that the workshops have become more international in nature. Participants now attend from all corners of the globe, and workshops have travelled around Europe – to Perugia (1987), Vienna (1988), Trento (1989), Toulouse (1990), Utrecht (1991), Munich (1992), Leuven (1993) and Exeter (1994).

The current workshop, hopefully, will be no exception. The invited speakers chosen for the meeting are all experts in their field, and all but one of their contributions are included in this volume. It is particularly appropriate that Murray Aitkin will deliver the opening paper of the workshop, as he was one of the invited speakers at the initial workshop. He will be presenting work on a general procedure for non-parametric maximum likelihood estimation of mixing distributions. Anthony Atkinson will introduce participants to optimal experimental design in generalised linear models, and Adelchi Azzalini will be speaking on new software tools in S-Plus and related practical problems in fitting models to repeated measurements. Last in the alphabet, but not least, James Booth will deliver work on bootstrap methods in generalised linear mixed models, illustrated by practical datasets collected in small spatial areas. This workshop, therefore, both continues with its GLM heritage, and looks forward to new developments in the analysis of repeated measures, mixed and random effect models.

There are also some forty contributed papers presented in this volume. They, together with the invited papers, are presented in alphabetic order of the name of the first author. These cover a wide variety of application areas, from epidemiology to education, and from agriculture to sociology. Many papers present analyses of real data sets, whereas others have a greater emphasis on theoretical and technical issues. All have been subject to a refereeing process by the Scientific Committee, who rejected approximately half of the papers submitted.

The Editors of this volume would like to thank all members of the Scientific Committee and other referees who again worked extremely hard in assessing the large number of papers submitted. The local organisers of the workshop listed below also deserve our thanks in advance, for what is always a busy and onerous task. Finally, we thank the authors for simplifying considerably the task of preparing this volume, both by submitting their papers in LaTeX in the correct format, by keeping strictly to the tight timescale, and for restricting their interesting ideas to eight pages.

GUHS, RH, BJF, GS-B (April 1995)

Scientific Committee: J. Engel (Eindhoven), L. Fahrmeir (Munich), A. de Falguerolles (Toulouse), A. Forcina (Perugia), B.J. Francis (Lancaster), P. Gherardini (Rome), R. Gilchrist (London), R. Hatzinger (Vienna), P. van der Heijden (Utrecht), J. Hinde (Exeter), E. Lesaffre (Leuven), B. Marx (Stanford), Ch.E. Minder (Berne), G.U.H. Seeber (Innsbruck), G. Tutz (Berlin).

Local Organising Committee: G. Marinell, G.U.H. Seeber, G. Steckel-Berger, Ch.M. Traweger, H. Ulmer, M. Völp (Innsbruck); R. Dittrich, R. Hatzinger (Vienna).

Table of Contents

NPML estimation of the mixing distribution in general statistical models with unobserved random effects

Murray Aitkin

ABSTRACT: General maximum likelihood computational methods have recently been described for longitudinal analysis and related problems using generalized linear models. These developments extend the standard methods of generalized linear modelling to deal with overdispersion and variance component structures caused by the presence of *unobserved random effects* in the models. The value of these methods is that they are not restricted by particular statistical model assumptions about the distribution of the random effects, which if incorrect might invalidate the conclusions. Despite this generality these methods are fully efficient, in the sense that the model is fitted by (nonparametric) maximum likelihood, rather than by approximate or inefficient methods. The computational implementation of the methods is straightforward in GLM packages like GLIM4 and S+.

An important feature of these computational methods is that they are applicable to a much broader class of models than the GLMs described above, and provide new general computational solutions to fitting a very wide range of models using *latent variables*.

KEYWORDS: Mixture; Overdispersion; Variance Components; Nonparametric Maximum Likelihood; EM Algorithm; Gauss-Newton Algorithm

1   Background

The analysis of longitudinal studies with normally distributed response variables is well-established, and maximum likelihood methods are widely available for modelling intra-class (equal) correlation structures of the repeated measures over time. (Maximum likelihood methods for autoregressive or other patterned correlation structures are also available, though much less widely.) The same methods apply to the analysis of *hierarchical* or *nested* studies, for example in school effectiveness studies which test pupils nested in schools. However for binary, categorical or count data measured repeatedly, the available statistical methods are much less satisfactory. The difficulty is that the generalizations of the standard exponential family (binomial, Poisson or gamma) models which allow for repeated measures by common random effects (equal correlation) do not have a closed analytic form for the likelihood function, and standard methods for normally distributed variables cannot be applied.

To avoid this difficulty, two different approximate procedures have been used. The first is to assume that the common random effects are normally distributed as in the normal variance component model, and to approximate the exponential family likelihood in one of several ways by a normal likelihood; this allows methods for normal variance component models to be applied approximately to non-normal distributions. This *conditional* approach has been applied in slightly different ways by McGilchrist (1994), Longford (1994 Chapter 8) and Goldstein (1991). The second is to assume that the *marginal* distribution of the response is from the exponential family and to model the covariance structure induced by the repeated measures in one of several reasonable ways (Liang and Zeger 1986, Diggle, Liang and Zeger 1994); the model is fitted by a generalized quasi-likelihood method (generalized estimating equations, GEE). This second marginal approach is inconsistent with the random effect model structure, since the marginal distribution of the response must be a mixed, not a pure, exponential family distribution. Thus the GEE approach actually fits a different model; there is considerable disagreement over whether the conditional or the marginal approach is preferable.

A disadvantage of *any* approach using a specified parametric form (e.g. normal) for the mixing distribution of the random effects is the possible sensitivity of the conclusions to this specification. The influential paper by Heckman and Singer (1984) showed substantial changes in parameter estimates in variance component models with quite small changes in mixing distribution specification; Davies (1987) showed similar effects. This difficulty can be avoided by *nonparametric maximum likelihood (NPML) estimation of the mixing distribution*, concurrently with the structural model parameters. In this approach the random effect distribution is unspecified, but remarkably can itself be estimated by ML (Kiefer and Wolfowitz 1956). The NPML estimate is well-known to be a discrete distribution on a finite number of mass-points (Kiefer and Wolfowitz 1956, Laird 1978, Lindsay 1983), but actually finding the NPML estimate is widely regarded as computationally intensive, the particular difficulty being the location of the mass-points. Current approaches use Gateaux or directional derivatives (Ezzet and Davies 1988, Follmann and Lambert 1989, Lesperance and Kalbfleisch, 1992, Böhning, Schlattman and Lindsay 1992). Barry, Francis and Davies (1989) remark:"[NPML] is not a simplification of the parametric approach as the identification of the number, location and masses of these points of support present formidable computational problems."

An important paper by Hinde and Wood (1987) addressed the computational issues of NPML estimation in the framework of the two-level logistic variance component model. They showed that quite generally, both the mass-point locations and the masses can be estimated straightforwardly by ML within the framework of a finite mixture of GLMs, allowing the straightforward full NPML estimation of the mixing distribution. The key to locating the mass-points is to incorporate them as a factor in an extended regression model obtained by conceptually duplicating the data for each of the mass-points. The masses are estimated by standard mixture ML methods. This process is repeated with the number of mass-points increasing (from 1) until the maximized likelihood stabilises at the NPMLE. Convergence of the EM algorithm in these NPML problems may be very slow, as information in the data about the mixing distribution may be very limited, but the algorithm is easily programmed, for example in GLIM4 or S-plus.

In a paper to appear in Statistics and Computing (Aitkin 1995a), I give the implementation of the EM algorithm using GLIM4 macros for the simpler case of overdispersed GLMs, which have a closely related but simpler structure (in the sense that each observation has its own random effect rather than sharing it with other observations), and discuss a wide range of applications which make clear the value and straightforwardness of the nonparametric ML approach. This paper also gives a simple new solution to a chronic problem with the EM algorithm, which does not provide standard errors for the ML estimates, only the estimates themselves (and the maximized likelihood). The lack of standard errors from the EM algorithm can be remedied in regression models by expoiting the asymptotic equivalence of the Wald and likelihood ratio tests for a parameter. Additional model fits are carried out omitting each explanatory variable in turn from the final model (Aitkin 1995a). The standard error for each parameter is then calculated as the absolute value of the parameter estimate divided by the square root of the deviance change on omitting the variable.

This provision of standard errors for the EM algorithm increases the usefulness of the EM approach; it does not require additional programming, but does require additional model fitting. A further paper in preparation (Aitkin 1995b) will give a similar analysis of a range of variance component applications: it is a particular strength of the NPML approach that overdispersion and variance component models can be fitted using essentially the same methods.

2   Computational Efficiency

A standard difficulty with the routine use of the EM algorithm is its slowness: this is notorious in mixture modelling problems. In a paper to appear in Statistics and Computing, Aitkin and Aitkin (1995) show that the Gauss-Newton (GN) algorithm using the Hessian matrix can be combined with EM in a hybrid algorithm which has the best properties of both: it converges reliably from any starting values like EM, but converges quadratically near the maximum like GN (instead of linearly or sub-linearly like EM). A computational study reported in the paper of the difficult two-component normal mixture with different variances shows that the hybrid algorithm saves a consistent 40% of computational time over EM, and may give dramatic improvements on data sets for which EM takes hundreds of iterations. The hybrid algorithm also provides standard errors for the estimated parameters based on the observed information.

This algorithm should perform very well on the overdispersed and variance component GLM applications above, since the unobserved structure is limited to the single extra random effect variable: the rest of the model structure is observable.

The general NPML computational approach is applicable to a much broader class of models than the GLM generalizations described above, and provides new general computational solutions to fitting a very wide range of models using *latent variables*. When implemented in efficient algorithms, it should provide new fast efficient methods for model fitting to a very wide range of problems. I now describe some of these extensions.

3   Model-fitting Extensions by NPML

3.1   NPML Kernel Density Estimation

Given a sample $y_1, ..., y_n$ from an unknown density $f(y)$, the kernel density estimate is

$$\tilde{f}(y) = \sum_{i=1}^{n} K(\frac{y - y_i}{h})/nh$$

where $K(u)$ is the kernel function (usually Gaussian), and h is the "smoothing parameter" or "bandwidth". So $\tilde{f}$ is a mixture of n Gaussian components with equal proportions $1/n$ located at the observed data values y_i. The common standard deviation h has to be estimated by some method (e.g. cross-validation); the choice of h is one of the principal issues in kernel density estimation.

Consider the continuous Gaussian mixture

$$f(y) = \int K(\frac{y - \theta}{h})\pi(\theta)d\theta/h$$

where $\pi(\theta)$ is an unknown mixing distribution on the mean θ. Then the NPMLE of $f(y)$ is

$$\hat{f}(y) = \sum_{k=1}^{K} \hat{\pi}_k K(\frac{y - \hat{\theta}_k}{\hat{h}})/\hat{h} ,$$

a K-component Gaussian mixture. So the NPMLE of f is a kernel density estimate, but with far fewer than n components which are also unequally weighted in general. Further, the common standard deviation is estimated by ML. From the NPML point of view, the usual kernel density estimate must be sub-optimal, as it has a lower likelihood over the parameter space of π_k, θ_k and h, and will be much more "wiggly" than appropriate because of the n components. This result can be generalized easily to mixing on both θ and h.

A related idea has been used by West (1993) in constructing kernel estimates for posterior densities from simulated draws from the posterior. With 10,000 observations the usual kernel density estimates are impractical, and West uses a much smaller number of components (around 500) obtained by agglomerative clustering of nearby points. By applying the above approach we can obtain the NPML estimate of the posterior density.

3.2 Extension of Overdispersion Models to Random Coefficient Regression Models

Overdispersion models can be viewed as random intercept models. These can be simply generalized to random slope models, for example in the Poisson regression model

$$Y_i | a_i, b_i, x_i \sim P(e^{a_i + b_i x_i})$$

with $a_i, b_i \sim \pi(a, b)$ where the mixing distribution $\pi(a, b)$ is specified. The NPML approach can be extended to the case where $\pi(a, b)$ is unknown, when the NPMLE of $\pi(a, b)$ is a K-point distribution in the (a, b) plane. For the normal regression model a further extension of this approach can allow σ^2 to vary also.

This generalization of NPML estimation for random coefficient models in the exponential family provides a fully efficient distribution-free alternative to generalized quasi-likelihood approaches or to approximate maximum likelihood approaches with normal random parameters, as in Longford (1993 Chapter 6) or Fahrmeir and Tutz (1994 Chapters 7 and 8). It can in principle be extended to p explanatory variables without the computational cost of p-dimensional numerical integration, since the number of mass-points required in p dimensions will usually be much less than Kp.

3.3 Latent Variable Models

The overdispersion and variance component models are only a special case of a very general class of models with *unobserved* or *latent variables*. Any such model can be generalized to one with an *arbitrary* distribution for the latent variables which is estimated by NPML. This generalization may in turn allow the generalization of the original model. One important example is the *normal factor model*, in which observed variables Y are regressed on unobserved factors Z. The sample mean $\bar{Y}$ and covariance matrix S are jointly sufficient statistics for the parameters in the normal factor model, which is critically linear and normal in the factors Z, in the sense that if either linearity of the regression of Y on Z or the normality of Z fails, then the model no longer has a closed-form likelihood, $\bar{Y}$ and S are not sufficient, and maximum likelihood model fitting becomes very complicated.

By allowing the distribution of Z to be arbitrary and estimating it by NPML, we can remove the linearity restriction and fit more general regression models, e.g. quadratic or response surface models. Computational methods much faster than EM are required here, as EM is notoriously slow in the usual normal factor model, where all the structure depends on the unobservable factors.

3.4   Incomplete Data Models

Incomplete data on explanatory variables is a routine difficulty of normal regression model fitting with social survey data. The EM algorithm is widely used for this problem, but in order to calculate in the E-step the conditional expectations of the sufficient statistics, a model is needed for the joint distribution of all the explanatory variables on which any observations are missing. This frequently amounts to having to specify a full joint model for all the variables, explanatory as well as response. For example, Little and Schluchter (1985) specify a full multivariate normal distribution for the continuous variables, response or explanatory, conditional on a multinomial distribution for all the categorical explanatory variables. Thus even simple regression problems require complex joint distribution specifications instead of just the specification of the distribution of the response, conditional on the explanatory variables.

The NPML estimate gives a new approach to this problem, since the NPML estimate of the joint distribution of the explanatory variables is just the empirical joint mass function. The conditional expectations of the sufficient statistics are then evaluated with respect to this joint distribution by finite summations without any model assumption, and the NPML estimate is itself updated in the E-step from the expected incomplete observations.

The NPML approach should be valuable, for both reducing bias and increasing precision, in the common case where randomly missing values are spread over all the variables, resulting in a substantial data loss when only "complete cases" are analysed using casewise deletion. The EM algorithm should converge rapidly in this case since the empirical distributions for the explanatory variables will be hardly changed by the imputation of a small number of observations on each variable.

3.5 Measurement Error in Explanatory Variables

Measurement error is an endemic problem in social and medical investigations of all kinds. It presents serious computational difficulties even in simple models. In the simplest example, we want to regress Y on an explanatory variable X, but X can be observed only with random measurement error. If Y is normal and the measurement error and the distribution of X are also normal, the model is unidentifiable if all parameters are unknown. Some assumption about the model variances is required to proceed at all.

Assuming normal measurement error but leaving the distribution of X unspecified, we can estimate it by NPML and also estimate the other model parameters; the unidentifiability problem in the full normal model is removed. (This is simply a consequence of the fact that normal mixtures of normals are not identifiable, but finite mixtures of normals are identifiable.)

3.6 "Smooth" Model-Fitting

The class of *generalized additive models* (GAMs, Hastie and Tibshirani 1990) allows the dependence of a response on an explanatory variable to be represented by a general smooth function, allowing local non-linearities for example in a generally linear logistic regression. Follman and Lambert (1989) noted that NPML estimation of a random intercept term would provide a similar local non-linearity and compared NPML estimation and GAM estimation on a binomial logit model. These comparisons can be extended to GLMs in which apparent local non-linearity in several variables can be treated by GAM or NPML approaches. This should also be possible with the closely related *break-point models*; these have a sudden change in a parameter value at some point in the observation sequence.

4 Conclusion

The extensions I have described give a clear picture of the power of the NPML approach. I have implemented the computational methods for both overdispersion and variance component models as EM algorithms in GLIM4; parameter standard errors can be obtained by the variable omission device described above. Thus a general EM method is available for these exponential family models which gives full ML fitting with standard errors.

The hybrid EM/GN algorithm for normal mixtures described by Aitkin and Aitkin (1995) improves the speed of convergence of EM while still guaranteeing convergence by starting with EM and then switching to GN when near the MLE; this also gives standard errors based on the observed information. The implementation of this algorithm should provide an efficient computational method for these models.

References

Aitkin, M. (1995a) A general maximum likelihood analysis of overdispersion in generalized linear models. *Statistics and Computing* **5**, to appear.

Aitkin, M. (1995b) A general maximum likelihood analysis of two-level variance component structure in generalized linear models. In preparation.

Aitkin, M. and Aitkin, I. (1995) A hybrid EM/Gauss-Newton algorithm for maximum likelihood in mixture distributions. *Statistics and Computing* **5**, to appear.

Aitkin, M., Anderson, D.A., Francis, B.J. and Hinde, J.P. (1989) *Statistical Modelling in GLIM*. Oxford: University Press.

Barry, J.T., Francis, B.J. and Davies, R.B. (1989) SABRE: software for the analysis of binary recurrent events. in *Statistical Modelling* New York: Springer-Verlag.

Böhning, D., Schlattman, P. and Lindsay, B. (1992) Computer-assisted analysis of mixtures (C.A.MAN): statistical algorithms. *Biometrics* **48**, 285-303.

Davies, R.B. (1987) Mass point methods for dealing with nuisance parameters in longitudinal studies. in *Longitudinal Data Analysis* (ed. R. Crouchley). Aldershot, Hants: Avebury.

Diggle, P.J., Liang, K.-Y. and Zeger, S.L. (1994) *The Analysis of Longitudinal Data*. Oxford: Clarendon Press.

Ezzet, F. and Davies, R.B. (1988) *A manual for MIXTURE*. Lancaster, UK: Centre for Applied Statistics.

Fahrmeir, L. and Tutz, G. (1994) *Multivariate Statistical Modelling Based on Generalized Linear Models*. New York: Springer-Verlag.

Follman, D.A. and Lambert, D. (1989) Generalizing logistic regression by nonparametric mixing. *J. Amer. Statist. Assoc.* **84**, 295-300.

Goldstein, H. (1991) Nonlinear multilevel models for discrete response data. *Biometrika* **78**, 45-51.

Hastie, T.J. and Tibshirani, R.J. (1990) *Generalized Additive Models*. London: Chapman and Hall.

Heckman, J.J. and Singer, B. (1984) A method for minimizing the impact of distributional assumptions in econometric models of duration. *Econometrica* **52**, 271-320.

Hinde, J.P. and Wood, A.T.A. (1987) Binomial variance component models with a non-parametric assumption concerning random effects. in *Longitudinal Data Analysis* (ed. R. Crouchley). Aldershot, Hants: Avebury.

Kiefer, J. and Wolfowitz, J. (1956) Consistency of the maximum likelihood estimator in the presence of infinitely many nuisance parameters. *Ann. Math. Statist.* **27**, 887-906.

Laird, N. M. (1978) Nonparametric maximum likelihood estimation of a mixing distribution. *J. Amer. Statist. Assoc.* **73**, 805-811.

Lesperance, M.L. and Kalbfleisch, J.D. (1992) An algorithm for computing the nonparametric MLE of a mixing distribution. *J. Amer. Statist. Assoc.* **87**, 120-126.

Liang, K.-Y. and Zeger, S.L. (1986) Longitudinal data analysis using generalized linear models. *Biometrika* **73**, 13-22.

Lindsay, B.G. (1983) The geometry of mixture likelihoods, part I: a general theory. *Ann. Statist.* **11**, 86-94.

Little, R.J.A. and Schluchter, M.D. (1985) Maximum likelihood estimation for mixed continuous and categorical data with missing values. *Biometrika* **72**, 497-512.

Longford, N.T. (1994) *Random Coefficient Models.* Oxford: Clarendon Press.

McGilchrist, C.A. (1994) Estimation in generalized mixed models. *J. Roy. Statist. Soc. B* **56**, 61-69.

West, M. (1993) Approximating posterior distributions by mixtures. *J. Roy. Statist. Soc. B* **55**, 409-422.

Some Topics in Optimum Experimental Design for Generalized Linear Models

Anthony C. Atkinson

ABSTRACT:
Optimum experimental designs for generalized linear models are found by applying the methods for normal theory regression models to the information matrix for weighted least squares. The weights are those in the iterative fitting of the model. Examples for logistic regression with two variables illustrate the differences between design for normal theory models and that for other GLMs.

KEYWORDS: Bayesian design; c-optimality; D-optimality; generalized linear model; logistic regression.

1 Introduction

The standard method of fitting generalized linear models, for example in GLIM, is by weighted least squares (McCullagh and Nelder, 1989). Optimum experimental designs for these models are found by applying design methods for normal theory regression models to the information matrix for weighted least squares, with the weights depending on the parameters of the linear predictor and on the link function.

Optimum experimental design is described in two recent books: Pukelsheim (1993), which emphasizes connections with convex programming, and Atkinson and Donev (1992) which stresses statistical aspects (designs for generalized linear models are discussed in their §22.5). An important paper on design for generalized linear models is Chaloner and Larntz (1989).

2 Optimality Theory

For the linear regression model $E(Y) = \beta^T f(x)$, where $f(x)$ is a $p \times 1$ vector of known functions of the k explanatory variables, the values of which can be specified by the experimenter, the information matrix for the least squares estimates of the parameters is proportional to $f(x)f^T(x)$. Experimental designs are sought which make large suitable functions of this matrix. The dependence of the information on the number of experimental trials n can be removed by standardization, working instead with the fractions of trials n_i/n at each x_i. More general mathematical results about designs are obtained on replacing the fractions by a measure ξ, yielding a 'continuous' design in which the n_i are no longer required to be integer. However practical designs do require that the n_i be integer.

For an experimental design represented as a measure ξ over the design region $\mathcal{X}$, the information matrix is

$$M(\xi) = \int_{\mathcal{X}} f(x)f^T(x)\xi(dx) = E_\xi f(x)f^T(x), \qquad (1)$$

which is not a function of β. Optimum design theory is concerned with minimization of the convex function $\Psi\{M(\xi)\}$. Because this is a well-behaved optimization problem, the optimality of any design ξ can be checked by using the directional derivative $\phi(x,\xi)$. This derivative measures the rate of change of $\Psi\{M(\xi)\}$ when the design ξ is perturbed by the addition of an infinitesimal design weight at x. It is thus a scalar. For the optimal design ξ^*, $\phi(x,\xi^*)$ is zero at the design points and positive elsewhere. The various equivalence theorems follow from this result.

One frequently used design criterion is D-optimality in which the generalized variance of the parameter estimates is minimized, when

$$\Psi\{M(\xi)\} = -\log|M(\xi)|,$$

equivalent to maximizing $|M(\xi)|$. Another frequently used criterion is c-optimality in which the variance of the linear combination $c^T\hat{\beta}$ is minimized, so that

$$\Psi\{M(\xi)\} = c^T M^{-1}(\xi)c.$$

The equivalence theorem for D-optimality relates to the variance of the prediction $\hat{y}(x)$ for the design ξ, standardized by taking the error variance as one. Let $d(x,\xi) = f^T(x)M^{-1}(\xi)f(x)$. Then, for D-optimality, the derivative $\phi(x,\xi) = p - d(x,\xi)$ and the maximum value, p, of $d(x,\xi)$ occurs at the points of support of the design.

Such equivalence theorems provide algorithms for the construction of designs, since values of x for which $d(x,\xi) > p$ are candidates for increased experimental effort. The theorems also provide methods of checking the optimality of designs. D-optimum designs for linear regression models depend on the model, although not on the value of β, and on the design region $\mathcal{X}$. The designs are typically supported at extreme points in the p-dimensional space of the $f(x)$, which, for second-order response surface models, translates into extremes and partial centroids of the lower dimensional space of $\mathcal{X}$.

3   Generalized Linear Models

The extension of optimum design theory to error distributions in the family of generalized linear models (McCullagh and Nelder, 1989) uses the information matrix

$$M(\xi, \theta) = E_\xi f(x) w(x, \theta) f^T(x),$$

where $w(x, \theta)$ is the iterative weight in the fitting algorithm. The greatest interest in the literature has been in logistic models for binary data, particularly with one explanatory variable. This linear logistic model is that

$$\log\{\mu/(1-\mu)\} = \eta = \alpha + \beta x \qquad (2)$$

where μ is the probability of death. The iterative weight, through which the optimum design depends on the parameters α and β, is $w = \mu(1-\mu)$. For a sufficiently large design region $\mathcal{X}$, the D-optimum design for $\alpha = 0, \beta = 1$ puts half the trials at $x = -1.543$ and half at $x = 1.543$. This is a special case of the result that p point D-optimum designs for models with p parameters put weight $1/p$ at each design point. As we shall see, D-optimum designs may have more than this minimum number of support points.

The c-optimum design for estimating the dose $\hat{x}$ at which the probability of success is π is found by maximising $c^T M^{-1}(\xi) c$ with c the vector of partial derivatives of $\hat{x}$ with respect to α and β. After some simplification this reduces to $c^T = (1 \quad \hat{x})$. For $\alpha = 0$ and $\beta = 1$ this design puts all the trials at $x = 0$, the value of $\hat{x}$.

Although designs for other values of the parameters can likewise be found numerically, design problems for a single x can often be solved in a canonical form, yielding a structure for the designs independent of the particular parameter values (Ford, Torsney and Wu, 1992). The D-optimum design in this problem puts the trials where the expected response is 0.176 and 0.824, the translation into the experimental variable x depending on the values of α and β.

4   Logistic Regression with Two Explanatory Variables

Although it was assumed in the previous section that the experimental region $\mathcal{X}$ was effectively unbounded, the design was constrained by the weight w to lie in a region in which μ was not too close to zero or one. But with more than one explanatory variable constraints on the region are necessary. For example, for the two variable model

$$\log\{\mu/(1-\mu)\} = \eta = \beta_0 + \beta_1 x_1 + \beta_2 x_2, \qquad (3)$$

with $\beta^T = (0, 1, 1)$, all points for which $x_1 + x_2 = 0$ yield a value of 0.5 for μ, however extreme the values of x. Some appreciation of the structure of optimum designs in such situations can be found by transformation of the design space in a manner similar to that of Ford *et al.* (1992). Because of the weighted least squares procedure, design for the model (3) is equivalent to design for the linear model

$$\eta = \beta_0 \sqrt{w} + \beta_1 \sqrt{w} x_1 + \beta_2 \sqrt{w} x_2, = \beta_0 z_0 + \beta_1 z_1 + \beta_2 z_2. \qquad (4)$$

Since $p = 3$, the induced design space $\mathcal{Z}$ is of dimension three. Three examples, projected onto z_1 and z_2 and so ignoring $z_0 = \sqrt{w}$, are given in Fig.1 for $\mathcal{X}$ the unit square. In Fig.1(a), $\beta^T = (0, 1, 1)$ and $\mathcal{X}$ is not much distorted. In Fig.1(b) $\beta^T = (0, 2, 2)$ so that for the corner of $\mathcal{X}$ for which $x_1 = x_2 = 1$, $\eta = 4$ and $\mu = 0.982$. This is well beyond the range for informative experiments and the projection of the induced design space appears to be folded over. As a result, experiments at extreme positions in $\mathcal{Z}$ will not necessarily be at extreme points in $\mathcal{X}$. The results in Fig.1(c) for $\beta^T = (2, 2, 2)$ are similar, but more extreme.

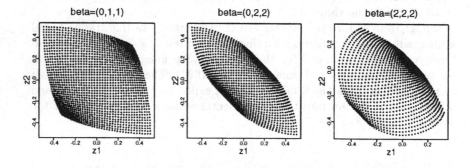

FIGURE 1. Induced design space $\mathcal{Z}$ for three sets of parameter values for the two variable logistic model

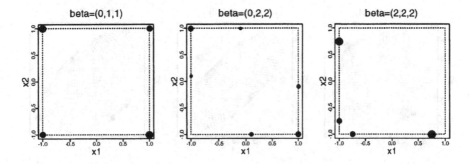

FIGURE 2. D-optimum designs for three sets of parameter values for the two variable logistic model: the design weights are proportional to the areas of the plotting symbols

D-optimum designs for these three sets of parameter values were found by searching over a grid of x values, with the optimum design measure ξ^* found by constrained numerical optimization. The resulting optimum designs are shown plotted in $\mathcal{X}$ in Fig.2 with symbol area proportional to design weight. In line with the earlier argument, the design in Fig.2(a) is a 2^2 factorial, although not quite equally weighted. For $\beta^T = (0, 2, 2)$ a six-point design is optimum, whereas Fig.2(c) shows an optimum four-point design when $\beta^T = (2, 2, 2)$. The dependence of these designs on the parameter values is clarified by the plots in $\mathcal{Z}$ space in Fig.3, in which the design points are seen to lie at or near vertices of the projection of the induced design region. In all designs some trials are at points with μ or $1 - \mu$ around 0.85. The remaining points are at $\mu = 0.5$

Despite the illumination shed by Fig.3, it is clear that canonical designs for two or more explanatory variables are likely to be difficult to find, even for simple design regions. Other design regions give rise, for example, to three-point optimum designs. Burridge and Sebastiani (1994) find canonical designs for one family of GLMs under restrictions on the values of the linear

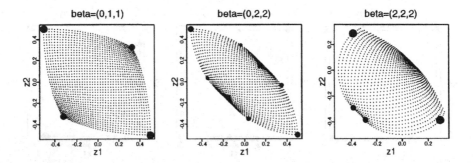

FIGURE 3. The designs of Fig.2 in the induced design space $\mathcal{Z}$

predictor.

5   Bayesian Design

The c-optimum design of §3 put all the trials at the estimated conditions of
the desired probability of success. This non-informative design results from
the use of point priors for the parameters, which give an over-precise specification of prior knowledge. Practical designs can be obtained which make
a more realistic use of prior information by maximising the expectation of
the design criterion over the prior distribution of θ. The extension of the
D-optimum criterion to incorporate prior information is to maximise

$$E_\theta \log |M(\xi,\theta)| = \int \log |M(\xi,\theta)| p(\theta) d\theta, \tag{5}$$

where $p(\theta)$ is the prior distribution of θ. The resulting design criterion is a
concave function for which a globally optimum design can be found by use
of the standard algorithms.
The extension of the condition for c-optimality to incorporate prior infor-

mation is similar, the optimum design minimising

$$E_\theta c^T(\theta) M^{-1}(\xi, \theta) c(\theta) = \int c^T(\theta) M^{-1}(\xi, \theta) c(\theta) p(\theta) d\theta. \qquad (6)$$

Optimum designs can be found by discretization of the prior $p(\theta)$, leading to maximization of a weighted sum of the criterion evaluated at each point. Examples for the model of §3 are given by Chaloner and Larntz (1989) which show how the number of design points increases with prior uncertainty. Atkinson, Demetrio and Zocchi (1995) use instead a Monte-Carlo method to sample from the prior distribution of θ. They find both D- and c-optimum Bayesian designs for a model with gender as a second factor in which the value of $\hat{x}$ is only implicitly defined.

Two last points on design for generalized linear models: one is that it has been assumed throughout this paper that the true model is known. Designs for discrimination between two generalized linear models are described by Ponce de Leon and Atkinson (1992), whereas Ponce de Leon and Atkinson (1993) use a parametric family of link functions to generate designs for the choice of link for binary data models. Finally, it is clear that widespread application of the methods of optimum experimental design requires software for finding the designs. Comments on these and other aspects of optimum design are given by Atkinson (1996).

References

Atkinson, A.C. (1996). The usefulness of optimum experimental designs (with discussion). *Journal of the Royal Statistical Society B*, **58**, (to appear).

Atkinson, A.C., C.G.B.Demetrio and S.Zocchi (1995). Optimum dose levels when males and females differ in response. *Applied Statistics*, **44**, 213–226.

Atkinson, A.C. and A.N.Donev (1992). *Optimum Experimental Designs* . Oxford university Press.

Burridge, J. and P.Sebastiani (1995). D-optimal designs for generalised linear models with variance proportional to the square of the mean. *Biometrika*, **81**, 295–304.

Chaloner, K. and K.Larntz (1987). Optimal Bayesian design applied to logistic regression experiments. *Journal of Statistical Planning and Inference*, **21**, 191–208.

Ford, I., B.Torsney and C.F.J.Wu (1997). The use of a canonical form in the construction of locally optimal designs for non-linear problems. *Journal of the Royal Statistical Society B*, **54**, 569–583.

McCullagh, P. and J.A.Nelder (1989). *Generalized Linear Models*. Chapman and Hall, London

Ponce de Leon, A.M. and A.C.Atkinson (1992). The design of experiments to discriminate between two rival generalized linear models. In: Fahrmeir, L., B.Francis, R.Gilchrist and G.Tutz (eds). *Advances in GLIM and Statistical Modelling: Proceedings of the GLIM92 Conference, Munich*, pp.159–164. Springer, New York.

Ponce de Leon, A.M. and A.C.Atkinson (1993). Designing optimal experiments for the choice of link function for a binary data model. In: Müller,W.G., H.P.Wynn and A.A.Zhigljavsky (eds). *Model-Oriented Data Analysis*, pp.25–36. Physica-Verlag, Heidelberg.

Pukelsheim, F. (1993). *Optimal Design of Experiments*. Wiley, New York

Autoregressive Modelling of Markov Chains

André Berchtold

ABSTRACT: The reduction of the number of parameters in high-order Markov chain already inspired several articles. In particular, Raftery (1985) proposed an autoregressive modelling which utilizes a same transition matrix for every lag. In this paper, we show that a model of the same type, but utilizing different matrices, gives best results and is not harder to estimate, even when the number of data is small.

KEYWORDS: Markov chain; General Transition Matrices (S-Matrices); General Autoregressive Model (GAM); Limit theorem; Quality of modelling

1 Introduction

Markov chains are utilized for a long time in a lot of domains including biology, physics, chemistry, music and econometrics... Unfortunately, as in most cases the number of available data is small, only the first-order transition matrix can be estimated, even when the data present a high-order dependence. The best solution for using high-order Markov chains seems to be the reduction of their number of parameters. Several propositions were made in this domain, especially by Pegram (1980), Logan (1981) and Raftery (1985). The most interesting of these models is the one proposed by Raftery, which is an autoregressive model utilizing a same one-order transition matrix for each lag. In this paper, we present a generalization of Raftery's model with the introduction of different types of transition matrices and we show that such a model gives best results and is not harder to estimate, especially when only a small number of data is available.

2 General Transition Matrices (S-Matrices)

Let $T = \{\ldots, t-3, t-2, t-1, t, t+1, t+2, \ldots\}$ be the set of time periods. Let P and F be two subsets of T. P and F define the matrix $S_{(P;F)}$ which contains the transition probabilities of any arrangement of the values of a variable X between these two subsets of periods. Any l-th order standard matrix can be rewritten as a S-matrix. This notion of General Transition Matrices is thus a generalization of standard transition matrices. For example :

$$S_{(t-l,\ldots,t-1;t-l+1,\ldots,t)} = Ql \text{ , the } l\text{-th order transition matrix}$$
$$S_{(t-l,\ldots,t-1;t)} = Rl \text{ , the reduced form of } Ql$$

Remark : The reduced form of a high-order transition matrix, defined by Pegram (1980), is a $m^l \times m$ matrix without structural zeros.

3   General Autoregressive Modelling (GAM)

Let $\{X_t; t \in N\}$ be a variable taking values in $V = \{1; \ldots; m\}$ and consider a l-th order Markov chain. Let $P = \{t-l, \ldots, t-1\}$ be the set of explanatory periods of the model. Let $P_{(h,g)}$ be subsets of P, where $h = \{1, \ldots, l-1\}$ represents the number of explanatory periods of the subset and $g = \{1, \ldots, C_l^h\}$ is the number of the h-explanatory periods subset of periods, C_l^h being the number of combinations of h elements among l. The principle of General Autoregressive Modelling is to consider the transitions between each of the subsets $P_{(h,g)}$ and the period t instead of the transitions between the two groups of periods $t - l, \ldots, t - 1$ and $t - l + 1, \ldots, t$. The model is then writing :

$$P(X_t = i_0 | X_{t-1} = i_1, \ldots, X_{t-l} = i_l) = \sum_{h=1}^{l-1} \sum_{g=1}^{C_l^h} \varphi_{(h,g)} s_{(P_{(h,g)};t)} \qquad (1)$$

where $s_{(P_{(h,g)};t)}$ is the transition probability between $\{X_{t-1} = i_1, \ldots, X_{t-l} = i_l\}$ and $X_t = i_0$ in the matrix $S_{(P_{(h,g)};t)}$, and where $\varphi_{(h,g)}$ are the autoregressive parameters. For all values of X_t, the GAM can be rewritten as :

$$\hat{\chi}_t' = \sum_{h=1}^{l-1} \sum_{g=1}^{C_l^h} \varphi_{(h,g)} \chi'_{(h,g)} S_{(P_{(h,g)};t)} \qquad (2)$$

where $S_{(P_{(h,g)};t)}$ are transition matrices, $\hat{\chi}_t'$ is the probability vector of the period t and $\chi'_{(h,g)}$ are selection vectors with dimensions :

$$\hat{\chi}_t' \; : \; 1 \times m \qquad \chi'_{(h,g)} \; : \; 1 \times m^h \qquad S_{(P_{(h,g)};t)} \; : \; m^h \times m$$

The GAM necessitates the estimation of two types of parameters : the S-matrices and the autoregressive parameters $\varphi_{(h,g)}$. The S-matrices are estimated from a serie of data. The $\varphi_{(h,g)}$ parameters are estimated by maximization of the following log-likelihood function under the constraints (4) and (5) :

$$L = \sum_{i_0=1}^{m} \cdots \sum_{i_l=1}^{m} n_{i_0,\ldots,i_l} \, log \left(\sum_{h=1}^{l-1} \sum_{g=1}^{C_l^h} \varphi_{(h,g)} \, S_{(P_{(h,g)};t)} \right) \qquad (3)$$

where $n_{i_0,\ldots,i_l}$ is the number of sequences of the form :

$$X_{t-l} = i_l \; ; \; \ldots \; ; \; X_{t-1} = i_1 \; ; \; X_t = i_0 \quad ; \quad \forall i_0,\ldots,i_l \in V$$

in the data. Since it must be a probability vector, $\hat{\chi}_t$ involves two restrictions :

$$1. \;\; \sum_{i=1}^{m} \hat{\chi}_t(i) = 1 \qquad 2. \;\; \hat{\chi}_t(i) \geq 0, \; \forall i$$

They are verified if we maximize (3) under the constraints :

$$\sum_{h=1}^{l-1} \sum_{g=1}^{C_l^h} \varphi_{(h,g)} = 1 \qquad\qquad (4)$$

$$\sum_{h=1}^{l-1} \sum_{g=1}^{C_l^h} \varphi_{(h,g)} S_{(P_{(h,g)};t)} \geq 0, \quad \forall i_0, i_1, \ldots, i_l \in V \qquad (5)$$

An alternative is to maximize (3) without the constraints (5) and to verify *a posteriori* that all results are probabilities.

4　Limit Theorem

In this section we demonstrate that the General Autoregressive Model has the same one-order limit distribution as the usual first-order Markov chain.

Proposition 4.1 *Let $\{X_t; t \in N\}$ be a variable taking values in $V = \{1,\ldots,m\}$ and suppose we have a serie of n consecutive data. Suppose that the first-order transition matrix (Q1) is regular with limit distribution $\pi = \{\pi_1,\ldots,\pi_m\}'$. Let $P = (P_1,\ldots,P_f)$ and $F = (t, P_1,\ldots,P_{f-1})$ be two subsets of periods defining a regular matrix $S_{(P;F)}$ with limit distribution ξ_S, and let $\omega = \{\omega_1,\ldots,\omega_m\}$ be its corresponding one-order limit distribution. Then :*

$$\lim_{n \to \infty} \pi = \lim_{n \to \infty} \omega$$

Sketch of the proof : As $Q1$ and $S_{(P;F)}$ are supposed regular, we have $\pi' = \pi' Q1$ and $\xi'_S = \xi'_S S_{(P;F)}$. Letting $\omega_{i_0} = \sum_{i_1,\ldots,i_{l-1}} \xi_{S(i_0,i_1,\ldots,i_{l-1})}$, $\forall i_0 \in V$, and applying the law of large numbers (Kemeny & Snell (1976)), we have :

$$\lim_{n \to \infty} \pi \;=\; p \tag{6}$$

$$\omega' \;=\; \xi'_S S_{(P;t)}$$

$$\lim_{n \to \infty} \omega \;=\; p \tag{7}$$

where $p = \{p_1, \ldots, p_m\}'$ is the probability vector of every value $1, \ldots, m$ in the data, and where $S_{(P;t)}$ is the reduced form of $S_{(P;F)}$. Finally, equalizing 6 and 7, we obtain :

$$\lim_{n \to \infty} \pi = \lim_{n \to \infty} \omega$$

$$\square$$

Theorem 4.1 *Limit Theorem for the General Autoregressive Model*
Let $\{X_t; t \in N\}$ be a variable taking values in $V = \{1, \ldots, m\}$ and consider the General Autoregressive Model. Suppose that all S-matrices of the model are regular, that the GAM produces only strictly positive probabilities and that π is the limit distribution of the first-order transition matrix. Then, as the number of data $n \to \infty$:

$$\lim_{t \to \infty} P(X_t = i_0 | X_1 = i_1, \ldots, X_l = i_l) = \pi_{i_0} \quad , \quad i_0, i_1, \ldots, i_l \in V$$

Sketch of the proof : Let $ARQl$ be a matrix with elements :

$$P(X_t = i_0, X_{t-1} = i_1, \ldots, X_{t-l+1} = i_{l-1} | X_{t-1} = j_1, \ldots, X_{t-l} = j_l)$$

$$= \begin{cases} \sum_{h=1}^{l-1} \sum_{g=1}^{C_l^h} \varphi_{(h,g)} S_{(P_{(h,g)};t)} & , \; if \; i_f = j_f \, , \; f = 1, \ldots, l-1 \\ \\ 0 & , \; otherwise \end{cases}$$

Since by hypothesis the GAM produces only strictly positive probabilities, all states are aperiodic, intercommunicate, and $ARQl$ is irreducible. Then, $ARQl$ is regular and has an unique limit distribution ξ satisfying $\xi' = \xi' ARQl$. Let $\omega = (\omega_1, \ldots, \omega_m)'$ be the one-order limit distribution corresponding to ξ. Then :

$$\xi' ARRl = \omega' \tag{8}$$

where $ARRl$ is the reduced form of $ARQl$. Let us rewrite $ARRl$ as :

$$ARRl = \sum_{h=1}^{l-1} \sum_{g=1}^{C_l^h} \varphi_{(h,g)} U_{(h,g)}$$

where $U_{(h,g)}$ are $m^l \times m$ matrices whose the probability at the intersection between the row defined by $\{X_{t-1} = i_1, \ldots, X_{t-l} = i_l\}$ and the column defined by $X_t = i_0$ is the corresponding probability in the matrix $S_{(P_{(h,g)};t)}$. Then, the k-th element of $\xi' U_{(g,h)}$ may be written :

$$\sum_{i_1,\ldots,i_l=1}^{m} \xi_{i_1,\ldots,i_l} \, s_{(P_{(h,g)};t)}$$

Summing first over the explanatory periods of $s_{(P_{(h,g)};t)}$ we obtain :

$$\sum_{\substack{i_f=1 \\ i_f \in P_{(h,g)}}}^{l} s_{(P_{(h,g)};t)} \sum_{\substack{i_f=1 \\ i_f \notin P_{(h,g)}}}^{l} \xi_{i_1,\ldots,i_l} = \sum_{\substack{i_f=1 \\ i_f \in P_{(h,g)}}}^{l} \xi_{S_{(P_{(h,g)};t)}} s_{(P_{(h,g)};t)}$$

where $\xi_{S_{(P_{(h,g)};t)}}$ is the limit distribution corresponding to $S_{(P_{(h,g)};t)}$. Then :

$$\xi' ARRl = \xi' \sum_{h=1}^{l-1} \sum_{g=1}^{C_l^h} \varphi_{(h,g)} U_{(g,h)} = \sum_{h=1}^{l-1} \sum_{g=1}^{C_l^h} \varphi_{(h,g)} \xi'_{S_{(P_{(h,g)};t)}} S_{(P_{(h,g)};t)}$$

And by application of proposition 4.1 :

$$\xi' ARRl = \sum_{h=1}^{l-1} \sum_{g=1}^{C_l^h} \varphi_{(h,g)} \pi' = \pi' \quad , \quad if \ n \ \to \infty \qquad (9)$$

Finally equalizing (8) and (9) we obtain :

$$\omega' = \pi'$$

$\square$

5   Measures of Quality

We present here two measures of the modelling quality of the GAM. The first is the mean deviation between a reference matrix and the matrix generated by the model. Formally, it is writing :

$$I_{qm}\left(\frac{ARSl}{Sl}\right) = \frac{1}{m^{(l+1)}} \sum_{i=1}^{m^l} \sum_{j=1}^{m^l} |e_{ij}|$$

where $E = [e_{ij}] = Sl - ARSl$ is the deviation matrix between Sl, the reduced form of the reference matrix, and $ARSl$ its modelling. This measure is comprised between 0 and $\frac{2}{m}$. The second measure of quality is the variance of deviations between Sl and $ARSl$. Formally, it is writing :

$$I_{qv}\left(\tfrac{ARSl}{Sl}\right) = \frac{1}{m^{(l+1)}} \sum_{i=1}^{m^l} \sum_{j=1}^{m^l} e_{ij}^2$$

This measure is comprised between 0 and $\frac{2}{m}$.

6 Numerical Illustration

In this section, we present a numerical illustration of the General Autoregressive Modelling. Let $\{X_t; t \in N\}$ be a variable taking values in $V = \{1; 2\}$ and consider the following third-order transition matrix :

$$X_t$$

X_{t-3}	X_{t-2}	X_{t-1}	1	1	1	1	2	2	2	2
1	1	1	0.90	0	0	0	0.10	0	0	0
2	1	1	0.10	0	0	0	0.90	0	0	0
1	2	1	0	0.65	0	0	0	0.35	0	0
2	2	1	0	0.35	0	0	0	0.65	0	0
1	1	2	0	0	0.90	0	0	0	0.10	0
2	1	2	0	0	0.10	0	0	0	0.90	0
1	2	2	0	0	0	0.65	0	0	0	0.35
2	2	2	0	0	0	0.35	0	0	0	0.65

$Q3 =$ (for rows 3–6 on the left)

We utilized $Q3$ to generate 5 sequences of 50 data each (A-E), for which we estimated the third-order transition matrix $(S_{(t-3,t-2,t-1;t)})$. We estimated six S-matrices of orders 1 and 2 too, namely :

1. $S_{(t-1;t)}$ 2. $S_{(t-2;t)}$ 3. $S_{(t-3;t)}$
4. $S_{(t-2,t-1;t)}$ 5. $S_{(t-3,t-1;t)}$ 6. $S_{(t-3,t-2;t)}$

Then, we estimated 17 GAMs using different arrangements of the S-matrices. The next two tables give complete results for the sequence B and summarize the best results we obtain for each other sequence. We denote φ_i the φ-parameter corresponding to the i-th S-matrix. The model 1 is the third-order Markov chain $(S_{(t-3,t-2,t-1;t)})$ calculated on the data. The model 2 is Raftery's model with matrix $S_{(t-1;t)}$. The last two columns of each table give the measures of quality I_{qm} and I_{qv} calculated between each of the modellings and $S3$, the reduced form of the reference matrix $Q3$.

The parameter φ_1 is generally close to zero. This indicates that the matrix $S_{(t-1;t)}$ has few influence on the resulting matrix, what is not surprising, since the generating matrix $Q3$ was constructed such that X_{t-1} has no influence on X_t. For every sequence, several GAMs have better I_{qm} and I_{qv} that the traditionnal third-order transition matrix. This is especially the case for the sequence C, for which the best models have a mean deviation

Models	φ_1	φ_2	φ_3	φ_4	φ_5	φ_6	$I_{qm}(\frac{ARS3}{S3})$	$I_{qv}(\frac{ARS3}{S3})$
B1	-	-	-	-	-	-	0.2099	0.0840
B2	0.8157	0.7724	-0.5881	-	-	-	0.2857	0.0991
B4	0.0071	0.0820	0.9109	-	-	-	**0.1250**	**0.0170**
B5	-0.0251	-0.1692	0.4021	0.7922	-	-	0.2162	0.0956
B6	-0.3108	0.3005	-1.0028	-	2.0131	-	0.1475	0.0268
B7	-0.0208	0.0170	0.0042	-	-	0.9996	0.1520	0.0455
B8	-0.0404	-0.1548	0.3057	0.7829	0.1066	-	0.2134	0.0949
B9	-0.0113	-0.0668	-0.1124	0.5304	-	0.6601	0.2274	0.0842
B12	-0.0043	-	-	-	-	1.0043	0.1501	0.0452
B13	-	0.0850	-	-	0.9150	-	0.1305	0.0216
B14	-	-	0.2741	0.7259	-	-	0.2435	0.1057
B15	-	-	-	0.7023	0.2977	-	0.2399	0.1055
B16	-	-	-	0.4097	-	0.5903	0.2001	0.0646
B17	-	-	-	-	-0.0304	1.0304	0.1540	0.0481

TABLE 1. Sequence B

Models	φ_1	φ_2	φ_3	φ_4	φ_5	φ_6	$I_{qm}(\frac{ARS3}{S3})$	$I_{qv}(\frac{ARS3}{S3})$
A1	-	-	-	-	-	-	0.1497	0.0316
A7	0.2094	-0.2092	-0.0220	-	-	1.0218	0.1261	**0.0217**
A17	-	-	-	-	0.4925	0.5075	**0.1148**	0.0235
C1	-	-	-	-	-	-	0.1558	0.0642
C7	0.4826	-0.4375	-0.3036	-	-	1.2585	0.0548	**0.0043**
C12	0.0484	-	-	-	-	0.9516	**0.0495**	0.0049
D1	-	-	-	-	-	-	0.1771	0.0432
D4	0.2490	-0.1455	0.8965	-	-	-	**0.1392**	**0.0274**
E1	-	-	-	-	-	-	0.1830	0.0791
E4	0.2133	-0.1088	0.8955	-	-	-	**0.1250**	**0.0171**
E6	-0.2193	0.2414	-0.1913	-	1.1692	-	**0.1250**	0.0284

TABLE 2. Sequences ACDE

three times lower as model 1 and a very small variance of deviations. On the other side, Raftery's model presents bad results: It is worse than the other autoregressive modellings and, more important, it is worse than the standard third-order matrix too. It is interesting to remark that model 4, which utilizes only one-order matrices is generally good. This can be explained by the fact that sequences of 50 data are too small to obtain good estimates of matrices of order two. This penalizes the models utilizing these matrices. Finally, we give here, for each sequence, the best modelling of $Q3$ (in reduced form) :

$$A17 = \begin{pmatrix} 0.9171 & 0.0829 \\ 0.2361 & 0.7639 \\ 0.8991 & 0.1009 \\ 0.4053 & 0.5947 \\ 0.9718 & 0.0282 \\ 0.1831 & 0.8169 \\ 0.9539 & 0.0461 \\ 0.3522 & 0.6478 \end{pmatrix} \quad B4 = \begin{pmatrix} 0.7970 & 0.2030 \\ 0.2456 & 0.7544 \\ 0.8221 & 0.1779 \\ 0.2707 & 0.7293 \\ 0.7990 & 0.2010 \\ 0.2476 & 0.7524 \\ 0.8242 & 0.1758 \\ 0.2728 & 0.7272 \end{pmatrix} \quad C12 = \begin{pmatrix} 0.7747 & 0.2253 \\ 0.0927 & 0.9073 \\ 0.6478 & 0.3522 \\ 0.2778 & 0.7222 \\ 0.7816 & 0.2184 \\ 0.0996 & 0.9004 \\ 0.6547 & 0.3453 \\ 0.2846 & 0.7154 \end{pmatrix}$$

26   André Berchtold

$$
D4 = \begin{pmatrix} 0.7383 & 0.2617 \\ 0.3778 & 0.6222 \\ 0.7560 & 0.2440 \\ 0.3955 & 0.6045 \\ 0.7042 & 0.2958 \\ 0.3436 & 0.6564 \\ 0.7219 & 0.2781 \\ 0.3613 & 0.6387 \end{pmatrix} \quad E4 = \begin{pmatrix} 0.7642 & 0.2358 \\ 0.2114 & 0.7886 \\ 0.7360 & 0.2640 \\ 0.1833 & 0.8167 \\ 0.8324 & 0.1676 \\ 0.2797 & 0.7203 \\ 0.8042 & 0.1958 \\ 0.2515 & 0.7485 \end{pmatrix}
$$

7   Conclusion

In this paper, we have presented a generalisation of Raftery's autoregressive model. The results we have obtained indicate that the integration of additional information to the first-order transition matrix leads to better modellings without estimation's problems. For example, in the numerical part of this paper, the utilisation of transition matrices between $t-2$ or $t-3$ and t improve significantly the results in comparison with Raftery's model. The GAM is especially well suited for high-order transition matrices estimation with few data. In that regard, it proved to be superior to both traditionnal high-order Markov chains and other model's approaches.

Acknowledgments: I would like to thank Professor Gilbert Ritschard, of the University of Geneva, for his helpful reading of this paper.

References

Berchtold, A. (1994). Modélisation autorégressive des chaînes de Markov d'ordre *l*. Cahiers du Département d'Econométrie, 94.07, Université de Genève.

Kemeny, J.G. & Snell, J.L. (1976). *Finite Markov Chains.* Springer-Verlag.

Logan, J.A. (1981). A Structural Model of the Higher-Order Markov Process Incorporating Reversion Effects. *The Journal of Mathematical Sociology,* Vol. 8, 75-89.

Pegram, G.G.S. (1980). An Autoregressive Model for Multilag Markov 12 Chains. *Journal of Applied Probability,* Vol. 17, 350-362.

Raftery, A.E. (1985). A Model for High-Order Markov Chains. *Journal of the Royal Statistical Society B,* Vol. 47, No 3, 528-539.

Raftery, G.G.S. & Tavaré, S. (1994). Estimation and Modelling Repeated Patterns in High Order Markov Chains with the Mixture Transition Distribution Model. *Applied Statistics,* Vol. 43, No 1, 179-199.

A Case–Study on Accuracy of Cytological Diagnosis

A. Biggeri
M. Bini

ABSTRACT: Confortini, Biggeri *et al.* (Acta Cytologica, 1993) designed a study to assess the reliability of cytological diagnoses of a centralized laboratory in the screening for Cervical tumours. A 100 slides set standard was read by 16 raters; the majority diagnosis has been defined as the modal rating for each slide, and the target diagnosis was known by histopathology and clinical follow–up.

In the present paper we analyse ratings from seven raters on a dichotomous classification (negative vs positive: i.e. active diagnostic investigations to be performed), using latent class models and log–linear models.

The results from latent variable modelling compared with those obtained considering majority and target diagnoses using log–linear models, show that the latent variable should be interpreted more as a sort of modal judgement (majority diagnosis) than a true diagnosis.

KEYWORDS: Latent Class Models; Agreement

1 Introduction

The evaluation of agreement among several judges has a growing role in routine quality control procedures to be applied in the context of medical diagnoses based on the classifications of cytological smears or histological sections.

Up to now mainly univariate analyses appeared in the field literature. Regression models have been proposed by several authors (for a review see Agresti, 1992) but only in a methodological frame.

The aim of the present study is to explore the practical implication of a particular class of regression models, i.e. latent class models proposed by Agresti and Lang (1993) which have a sensible interpretation in term of an underlying (latent) classification. Our study will compare the results from applying latent class models and the results obtained using the *true* status or a conventional diagnosis based on the modal ratings.

We believe that a proper understanding of regression models will enhance their application in subjective research fields.

Section 2 introduces the data set to be analysed, section 3 reviews the latent class models used and section 4 presents and discusses the results obtained.

2   Materials

Confortini, Biggeri *et al.* (Acta Cytologica, 1993) designed a study to assess the reliability of cytological diagnoses of a centralized laboratory in the screening for tumours of Cervix Uteri. Here we analyse ratings from seven raters on a dichotomous classification (negative vs positive: i.e. active diagnostic investigations to be performed).

A 100 slides set standard had been assembled from the archive of the Screening Programme and from the Tuscany Cancer Registry. For each woman a complete follow-up history was available. Therefore the resercher has been able to assign a unique diagnosis to each slide, which has been assumed as representative of the *true* status.

The set standard was originally read by 16 raters and a majority diagnosis has been defined as the modal rating for each slide, while the target diagnosis was known by histopathology and clinical follow–up, as explained before.

3   Methods

Some latent class models have been proposed and applied by Agresti and Lang (1993) to explain agreement among raters. These models are appealing because of the simple interpretation of the latent class in term of "true" underlying classification, and the ease implementation on a set of multiple ratings.

These models are based on two fundamental assumptions: the *local independence* which states that within a given class of latent variable the raters choices are independent, and the *no interaction* (Darroch and McCloud, 1986) between object signals and differences in perception of the rater.

To re–examine these assumptions and models, it suffices to consider the simple case of two raters, A and B, who classify a sample of S objects on a same categorical scale with I categories.

The first assumption is justified by study design. For each given object the raters expressed their choice independently and blindly each other. Then the joint probability for an object s of being classified in the category h by rater A and i by rater B is obtained simply multiplying the probability of being classified by each rater separately:

$$\pi_{shi} = \phi_{sAh}\phi_{sBi}$$

where $s = 1, 2, \ldots, S$ and $h, i = 1, 2, \ldots, I$ denote the objects and the classification levels respectively.

When there is an unobserved categorical variable Y, with L levels, which differentiates the objects w.r.t. the classification adopted, those probabilities become:

$$\rho_{lri} = \phi_{sri}$$

where $l = 1, 2, \ldots, L$ and r denotes the rater.

As a consequence of this relation the local independence "moves" from a given object s to a given class l of the latent variable Y. Moreover, the probability π_{hi} of classifying an object ramdomly sampled from the population, which is defined as

$$\pi_{hi} = S^{-1} \sum_s^S \phi_{sAh} \phi_{sBi}$$

is equal to the marginal probability with respect to the variable Y.

The second assumption considered, states that the probability ϕ_{sri} can be simply expressed as a function of the product of the signal emitted by the object s about his belonging to category i, β_{si}, times the perception of rater r about the category i, ψ_{ri}:

$$\phi_{sri} = \theta_{sr} \beta_{si} \psi_{ri}$$

where θ_{sr} is a normalisation parameter for $\phi_{sr+} = 1$.

Indeed Darroch and McCloud commented as follows:

> ... the correct category for an object exists only in the eye of the beholder. Only partially of course, because the signals sent out by the object h about its degree of conformity to each category are received by all expert observers and differ, in general, from those sent out by object h'.

Substitution of ϕ_{sri} in the formula for π_{hi} leads to a *quasi–symmetric* table of probabilities.

Let $m_{hi} = n\pi_{hi}$ denote the expected frequencies for the observed table of ratings by A and B. These frequencies could be modelled as follows:

$$\log m_{hi} = \log \sum_l m_{hil} = \mu + \lambda_h^A + \lambda_i^B + \log[\sum_l \exp(\lambda_l^Y + \lambda_{hl}^{AY} + \lambda_{il}^{BY})]$$

where $\lambda(.)$ are the parameters for main effects and interactions between each rater and latent variable Y.

Agresti and Lang called this model *ordinary latent class*.

To obtain the quasi–symmetry model (Caussinus, 1965) it is sufficient to impose the condition of equality between interaction parameters, for each given level of the latent variable.
In the case of ordered categories the linear–by–linear quasi–symmetry model is defined substituting the interactions parameters with only one coefficient for the linear–by–linear interaction:

$$\log m_{hi} = \mu + \lambda_h^A + \lambda_i^B + \log[\sum_l \exp(\lambda_l^Y + \beta^{AY} x_h y_l + \beta^{BY} x_i y_l)]$$

where x and y are scores for observed and latent classes respectively.
Maximum likelihood estimates of the parameters in the latent class models can be obtained using a variation of the EM algorithm (Goodman, 1974). Arbitrary starting values satisfying the marginal observed counts, are used in the maximization step to obtain expected frequencies. These are used in the E-step to proportionally update the starting values of the previous cycle. In the M-step the iterative reweighted least squares method could be used. The EM procedure ends when the difference between the deviance values of two consecutive iterations is less than a small pre-specified arbitrary value (i.e. 0.000001).
We use the previous described latent class models to analyse the data described before. We consider that the latent variable can have at most three levels.

4   Results

The results of fitting the latent class models are shown in table I. Only the independence model badly fits the data. The quasi–symmetry model seems to provide an adequate fit, supporting the assumptions in the Darroch and McCloud conceptualization. The latent class models with three categories also behave well. The quasi–symmetric 3-level latent variable model provides the best fit and is more parsimonious.
The conditional probabilities of a positive diagnosis given the latent class are estimated for each rater using the fitted values from the quasi–symmetry 3-level latent class model (table II).
These probabilities are compared with those obtained cross–tabulating the data by raters by majority or target diagnosis (table II). We can notice that the conditional probabilities from the latent variable model are consistent with those obtained grouping the diagnostic categories 1-7 (negative and unclassified); 2-3-4 (benign lesion, mild displasia and moderate displasia); and 5-6 (severe displasia and carcinoma). Moreover, the ranking of the raters obtained from the latent variable modelling is coherent with that from the cross–tabulation by the majority diagnosis but not by the target diagnosis.

Table III reports the log Odds-ratios for the proportion of positive diagnoses by each pair of raters. These has been computed as the differences between the main effect estimates from the latent class model and, for matter of comparison, from the cross–tabulation raters by majority or target diagnosis. The results are consistent and predictable since they are based on the same marginal counts.

5   Conclusions

The results in table II have important consequences on both the usefulness and the interpretation of latent variable models for rater agreement. Indeed the latent variable should not be interpreted as a proxy of a true diagnosis. The latent class parameters are estimated in such a way that all the dependence among the raters displayed by the marginal observed table, is absorbed. The association among ratings is dependent on category distinguishability. The latent variable levels correspond therefore to equally distinguishable objects, but not truly different.

The object property of being recognizeable is function of the true status and moreover of the shared knowledge of that status. The latent class conditional probabilities of table II resemble those obtained from the majority diagnosis, i.e the modal judgement.

The consequences of this finding are two:

1. using latent class models cannot substitute an analysis of accuracy based on a gold–standard; indeed it is a powerful tool to investigate both agreement and marginal homogeneity among raters;

2. a simple univariate analysis which mimics latent class modelling could be based on majority or modal diagnosis.

Acknowledgments: We are grateful to Dr. Corrado Lagazio for having provided the implementation in Glim-4 of the EM algorithm used in the analysis.

TABLE I: Fitting results of latent class models

n. class latent	model	Likelihood-ratio statistic	Degrees of freedom
1	indipendence	448.76	120
1	quasi–symmetry	74.717	114
2	ordinary latent class	93.26	112
2	quasi–symmetry latent class	102.2	118
3	ordinary latent class	62.07	104
3	quasi–symmetry latent class	76.84	116
3	$l \times l$ quasi–symmetry latent class	84.63	117

TABLE II:

a) Conditional probabilities given latent variable

Rater	Latent classes		
	$L=1$	$L=2$	$L=3$
A	0.020	0.679	0.968
B	0.009	0.504	0.937
C	0.008	0.470	0.930
D	0.007	0.436	0.919
E	0.002	0.187	0.770
F	0.007	0.436	0.919
G	0.015	0.610	0.960

b) Conditional probabilities given Target diagnosis and Majority diagnosis

Rater	Target Diagnosis			Majority Diagnosis		
	1-7	2-3-4	5-6	1-7	2-3-4	5-6
A	0.01250	0.4038	0.8382	0.01250	0.3269	0.8971
B	0.01250	0.05769	0.9559	0.01250	0.1731	0.8676
C	0.01250	0.2500	0.7794	0.03750	0.2115	0.7794
D	0.01250	0.2500	0.7500	0.01250	0.1346	0.8382
E	0.01250	0.1346	0.5735	0.01250	0.01923	0.6618
F	0.01250	0.1346	0.8382	0.01250	0.09615	0.8676
G	0.01250	0.3269	0.8382	0.01250	0.2885	0.8676

TABLE III:

a) Differences between estimates of the parameters of quasi–symmetry model with $L = 3$

		A	B	C	D	E	F	G
Δ	A		-0.733	-0.871	-1.008	-2.223	-1.008	-0.304
	B			-0.138	-0.275	-1.49	-0.275	0.429
	C				-0.137	-1.352	-0.137	0.567
	D					-1.215	0	0.704
	E						1.215	1.919
	F							0.704

b) Target diagnosis: marginal homogeneity

		A	B	C	D	E	F	G
Δ	A		-0.480	-0.579	-0.679	-1.572	-0.679	-0.188
	B			-0.099	-0.199	-1.092	-0.199	0.292
	C				-0.100	-0.993	-0.100	0.391
	D					-0.893	0	0.491
	E						0.893	1.384
	F							0.491

c) Majority diagnosis: marginal homogeneity

		A	B	C	D	E	F	G
Δ	A		-0.560	-0.675	-0.789	-1.777	-0.789	-0.219
	B			-0.115	-0.229	-1.217	-0.229	0.341
	C				-0.114	-1.102	-0.114	0.456
	D					-0.988	0	0.570
	E						0.988	1.558
	F							0.570

References

Agresti, A. (1992). Modeling Patterns of Agreement and Disagreement. *Statistical Methods in Medical Research*, **1**, 201-218

Agresti, A., Lang, J.B. (1993). Quasi-Symmetric Latent Class Models, with Application to Rater Agreement. *Biometrics*, **49**, 131-139

Caussinus, H. (1965). Contribution à l'Analyse Statistique des Tableaux de Correlation. *Annales de la Faculté des Sciences de l'Université de Toulose*, **29**, 77-182

Confortini, M., Biggeri, A., Cariaggi, M.P., *et al.* (1993). Intralaboratory Reproducibility in Cervical Cytology. *Acta Cytologica*, **37**, 49-54

Darroch, J.N., McCloud, P.I. (1986). Category Distinguishability and Observer Agreement. *Australian Journal of Statistics*, **28**, 371-388

Goodman, L.A. (1974). Exploratory Latent Structure Analysis Using both Identifiable and Unidentifiable Models. *Biometrika*, **61**, 215-231

Dynamics and Correlated Responses in Longitudinal Count Data Models

R.Blundell
R.Griffith
F.Windmeijer

ABSTRACT: The aim of this paper is to examine the properties of dynamic count data models. We propose the use of a linear feedback model (LFM), based on the Integer Valued Autoregressive (INAR) class, for longitudinal data applications with weakly exogenous regressors. In these models the conditional mean is modelled linearly in the history of the process with a log-link in the exogenous variables. We explore the quasi-differencing GMM approach to eliminating unobserved heterogeneity. These ideas are illustrated using data on US R&D expenditures and patents.

KEYWORDS: Longitudinal Count Data; Integer Valued Autoregressive Process; Multiplicative Individual Effects; Generalized Method of Moments; Patents and R&D.

1   Introduction

In many economic applications to longitudinal, or panel, data the variable of interest is a count process - for example, the number of job applications, visits to the doctor, patent applications, or technological innovations made in a period of time. It is often the case that the process under study is inherently dynamic so that the history of the count process itself is an important determinant of current outcomes. Like other panel data applications it is also likely that unobserved heterogeneity, or individual effects, induce persistently different counts across individuals or firms. Distinguishing between dynamics and individual effects is made difficult in panel data by the correlation between the error process and the lagged dependent variable generated by the presence of individual effects. In linear models the within group estimator and the first differencing GMM estimator essentially eliminate the individual effect. In count data models, where a natural non-linearity is produced by the positive and discrete nature of the data, the presence of both individual effects and dynamics is particularly difficult

to deal with. Also, due to the strong likelihood in economic applications
of feedback from the errors to the explanatory variables, certain estimators like the GEE estimator of Liang and Zeger (1986), with a general
(non-independent) working variance, will be inconsistent.

In this paper the properties of several dynamic models for count data are
examined. Typically, where dynamics have been thought to be important,
lagged values of explanatory variables have been included in the exponential
mean function. Little research has looked at the properties of models that
include the *lagged dependent* variable itself in the exponential mean function. We argue that this specification is less interpretable and the stability
properties are much less robust to the specific form of the data process.
Instead, we propose use of a linear feedback model (LFM).

The LFM is derived from the Integer Valued Autoregressive (INAR) model
which has foundations in the generalisation of the Poisson model to the
ARMA case developed by Al-Osh and Alzaid (1987). We argue that the
LFM specification is particularly well suited to economic applications. In
the LFM model the expectation of an integer dependent variable - the
count variable - is modelled linearly in the history of the count process.
This mirrors the standard *linear* autoregressive model. However, since the
process must remain non-negative, only positive autoregressive processes
are permitted. The remaining components of the mean specification, which
include the observable and unobservable explanatory factors, are multiplicative as in the standard Poisson count specification which guarantees
a positive conditional mean. In this LFM specification it is easy to understand the dynamics and to assess the stability properties of the estimated
model.

In order to control for unobserved heterogeneity, we apply quasi-differenced
GMM estimators analogous to those for linear panel data models. To investigate our proposed model and estimators, an application is made to the
well known dataset of US firms' R&D expenditure and patenting activity
used in Hall *et al.* (1986) and Hausman *et al.* (1984).

2   Feedback and Correlated Individual Effects

In many economic applications, including the patents-R&D relationship we
consider, estimation of (dynamic) models is complicated by the fact that
explanatory variables are not always strictly exogenous. Instead, there may
be *simultaneity* between regressors and current shocks, or *feedback* from the
current shocks to future values of the regressors. For the simple linear model
specification

$$y_{it} = x'_{it}\beta + u_{it}$$

simultaneity is formalized by $E[u_{it}|x_{it}] \neq 0$, whereas feedback, or weak
exogeneity, which occurs when the regressors are predetermined implies

that $E[u_{it}|x_{iT}, ..., x_{i1}] \neq E[u_{it}|x_{it}, ..., x_{i1}]$. An example of a predetermined or weakly exogenous variable is the lagged dependent variable, y_{it-1}.

A commonly used estimator in panel data applications is the GEE estimator (Liang and Zeger (1986)) which is based on

$$E\left[\sum_i D_i' V_i^{-1}(y_i - \mu_i)\right] = 0, \tag{1}$$

where $y_i = \{y_{it}\}$; $\mu_i = \{\mu_{it}\}$; $D_i = \{\partial\mu_i/\partial\beta\}$; and V_i is the specified (or working) variance matrix of y_i. When the regressors in a panel data model are weakly exogenous, consistent estimates will only be obtained by specifying a diagonal working variance, as otherwise condition (1) is not satisfied.

A further complication is that in most economic panel data applications the individual effects are correlated with the explanatory variables, in which case the random effects (GLS) estimator is inconsistent. Use of the "fixed effects" or dummy variable (within groups) estimator is then the usual practice. However, when the regressors are predetermined, the within groups transformation (*i.e.* taking deviations from individual means) results in a bias due to the correlation of the past shocks with current regressors. A standard solution for the linear model is to take first differences and estimate by the Generalized Method of Moments (GMM). For the first order autoregressive case, the model specification is given by

$$y_{it} = \alpha y_{it-1} + \eta_i + u_{it},$$

and the moment conditions $E[(u_{it} - u_{it-1})|y_{it-2}, ..., y_{i1}] = 0$ can be used under certain regularity conditions.

3   Dynamic Specifications

Let y_{it} denote the discrete variable to be explained and for subject i, $i = 1, ..., N$, at time t, $t = 1, ..., T$; and let x_{it} denote a vector of explanatory variables. The exponential - or log-link - model of the form

$$E(y_{it}|x_{it}) = \exp(x_{it}'\beta), \tag{2}$$

is commonly used to model such count data. A straightforward way of incorporating dynamics in model (2) is to include lags of the explanatory variables:

$$E(y_{it}|x_{it}) = \exp(\sum_{j=0}^{p} x_{it-j}'\beta_j),$$

a model we label exponential distributed lag (EDL) model. The short run effect in this model is given by $\beta_0 E(y_{it})$ and the long run effect by $(\sum_{j=0}^{p} \beta_j)E(y_i)$.

An alternative specification that incorporates the entire history of the process[1] is to include the lagged dependent variable in the exponential mean function:

$$E(y_{it}) = \exp(\alpha y_{it-1} + x'_{it}\beta).$$

This model we call the exponential feedback model (EFM). The EFM has as a disadvantage that it is explosive for $\alpha > 0$ and therefore has no long run interpretation. Furthermore, in panel data α is very difficult to interpret due to the fact that the correlation between y_{it} and y_{it-1} is equal to $\alpha E(y_i)$. Where there are multiple time series with different means, α will be some weighted value. In particular, individuals with a high mean will have a dominant downward effect on the coefficient estimate.

A different specification, based on the INAR process,[2] is the linear feedback model (LFM):

$$E(y_{it}) = \gamma y_{it-1} + \exp(x'_{it}\beta),$$

with $\gamma > 0$. The correlation over time in this model is given by

$$corr(y_{it}, y_{it-j}) = \gamma^j,$$

and the short run and long run dynamics are given by $\beta(1 - \gamma)E(y_{it})$ and $\beta E(y_i)$ respectively. The stability and interpretability properties of the LFM model make it the favoured model. It also has an advantage over dynamic specifications like the one proposed by Zeger and Qaqish (1988), which includes $\ln(y_{it-1})$ in the exponential mean function, in that it avoids the need to transform zero values of the lagged dependent variable.

4   Individual Effects

For count data models unobserved heterogeneity or individual effects are normally modelled multiplicatively. For the LFM this results in

$$y_{it} = \gamma y_{it-1} + \exp(x'_{it}\beta)\nu_i u_{it}.$$

One way[3] to control for fixed effects (Wooldridge (1991), Chamberlain (1992)) is to quasi-difference the model as

$$r_{it}(\theta) = \frac{y_{it} - \gamma y_{it-1}}{\mu_{it}} - \frac{y_{it-1} - \gamma y_{it-2}}{\mu_{it-1}} = \nu_i(u_{it} - u_{it-1}),$$

---

[1]See also Fahrmeir and Tutz (1994) for a general discussion of specification and estimation of GLM time series models.

[2]See Al-Osh and Alzaid (1987), McKenzie (1988), Ronning and Jung (1992), Alzaid and Al-Osh (1993), and Brännäs (1994).

[3]For another approach that utilizes pre-sample information to correct for unobserved heterogeneity see Blundell et al. (1995).

with $\theta = (\gamma, \beta)$ and $\mu_{it} = \exp(x'_{it}\beta)$.

Defining $z_{it} = (y_{it-1}, x_{it})$, under weak exogeneity and the assumption

$$E(u_{it}|z_{i1}, ..., z_{it}) = 1,$$

the conditional mean of $r_{it}(\theta)$ is given by

$$E\left[r_{it}(\theta)|z_{i1}, ..., z_{it-1}\right] = E\left[r_{it}(\theta)|\nu_i, z_{i1}, ..., z_{it-1}\right] \times E\left[\nu_i|z_{i1}, ..., z_{it-1}\right] = 0$$

using the law of iterated expectations. This provides an orthogonality restriction that can be used to consistently estimate the model parameters θ by GMM, minimizing

$$\left(\frac{1}{N}\sum_{i=1}^{N} r'_i(\theta)Z_i\right) W_N \left(\frac{1}{N}\sum_{i=1}^{N} Z'_i r_i(\theta)\right),$$

with general weight matrix that allows for heteroscedasticity and correlation over time:

$$W_N = \left[\sum_{i=1}^{N} Z'_i \widetilde{r}_i \widetilde{r}_i Z_i\right]^{-1},$$

with $\widetilde{r}_i$ some initial (first round) estimate.

5   Application

The data we use to examine the properties of the alternative dynamic specifications is the well known patent-R&D panel data set of Hausman *et al.* (1984). It contains information on R&D expenditures and a count of patents applied for and subsequently granted for large US firms between 1972 and 1979. Work on this data has focused on the dynamic relationship between R&D and patents and has found little evidence of dynamic feedback. The highly autoregressive nature of the R&D data makes it difficult to identify the structure of dynamic feedback from the exponential distributed lag model which is typically used.

Results for the EDL and LFM specifications are given in Table 1. In column (1) the estimator for the EDL is a Quasi-Likelihood, or GEE estimator under the independence working assumption, not correcting for fixed effects. This estimator is identical to the GMM estimator when the instruments are the same as the RHS variables. Columns (2) and (3) give the results for the EDL and LFM quasi differenced GMM estimators.[4]

---

[4] Hausman *et al.* (1984) estimated the EDL model controling for individual effects by conditioning on the sufficient statistic, assuming R&D to be strictly exogenous. These results indicated a strong current effect of R&D with little impact through lagged R&D.

TABLE 1. The Estimated Patent Equation

Model Years	(1) EDL 74-79	(2) EDL 75-79	(3) LFM 74-79
Pat$_{-1}$			0.5642 *0.2693*
Pat$_{-2}$			0.2068 *0.0749*
ln R&D	0.3011 *0.1521*	0.6302 *0.3335*	0.8149 *0.3172*
ln R&D$_{-1}$	0.0262 *0.0828*	0.0422 *0.1480*	
ln R&D$_{-2}$	0.4114 *0.1614*	0.0112 *0.0993*	
v_1	3.2128	-3.3315	-1.7428
v_2	3.1990	0.7654	0.3516

The sample contains 446 firms. Standard errors are shown in italics and allow for a general covariance matrix per firm over time. All models include year dummies. Columns (2) and (3) are estimated using the quasi-differenced estimator. In column (1) the RHS variables are treated as weakly exogenous and therefore satisfy the conditions for instruments. In column (2) ln R&D$_{-1}$, ln R&D$_{-2}$, ln R&D$_{-3}$ and in column (3) Pat$_{-2}$, Pat$_{-3}$ and ln R&D$_{-1}$ are used as instruments. v_1 and v_2 are the Arellano and Bond (1991) tests for first and second order serial correlation which are asymptotically $N(0,1)$ distributed under the null of no serial correlation.

The estimates of the EDL model without taking account of the individual effects are shown in column (1). These suggest a U shape and indicate truncation bias. The tests indicate strong serial correlation, possibly due to unobserved heterogeneity. When the individual effects are differenced out in column (2), the U shape and truncation bias disappear, and there seems evidence of only contemporaneous correlation between patents and R&D. The LFM as shown in column (3) indicates a much smaller short run effect but a slightly larger long run impact of R&D on firms' patenting activity.

6 Conclusions

In this paper the use of the linear feedback model based on the INAR model for panel data applications is proposed. We show that the LFM specification is particularly well suited to economic applications. In the LFM model the expectation of an integer dependent variable - the count variable - is modelled linearly in the history of the count process. This mirrors the standard *linear* autoregressive model. However, since the process must remain non-negative, only positive autoregressive processes are permitted. The remaining components of the mean specification, which include the observable and unobservable explanatory factors, is multiplicative mirroring the standard Poisson count specification which guarantees a positive conditional mean.

In this LFM specification dynamics and stability are easy to assess. Following on from the work of Chamberlain (1992) and Wooldridge (1991) we are also able to develop quasi-differencing GMM estimators analogous to those for linear panel data models. To investigate our proposed model an application is made to the well known dataset of US firms R&D expenditure and patenting activity used in Hausman *et al.* (1984). Work on this data has focused on the dynamic relationship between R&D and patents and has found little evidence of dynamic feedback. We show, using the alternative LFM specification, there is strong evidence of a slow adjustment of successful patent outcomes to R&D investment.

Acknowledgments: We are grateful to Manuel Arellano, Andrew Chesher, Bo Honoré, Ekaterina Kyriazidou, Neil Shephard and seminar participants at Nuffield and Aarhus for helpful comments. This study is part of the program of research of the ESRC Analysis of Large and Complex Data Sets at University College London and the ESRC Centre for the Microeconomic Analysis of Fiscal Policy at the Institute for Fiscal Studies. The financial support of the ESRC is gratefully acknowledged. We are grateful to Bronwyn Hall for providing the data on patents and R&D used in this study. The usual disclaimer applies.

References

Al-Osh, M.A., and A.A. Alzaid (1987). First-Order Integer Valued Autoregressive (INAR(1)) Process. *Journal of Time Series Analysis*, **8**, 261-275

Alzaid, A.A., and M.A. Al-Osh (1993). Some Autoregressive Moving Average Processes with Generalised Poisson Marginal Distributions. *Annals of the Institute of Mathematical Statistics*, **45**, 223-232

Arellano, M., and S. Bond (1991). Some Tests of Specification for Panel Data: Monte Carlo Evidence and an Application to Employment Equations. *Review of Economic Studies*, **58**, 277-98

Blundell, R., R. Griffith and F. Windmeijer (1995). Individual Effects and Dynamics in Count Data Models, Discussion Paper 95-03, Dept. of

Economics, University College London

Brännäs, K. (1994). Estimation and Testing in Integer Valued AR(1) Models. Umeå Economic Studies Paper No. 335, February

Chamberlain, G. (1992). Comment: Sequential Moment Restrictions in Panel Data. *Journal of Business & Economic Statistics*, **10**, 20-26

Fahrmeir, L., and G. Tutz (1994). *Multivariate Statistical Modelling Based on Generalized Linear Models*. Springer Series in Statistics, Springer-Verlag, New York

Hall, B., Z. Griliches and J. Hausman (1986). Patents and R and D: Is there a lag? *International Economic Review*, **27**, 265-283

Hausman, J., B. Hall, and Z. Griliches (1984). Econometric Models for Count Data and an Application to the Patents-R&D Relationship. *Econometrica*, **52**, 909-938

Liang, K.-Y., and S.L. Zeger (1986). Longitudinal Data Analysis using Generalized Linear Models. *Biometrika*, **73**, 13-22

McKenzie, E. (1988). Some ARMA Models for Dependent Sequences of Poisson Counts. *Advances in Applied Probability*, **20**, 822-835

Ronning, G., and R.C. Jung (1992). Estimation of a First Order Autoregressive Process with Poisson Marginals for Count Data. In: Fahrmeir, L., B. Francis, R. Gilchrist, and G. Tutz (eds). *Advances in GLIM and Statistical Modelling. Lecture Notes in Statistics*, **78**, 188-194, Springer-Verlag, Berlin

Wooldridge, J.M. (1991). Multiplicative Panel Data Models without the Strict Exogeneity Assumption. MIT Working Paper No. 574, March

Zeger, S.L. and B. Qaqish (1988). Markov Regression Models for Time Series: A Quasi-Likelihood Approach. *Biometrics*, **44**, 1019-1031

Bootstrap Methods for Generalized Linear Mixed Models With Applications to Small Area Estimation

James Booth

ABSTRACT: Generalized linear mixed models (GLMMs) provide a unified framework for analyzing relationships between binary, count or continuous response variables and predictors with either fixed or random effects. Recent advances in approximate fitting procedures and Markov Chain Monte Carlo techniques, as well as the widespread availability of high speed computers suggest that GLMM software will soon be a standard feature of many statistical packages. Although the difficulty of fitting of GLMMs has to a large extent been overcome, there are still many unresolved problems, particularly with regards to inference. For example, analytical formulas for standard errors and confidence intervals for linear combinations of fixed and random effects are often unreliable or not available, even in the classical case with normal errors. In this paper we propose the use of the parametric bootstrap as a practical tool for addressing problems associated with inference from GLMMs. The power of the bootstrap approach is illustrated in two small area estimation examples. In the first example, it is shown that the bootstrap reproduces complicated analytical formulas for the standard errors of estimates of small area means based on a normal theory mixed linear model. In the second example, involving a logistic-normal model, the bootstrap produces sensible estimates for standard errors, even though no analytical formulas are available.

KEYWORDS: Analytical Correction; Empirical BLUP; Mixed Model; Monte Carlo Approximation; Prediction Interval; Standard Error.

1 Introduction

In this paper we suggest parametric bootstrap methods for inference from generalized linear mixed models (GLMMs). We show that the bootstrap provides automatic frequentist solutions to problems associated with inference from GLMMs that are otherwise either analytically intractable or

requiring of complicated analytical methods. The bootstrap approach is
illustrated in two examples in which GLMMs are used to obtain shrinkage
estimates of small area means. A related application of the bootstrap in an
empirical Bayes framework, based on an idea outlined by Efron (1987), is
considered by Carlin and Gelfand (1991).

The first example involves some well known data from Battese, Harter
and Fuller (1988) on the area used for growing corn and soybeans in 12
Iowa counties. These authors proposed a mixed effects normal theory linear model for estimating the county means and derive analytical corrections
for naive estimates of standard errors to take account of sampling variability in the variance component estimates. It turns out that application of
the parametric bootstrap in this problem reproduces the results of Battese,
Harter and Fuller (BHF) almost exactly. That is, the bootstrap essentially
performs the analytical corrections automatically without any prior knowledge of their existence.

The second example involves data from 13 counties in north central Florida
on the number of births to underage mothers and adult mothers during
the three year period, 1989-91. We propose the use of a logistic-normal
GLMM for estimating county specific rates of births to underage mothers.
As in the Iowa farm data example, ignoring the sampling variability of the
estimated random effects variance can lead to underestimation of standard
errors of the linear predictor and hence inaccurate confidence intervals. In
this problem, the parametric bootstrap method appropriately adjusts the
standard errors even though no analytical correction is available.

An outline of the rest of the paper is as follows. In Section 2 we define the
GLMM and discuss fitting strategies. A brief description of the parametric
bootstrap and its application in the GLMM context is given in Section 3.
Application to small area estimation problems is considered in Section 4.

2   Notation and definition of GLMM

Let $\mathbf{x}_i$ $(p \times 1)$ and $\mathbf{z}_i$ $(q \times 1)$ be known covariate values associated with the
ith response, y_i, $i = 1, \ldots, n$. The linear predictor associated with the ith
response is then defined by

$$\eta_i = \mathbf{x}_i^t \alpha + \mathbf{z}_i^t \mathbf{b}, \qquad (1)$$

where α is a p-vector of fixed effects which are typically unknown and $\mathbf{b}$ is
a q-vector of random effects.

We shall assume in what follows that the marginal distribution of $\mathbf{b}$ is multivariate normal with zero mean and covariance matrix D. Often, the matrix
D is assumed to have a particular structure determined by a parameter γ.
In this case we sometimes write, $D = D(\gamma)$, to emphasize this dependence.
A GLMM for the response vector $\mathbf{y} = (y_1, \ldots, y_n)^t$ is now completely spec-

ified by assuming that the conditional distribution of **y** given **b** is described by a generalized linear model (McCullagh and Nelder, 1989).

Let μ_i denote the conditional expectation of y_i given **b** and suppose that μ_i is related to the covariates, $\mathbf{x}_i$ and $\mathbf{z}_i$ by a link function g to the linear predictor in 1. That is

$$g(\mu_i) = \eta_i . \tag{2}$$

Then the n responses are assumed to be conditionally independent given **b** with exponential family densities of the form

$$f(y_i|\phi, \alpha, \mathbf{b}) = c(y_i, \phi a_i) \exp\left[\frac{1}{\phi a_i}\{\theta_i y_i - \kappa(\theta_i)\}\right], \tag{3}$$

where $\mu_i = \kappa'(\theta_i)$, the a_i's are known weights and ϕ is a scalar parameter which may be known or unknown.

In principle estimates of the unknown parameters can be obtained directly from the likelihood function

$$
\begin{aligned}
l(\alpha, \phi, \gamma) \quad &\propto \quad \int f(\mathbf{y}, \mathbf{b}|\alpha, \phi, \gamma) d\mathbf{b} \\
&= \quad \int \left\{\prod_{i=1}^{n} c(y_i, \phi a_i)\right\} |2\pi D(\gamma)|^{-1/2} \\
&\quad \times \exp\left[\sum_{i=1}^{n} \frac{1}{\phi a_i}\{\theta_i y_i - \kappa(\theta_i)\} - \frac{1}{2}\mathbf{b}^t D(\gamma)^{-1}\mathbf{b}\right] d\mathbf{b} . \tag{4}
\end{aligned}
$$

Unfortunately, a closed form expression for the integral in 4 is only available when the conditional density in 3 is itself normal. In some simple cases it is feasible to obtain maximum likelihood estimates by evaluating the integral in 4 numerically and/or by using the EM algorithm (eg. Anderson and Hinde, 1988). In general, the intractability of the likelihood function 4 has been a major obstacle to the use of GLMMs in practice. However, recent theoretical developments, such as the approximate fitting procedures of Breslow and Clayton (1993) and McGilchrist (1994) as well as Markov Chain Monte Carlo (MC^2) techniques (eg. Zeger and Karim, 1991), have made the use of GLMMs a practical proposition.

The approaches of Breslow and Clayton and McGilchrist, mentioned above, are both rooted in the following idea which can be traced back to Henderson (1950) in the normal theory mixed linear model case. Suppose that the parameters ϕ and γ are known. Then, maximizing the joint density $f(\mathbf{y}, \mathbf{b}|\alpha, \phi, \gamma)$ in 4 with respect to $(\alpha, \mathbf{b})$ leads to the estimating equations

$$\sum_{i=1}^{n} \frac{(y_i - \mu_i)\mathbf{x}_i}{\phi a_i v(\mu_i) g'(\mu_i)} = \mathbf{0} \tag{5}$$

and

$$\sum_{i=1}^{n} \frac{(y_i - \mu_i)\mathbf{z}_i}{\phi a_i v(\mu_i) g'(\mu_i)} = D^{-1}\mathbf{b} \tag{6}$$

for the fixed effects α and the random effects $\mathbf{b}$ respectively, where $v(\mu_i) = \kappa''(\theta_i)$ is the variance function of the generalized linear model conditional on $\mathbf{b}$. In the normal theory case the solution to equations 5 and 6 is given by the generalized least squares (GLS) estimate of α and the best linear unbiased predictor (BLUP) of $\mathbf{b}$. See Robinson (1993) for a review of best linear unbiased prediction and related techniques. In general, equations 5 and 6 can be solved by iteratively reweighted least squares (IRLS). When ϕ and/or γ are unknown, Breslow and Clayton suggest approximate REML and other techniques for estimating the additional parameters.

3   Parametric Bootstrap

The parametric bootstrap, as described by Efron (1982), is defined as follows. Suppose that the random vector $\mathbf{u}$ comes from a distribution in a parametric family $\{F_\lambda : \lambda \in \Lambda\}$, where λ is unknown. Let $\hat{\lambda}$ denote an estimate of λ based on $\mathbf{u}$. Suppose that we are interested in estimating a functional $t(F_\lambda)$ of the unknown distribution such as the standard error of a particular component of $\hat{\lambda}$. Then, a bootstrap estimate of $t(F_\lambda)$ is simply $t(F_{\hat{\lambda}})$.

In the context of GLMMs an important problem is the estimation of "standard errors" for the fitted linear predictor $\hat{\eta} = \mathbf{x}^t\hat{\alpha} + \mathbf{z}^t\hat{\mathbf{b}}$ at a particular set of covariate values $\mathbf{x}$ and $\mathbf{z}$. Here we define standard error to mean the square root of the mean squared error of prediction

$$\nu^2 = E\left\{[\hat{\eta} - \eta]^2\right\} = E\left\{\left[\mathbf{x}^t(\hat{\alpha} - \alpha) + \mathbf{z}^t(\hat{\mathbf{b}} - \mathbf{b})\right]^2\right\} \qquad (7)$$

The expectation in 7 is completely determined by the joint distribution, F_λ of $(\mathbf{y}, \mathbf{b})$ defined in Section 2, where $\lambda = (\alpha, \phi, \gamma)$. The bootstrap estimate of ν^2 is therefore

$$\hat{\nu}^2 = E^*\left\{[\hat{\eta}^* - \eta^*]^2\right\} = E^*\left\{\left[\mathbf{x}^t(\hat{\alpha}^* - \hat{\alpha}) + \mathbf{z}^t(\hat{\mathbf{b}}^* - \mathbf{b}^*)\right]^2\right\}, \qquad (8)$$

where the "*" notation indicates that the expected value is taken with respect to the fitted distribution with parameter $\hat{\lambda} = (\hat{\alpha}, \hat{\phi}, \hat{\gamma})$; i.e. $(\mathbf{y}^*, \mathbf{b}^*) \sim F_{\hat{\lambda}}$. We propose here that approximate confidence (or prediction) intervals for a monotone function $h(\eta)$ be obtained as $h(\hat{\eta} \pm z\hat{\nu})$, as in the fixed effects case.

A naive approximation to ν^2 is obtained by ignoring the sampling variability of $\hat{\gamma}$. Suppose that ϕ is known and let $\hat{\eta}(\gamma)$ denote the fitted linear predictor as a function of the "true" variance component vector γ. Let X denote the $(n \times p)$ matrix with ith row $\mathbf{x}_i^t$ and Z the $(n \times q)$ matrix with ith row $\mathbf{z}_i^t$. In addition, let $w_i = 1/\{\phi a_i v(\mu_i)g'(\mu_i)^2\}$ and $V = W^{-1} + Z^t DZ$, where $W = \text{diag}(w_i)$. Then

$$\nu^2 \approx E\left\{(\hat{\eta}(\gamma) - \eta)^2\right\}$$

$$\approx \mathbf{z}^t (D - DZ^t \hat{V}^{-1} ZD) \mathbf{z}$$
$$+ (\mathbf{x} - X^t \hat{V}^{-1} ZD\mathbf{z})^t (X^t \hat{V}^{-1} X)^{-1} (\mathbf{x} - X^t \hat{V}^{-1} ZD\mathbf{z}) , \quad (9)$$

where $\hat{V}$ is the matrix V evaluated at $\hat{\mu}$. The motivation for the second approximation in 9 is that it is exact in the normal theory case with identity link, where $w_i = 1/(\phi a_i)$ does not depend on μ_i and hence $\hat{\eta}(\gamma)$ is linear in the components of $\mathbf{y}$.

Battese et al. (1988) obtain a better analytical approximation to 7 in the normal theory, identity link case. Details of their approximation are given in Fuller and Harter (1987). The approach is closely related to that of Kackar and Harville (1984) in which $\hat{\eta} = \hat{\eta}(\hat{\gamma})$ is expanded in a first-order Taylor series about γ. The resulting linear approximation to $\hat{\eta} - \eta$ is then used to obtain an expansion for ν^2. We note that both approximations to ν^2 are functions of the unknown parameter vector γ (and ϕ, if it is unknown). The variance estimates actually used in practice are the same functions evaluated at $\hat{\gamma}$ (and $\hat{\phi}$); that is, the corresponding approximations to the' bootstrap estimate $\hat{\nu}^2$ in 8.

An alternative to analytical approximation of $\hat{\nu}^2$ in 8 is Monte Carlo approximation based on "resampling". This approach automatically takes account of the sampling variability in the estimated variance component vector $\hat{\gamma}$ and avoids the need for complicated expansions. Let $(\mathbf{y}_1^*, \mathbf{b}_1^*), \ldots,$ $(\mathbf{y}_R^*, \mathbf{b}_R^*)$ denote i.i.d. vectors generated from the fitted model $F_{\hat{\lambda}}$. That is, for each $r = 1, \ldots, R$, $\mathbf{b}_r^* \sim N(0, D(\hat{\gamma}))$ and $\mathbf{y}_r^*$ is a "resample" consisting of n independent data values generated from the model 3 with ϕ, α and $\mathbf{b}$ replaced by $\hat{\phi}$, $\hat{\alpha}$ and $\hat{\mathbf{b}}_r^*$ respectively. Then a Monte Carlo approximation to $\hat{\nu}^2$ is given by

$$\hat{\nu}^2 \approx \frac{1}{(R-1)} \sum_{r=1}^{R} (\hat{\eta}_r^* - \eta_r^*)^2 , \quad (10)$$

where $\eta_r^* = \mathbf{x}^t \hat{\alpha} + \mathbf{z}^t \mathbf{b}_r^*$ and $\hat{\eta}_r^*$ is the version of $\hat{\eta}$ computed using the rth resample $\mathbf{y}_r^*$ rather than the actual sample $\mathbf{y}$. It can easily be verified that the right side of 10 converges to the left side with probability one as $R \to \infty$ conditional on the sample.

4   Application to small area estimation

In this section we illustrate the use of the bootstrap in two small area estimation problems. For a recent review of small area estimation methodology, see Ghosh and Rao (1994).

Suppose that the response values come from k different small areas or groups. Let y_{ij}, $j = 1, \ldots, n_i$, denote the jth response in the ith small area and let $\mathbf{x}_{ij}$ $(p \times 1)$ be a vector of known covariates associated with the (i, j)th response. In practice, the value of $n = \sum_{i=1}^{k} n_i$ is often chosen so that characteristics of the overall population can be estimated accurately. If

k is large, the number of responses per small area may be very small, even zero in some cases if the sampling scheme allows it. Thus, direct estimates of individual small area characteristics are likely to be extremely unreliable. One way of dealing with this problem is to "borrow strength" from other small areas using a GLMM with linear predictor

$$\eta_{ij} = \mathbf{x}_{ij}^t \alpha + b_i \tag{11}$$

for the (i,j)th response, where the b_i is a random effect associated with the small area i. It is usually assumed that the b_i's are i.i.d. normal with mean zero and variance σ_b^2. That is, $q = k$ and $D = \sigma_b^2 I$ in the notation of Section 2, where I is the identity matrix.

EXAMPLE 1: Battese, Harter and Fuller (1988) use a GLMM with linear predictor of the form given in 11 and normal errors to estimate the area used for growing corn in 12 Iowa counties. Their response values are precise areas based on surveying a small fraction of "segments" in each county. The sample sizes, n_i, range from 1 to 5 with the total sample size being $n = 36$. In addition to the survey measurements, information on crop areas for the entire region is available from satellite images. The satellite estimates of corn and soybean crop areas are used as covariates in the GLMM. Thus, the complete model considered by Battese et al. is

$$y_{ij} = \mathbf{x}_{ij}^t \alpha + b_i + \epsilon_{ij}, \tag{12}$$

where the ϵ_{ij}'s are i.i.d. normal errors with variance σ_ϵ^2.
According to the model, the area used for growing corn in the ith county (measured in hectares per segment) is given by

$$\bar{y}_{i(p)} = \bar{\mathbf{x}}_{i(p)}^t \alpha + b_i + \bar{\epsilon}_{i(p)}, \tag{13}$$

where (p) indicates that the average is taken over all segments in the county. Since the number of segments per county is several hundred or more, the average error term $\bar{\epsilon}_{i(p)}$ is negligible and hence corn area estimates are given by

$$\hat{\bar{y}}_{i(p)} = \bar{\mathbf{x}}_{i(p)}^t \hat{\alpha} + \hat{b}_i, \tag{14}$$

where $\hat{\alpha}$ and $\hat{b}_i$ denote estimates of fixed and random effects respectively. BHF propose a two-stage procedure for fitting their model in which the variance components, σ_b^2 and σ_ϵ^2, are first estimated by a method of moments, and then "empirical" GLS and BLUP estimates are obtained for α and $\mathbf{b}$ respectively using the estimated variance components in equations 5 and 6.
The "naive" and BHF analytical approximations together with Monte Carlo approximations to the bootstrap standard errors for corn area estimates, based on $R = 2000$ independent resamples, are displayed in Table 1.

County	Sampled Segments	Predicted Hectares	Naive S.E.	BHF S.E.	Bootstrap S.E.
Cerro Gordo	1	122.2	8.8	9.6	9.7
Hamilton	1	126.3	8.7	9.5	9.7
Worth	1	106.2	8.7	9.3	9.4
Humboldt	2	108.0	7.3	8.1	8.1
Franklin	3	145.0	6.1	6.5	6.4
Pocahontas	3	112.6	6.2	6.6	6.5
Winnebago	3	112.4	6.1	6.6	6.6
Wright	3	122.1	6.2	6.7	6.6
Webster	4	115.8	5.5	5.8	5.8
Hancock	5	124.3	5.1	5.3	5.4
Kossuth	5	106.3	5.0	5.2	5.2
Hardin	5	143.6	5.2	5.7	5.6

TABLE 1. Comparison of Analytical and Bootstrap Standard Errors for the Crop Area Estimates in 12 Iowa Counties

EXAMPLE 2: The first two columns in Table 2 concern teen pregnancy rates in $k = 13$ north central Florida counties during the 3 year period 1989-91 (Gainesville Sun, April 30, 1994). Notice that the total number of births in some of the counties is small and hence the corresponding direct rate estimates are extremely unreliable.

Let n_i denote the total number of births in county i during the reporting period and let y_i denote the number involving "underage" mothers (i.e. women 17 or under). We suppose that, conditional on n_i and a county specific effect b_i, $y_i \sim \text{binomial}(n_i, \pi_i)$, where

$$\eta_i = \text{logit}(\pi_i) = \alpha + b_i, \qquad (15)$$

for $i = 1, \ldots, k$.

If the county specific effects, b_i, $i = 1, \ldots, k$, are fixed and all equal to zero, then model 15 implies a common "underage" pregnancy rate for all 13 counties. However, with a deviance of 89.9 on 12 degrees of freedom, this model is not supported by the data. On the other hand, if the b_i's are arbitrary fixed effects, then the model is saturated and the estimates of the county specific rates are the same as those given by the newspaper. As a compromise we propose modeling the county specific effects as i.i.d. normal random variables with mean zero.

Estimates of the underlying "underage" pregnancy rates per 1000 births for the 13 counties based on the logistic-normal model were obtained using the EM algorithm (Anderson and Hinde, 1988). These are displayed in Table 2 alongside the direct estimates. Notice that the estimates obtained using the random effects model tend to shrink the direct estimates towards the

County	Total No. per Year	Direct Rate	Direct 95% C.I.	Model Rate	Naive 95% C.I.	Bootstrap 95% C.I.
Alachua	2848	32.2	28.6,36.1	32.4	28.9,36.3	29.0,36.1
Bradford	344	48.1	36.9,63.4	46.5	36.1,59.8	35.3,61.0
Clay	1617	22.7	18.8,27.3	23.6	19.9,28.1	20.4,27.4
Columbia	688	50.4	41.7,60.7	49.1	41.0,58.9	39.5,61.0
Dixie	160	43.8	28.7,66.2	41.8	29.2,59.5	28.4,61.1
Gilchrist	133	20.0	10.1,39.6	27.8	18.0,42.7	18.1,42.5
Hamilton	171	79.8	59.4,107.	68.4	51.4,90.6	46.4,99.9
Lafayette	66	35.5	16.9,72.3	36.9	22.9,59.2	22.6,59.9
Levy	350	28.6	20.0,40.6	30.6	22.6,41.3	22.6,41.3
Marion	2753	29.4	26.0,33.3	29.7	26.3,33.5	26.6,33.2
Putman	982	43.8	37.0,51.8	43.4	36.8,51.0	36.2,51.8
Suwannee	351	36.1	26.4,49.2	36.5	27.6,48.2	27.1,49.0
Union	135	54.3	36.0,81.1	48.4	33.8,68.9	32.6,71.4

TABLE 2. Teen Pregnancy in North Central Florida. (Estimated rates of babies born to mothers 17 and under per 1000 births, based on data from 1989-91.)

overall rate, 33.9, for the 13 counties. The amount of shrinkage is greatest in the rural counties where the total number of births is small. Notice, in particular, the results for Hamilton county which had by far the highest direct rate.

Also displayed are 95% confidence intervals based on both the direct and GLMM approaches. In the latter case, intervals are constructed using both the naive estimates of standard errors for the linear predictors as given in 9 and approximate bootstrap standard errors based on $R = 1000$ resamples from the fitted model ($\hat{\alpha} = -3.23$, $\hat{\sigma}_b = 0.324$). Again, the results for Hamilton county are the most striking, with the naive interval being much shorter than its bootstrap counterpart. To the best of our knowledge, no tractable analytical correction to the naive standard errors is available in this problem.

Acknowledgments: The author would like to thank Brett Presnell for his suggestions and for reading the manuscript prior to submission.

References

Anderson, D.A. and Hinde, J.P. (1988). Random effects in generalized linear models and the EM algorithm. *Communications in Statistics: Theory and Methods*, **17**(11), 3847-3856.

Battese, G.E., Harter, R.M. and Fuller, W.A. (1988). An error-components model for prediction of county crop areas using survey and satellite data. *Journal of the American Statistical Association*, **83**, 28-36.

Breslow, N.E. and Clayton, D.G. (1993). Approximate inference in generalized linear mixed models. *Journal of the American Statistical Association*, **88**, 9-25.

Carlin, B.P. and Gelfand, A.E. (1991). A sample reuse method for accurate parametric empirical Bayes confidence intervals. *Journal of the Royal Statistical Society*, **B 53**, 189-200.

Henderson, C.R. (1950). Estimation of genetic parameters (abstract). *Annals of Mathematical Statistics*, **21**, 309-310.

Efron, B. (1982). *The jackknife, the bootstrap and other resampling plans.* Volume 38, CBMS-NSF Regional Conference Series in Applied Mathematics. SIAM.

Efron, B. (1987). Discussion of "Empirical Bayes confidence intervals based on bootstrap samples", *Journal of the American Statistical Association*, **82**, 754.

Fuller, W.A. and Harter, R.M. (1987). The multivariate components of variance model for small area estimation. *Small Area Statistics: An International Symposium*. Eds. R. Platek, J.N.K. Rao, C.E. Sarndal, and M.P. Singh. New York: John Wiley, 103-123.

Ghosh, M. and Rao, J.N.K. (1994). Small area estimation: an appraisal (with discussion). *Statistical Science* **9**, 55-93.

Kackar, R.N. and Harville, D.A. (1984). Approximations for standard errors of estimators of fixed and random effects in mixed linear models. *Journal of the American Statistical Association*, **79**, 853-862.

McCullagh, P. and Nelder, J.A. (1989). *Generalized Linear Models*, 2nd Edition. Chapman and Hall.

McGilchrist, C.A. (1994). Estimation in generalized mixed models. *Journal of the Royal Statistical Society* **B 56**, 61-70.

Robinson, G.K. (1993). That BLUP is a good thing: The estimation of random effects (with discussion). *Statistical Science* **6**, 15-51.

Zeger, S.L. and Karim, M.R. (1991). Generalized linear models with random effects; a Gibb's sampling approach. *Journal of the American Statistical Association* **86**, 79-86.

Confidence Intervals for Threshold Parameters

R.C.H. Cheng and W.B. Liu

ABSTRACT: The calculation of a confidence interval for an unknown threshold is sometimes a non-standard problem. We use a modification of a score statistic initially suggested by Smith for hypothesis testing. The limiting distribution follows a stable law. We use the Mellin transform to obtain explicit formulas for the parameters defining this distribution. We show that if the distribution is to remain non-degenerate over the range of possible parameters values then a special standardization has to be used. We also so show how the score statistic can be modified so that it converges to just a single distribution. This simplifies the tabulation required for practical confidence interval calculations.

KEYWORDS: Stable-Law Distributions; Mellin Transform

1 Introduction

Shifted location models are generalizations of simpler models obtained by adding an offset threshold or shifted location which translates the distribution to an arbitrary origin. Such models are useful when there is reason to believe that an observed response cannot occur below some minimum, unknown threshold. Such models have been considered in reliability studies as an extension of the Weibull 'weakest link' model; in operational research problems where service times cannot fall below some minimum; and in physical response problems where some minimum response time is required. It is well known that the problem of estimating such a shifted location can be non-regular in the sense of not satisfying standard asymptotic normality properties, and the difficulties encountered in using maximum likelihood methods have been discussed by many authors; see Cheng and Traylor (1995), for a recent review. Various methods have been proposed for handling the difficulty and satisfactory methods of estimation do now exist. A problem which has not received so much attention is the construction of confidence intervals for such estimators, and this is the subject of the present paper.

In Section 2 we formulate the problem. In Section 3 we discuss the calculation of confidence intervals. The limiting distributions of the statistics used are derived in Sections 4 and 5. Section 6 gives a numerical example.

2   Threshold Estimators

We consider the estimation of a threshold parameter in continuous distributions. We denote the probability density function (pdf) by $f(x, \theta)$ where $\theta = (\mu, \phi)$ with μ the unknown threshold. We shall assume $\phi = (\alpha, \sigma)$ so that three parameters are unknown (though this is not a rigid restriction) and that the density has the form

$$f(x, \theta) = \begin{array}{ll} (x - \mu)^{\alpha-1} g[(x - \mu)/\sigma, \alpha] & if \ \ x > \mu, \\ 0 & if \ \ x < \mu, \end{array} \tag{1}$$

where g tends to a finite quantity $c(\phi) > 0$ independent of μ, as $x \to \mu$. When $\alpha > 1$, $f(x, \theta) \to 0$ as $x \to \alpha$, but $f(x, \theta)$ is J-shaped when $\alpha < 1$. The unknown true parameters will be denoted by $\theta_0 = (\mu_0, \alpha_0, \sigma_0)$. The formulation covers the three-parameter gamma pdf:

$$f_G(x, \theta) = \sigma^{-\alpha} \frac{(x - \mu)^{\alpha-1}}{\Gamma(\alpha)} \exp\left[-\frac{(x - \mu)}{\sigma}\right], \qquad x > \mu, \tag{2}$$

and the three-parameter Weibull pdf:

$$f_W(x, \theta) = \alpha \frac{(x - \mu)^{\alpha-1}}{\sigma^{\alpha}} \exp\left[-\left(\frac{x - \mu}{\sigma}\right)^{\alpha}\right], \qquad x > \mu. \tag{3}$$

The behaviour of the estimators is determined by the degree of contact of $f(x, \theta)$ with the x-axis as $x \to \mu$. The basic facts are:
(i) When $\alpha_0 > 2$ the problem is regular with the maximum likelihood (ML) estimate $\hat{\theta}$ satisfying the usual asymptotic normality properties.
(ii) When $\alpha_0 = 2$ then the ML estimate has the property:

$$\hat{\mu} - \mu_0 = O_p\left(\frac{1}{(n \log n)^{\frac{1}{2}}}\right)$$

while $\hat{\phi}$ has standard asymptotic normality properties with $\mu = \hat{\mu}$ treated as being known as far as the distributional properties of $\hat{\phi}$ are concerned.
(iii) When $\alpha_0 < 2$ then there exist estimates (the ML estimate only exists when $\alpha_0 > 1$) with the property:

$$\hat{\mu} - \mu_0 = O_p\left(\frac{1}{n^{\frac{1}{\alpha_0}}}\right),$$

while $\hat{\phi}$ has standard asymptotic normality properties with again $\mu = \hat{\mu}$ treated as being known.

Therefore, when $\alpha_0 > 2$, confidence intervals can be constructed, at least asymptotically, for all parameters, using standard normal theory. When $\alpha_0 < 2$ then ϕ can be treated by standard means. The problem thus reduces to finding a confidence interval for μ_0 in the case $\alpha_0 < 2$.

3 Confidence Intervals

Smith (1985) suggested the using the score statistic

$$\frac{\partial L}{\partial \mu} = n^{-1} \sum_{i=1}^{n} \frac{\partial}{\partial \mu} \log f(x_i, \theta) \tag{4}$$

to test hypotheses about μ_0. In the regular case $\alpha_0 > 2$, use of (4) is equivalent to use of the Wald statistic or the likelihood ratio. In the non-regular case it is not clear how efficient the score statistic is. However it does afford a tractable approach.

Smith (1985) showed that if $1 < \alpha_0 < 2$ or if $\alpha_0 < 1$ then

(i) $\partial L/\partial \mu \mid_{\theta_0} = n^{-1}(\alpha_0 - 1)T(\alpha_0) + R$, where $T(\mu_0) = \sum(x_i - \mu_0)^{-1}$, and where, asymptotically, R can be neglected,

(ii) $\partial L/\partial \mu \mid_{\theta_0}$ and $T(\mu_0)$, when appropriately standardized, have the same limiting non-degenerate stable-law distribution.

(iii) This limit distribution is unchanged if ϕ_0 is replaced by an $\sqrt{n}$-consistent estimator.

$T(\mu_0)$ and its generalization which we take in the standardized form:

$$W_n(\theta, \lambda) = a_n^{-1} \sum Y_i^{-\lambda} - nb_n \tag{5}$$

where

$$Y_i = (X_i - \mu)/\sigma, \tag{6}$$

and $\lambda > 0$ is a quantity at our disposal, can be used to construct a confidence interval for μ_0. (The notation indicates that W_n depends on θ, as a_n and b_n will in general depend on θ.) For a given confidence level, p say, we take appropriate percentage points, w_- and w_+, say, so that

$$\Pr(w_- < W_n < w_+) = 1 - p. \tag{7}$$

Then we replace ϕ by a suitable estimator $\tilde{\phi}$ and solve $W_n = w_-$ and $W_n = w_+$, ie:

$$\sum [(x_i - \mu_\pm)/\sigma]^{-\lambda} = a_n(w_\pm + nb_n)$$

to give limits μ_- and μ_+ for a $p100\%$ confidence limit for μ_0:

$$\mu_- < \mu_0 < \mu_+. \tag{8}$$

Smith suggested estimating μ_0 using $\hat{\mu} = X_{(1)}$, the smallest observation, and estimating the other parameters from the remaining observations with

μ set equal to $\hat{\mu} = X_{(1)}$. An awkwardness of this approach is that we then need to make the probability statement (7) one-sided with the confidence interval taking the form

$$\mu_- < \mu_0 < X_{(1)} \qquad (9)$$

where the point estimate is at one end of the interval. A possibly more satisfactory method is to use a spacings based estimator (Cheng and Amin, 1983) where the estimator is bounded away from $X_{(1)}$.

For the method to be usable we need to find coefficients a_n and b_n for which W_n has a tractable as well as non-degenerate limiting stable distribution; and we need to identify the distribution precisely. The parameter λ is at our disposal and we consider the cases $\lambda = 1$ and $\lambda = \alpha$. The latter case is interesting because the limiting distribution becomes independent of α. This allows confidence intervals to be calculated based on the tabulation of just one distribution.

Another possibility is the limit $\lambda \to \infty$ which in essence picks out just the smallest observation

$$W_n(\theta, \infty) = a_n[(X_{(1)} - \mu)/\sigma]^{-1} - nb_n.$$

In certain situations where $\alpha = 1$, $X_{(1)}$ is sufficient for μ. Thus a confidence interval based on (the asymptotically Weibull distributed) $X_{(1)}$ alone may be reasonably efficient as well as being considerably easier to calculate than one based on $W_n(\theta, \lambda)$. We do not discuss this case further here.

Smith gives some results for the case $\lambda = 1$ but does not identify the coefficients nor the distribution explicitly. One method is to follow Feller's seminal approach using the theory of slowly varying functions A succinct summary of this method is given by Chow and Teugels (1978). However the method is quite difficult to apply in the case $\lambda = \alpha$. We give an alternative method based on the Mellin transformwhich covers both cases.

4   Stable Limits of Sums of Reciprocals

4.1   Case $\lambda = 1$ ($\alpha < 2$)

The two key parameters in (1) are the threshold μ and the power parameter α. It will be convenient to regard the other parameters as being known. In all the cases that we consider the distributional results are not altered by substituting $\sqrt{n}$-consistent estimates for these unknown values. W_n, (5), becomes

$$W_n(\theta, 1) = a_n^{-1} \sum Y_i - nb_n.$$

We obtain its distribution from its moment generating function. This is

$$E(e^{-sW_n}) = \{ \int_{-\infty}^{\infty} e^{-s(\frac{y}{a_n} - b_n)} f_Y(y) dy \}^n. \qquad (10)$$

The integral can be written as

$$\int_{-\infty}^{\infty} e^{-s(\frac{y}{a_n}-b_n)} f_Y(y)dy = e^{sb_n} L(s),\qquad(11)$$

where

$$L(s) = \int_{-\infty}^{\infty} e^{-\frac{sy}{a_n}} f_Y(y)dy.$$

We now take the Mellin transform of $L(s)$:

$$M_L(p) = \int_0^{\infty} s^{p-1} L(s)ds = a_n^p \Gamma(p)\nu_Y(p),\qquad(12)$$

say, where $\nu_Y(p) = \int y^{-p} f_Y(y)dy$. This allows a series expansion of $L(s)$ to be obtained using the Mellin inversion formula

$$L(s) = \frac{1}{2\pi i}\int_{c-i\infty}^{c+i\infty} (\frac{s}{a_n})^{-p}\Gamma(p)\nu_Y(p)dp \qquad c > 0.\qquad(13)$$

The asymptotic form of $\Gamma(p)\nu_Y(p)$ for $|p| \to \infty$ allows the integral to be closed to the left with a Bromwich contour. The terms in the expansion of $L(s)$ thus come from the residues of the integrand and do not correspond simply to integer powers.

Example 1. Weibull distribution. Here $\nu_Y(p) = \Gamma(1 + \frac{p}{\alpha})$. The integrand in (13) has poles at $p = -j$ and $p = -\alpha(j+1)$, $j = 0, 1, 2, \ldots$ For the case $\alpha < 1$ the two leading residues corresponding to the poles at $p = 0$ and $p = -\alpha$ are $R_0 = 1$ and $R_1 = -(s/a_n)^{\alpha}\Gamma(1 - \alpha)$, and we have

$$L(s) = 1 - \Gamma(1-\alpha)\left(\frac{s}{a_n}\right)^{\alpha} + O(s).$$

Substituting of this into (11) and (10) we find that if $a_n = [n\Gamma(1-\alpha)]^{1/\alpha}$, $b_n = 0$, then the generating function has the limit

$$\lim_{n\to\infty} E(e^{-sW_n}) = e^{-s^{\alpha}}\qquad(14)$$

For the case $\alpha > 1$, a similar calculation shows that if $a_n = [n\Gamma(2-\alpha)/(\alpha-1)]^{1/\alpha}$, $b_n = a_n^{-1}\Gamma(1 - \frac{1}{\alpha})$ then

$$\lim_{n\to\infty} E(e^{-sW_n}) = e^{s^{\alpha}}.\qquad(15)$$

Example 2. Gamma distribution. If $\alpha < 1$, $a_n = [n\Gamma(1-\alpha)/\Gamma(1+\alpha)]^{1/\alpha}$ and $b_n = 0$, we obtain again the limit (14). If $\alpha > 1$, $a_n = \{n\Gamma(2-\alpha)/[(\alpha-1)\Gamma(1+\alpha)]\}^{1/\alpha}$ and $b_n = a_n^{-1}\Gamma(1-\alpha)/\Gamma(\alpha)$ we obtain the limit (15).

For numerical calculations we need percentage points of these stable distributions, as defined in (7). The best method of calculating percentage points is to use the integral representations of the cdf found by Zolotarev

(1964) and presented by Lukacs (Theorem 5.9.3, 1970). Zolotarev writes, $\varphi(t)$ the characteristic function of stable law distributions as

$$\log \varphi(t) = \begin{cases} itd - c\,|t|^{\alpha}\exp\{-i\frac{\pi}{2}\beta K(\alpha)sgn(t)\} & if\ \alpha \neq 1 \\ itd - c\,|t|\,\{\frac{\pi}{2} + i\beta sgn(t)\log|t|\} & if\ \alpha = 1 \end{cases} \tag{16}$$

where $K(\alpha) = 1 - |1-\alpha|$, $0 < \alpha \leq 2$, $-1 \leq \beta \leq 1$, $c > 0$. Expression (14) is the special case where $\alpha < 1$, obtained by setting $d = 0$, $c = 1$, $\beta = 1$, with $s = -it$. Similarly (15) is the special case where $\alpha > 1$, obtained by setting $d = 0$, $c = 1$, $\beta = -1$, with $s = -it$.

4.2   Case $\lambda = \alpha$ (with $\alpha < 2$), or $\alpha= 1$

These two cases give rise to the same analysis, which is complicated by the fact that the leading poles of the integrand (13) are at $p = 0$ and $p = -1$, with the second being a double pole. We illustrate with the examples of the Weibull and gamma distributions. We find after some calculation that

$$\begin{aligned} L(s) &= \tfrac{1}{2\pi i} \int_{c-i\infty}^{c+i\infty} \Gamma^{-1}(A)(\tfrac{s}{a_n})^{-p}\Gamma(p)\Gamma[A(1+p)]dp \\ &= 1 + \tfrac{s}{\Gamma(1+A)a_n}[\log(s/a_n) + AE + E - 1] + \dots \end{aligned}$$

where E is Euler's constant, $A = 1$ in the Weibull case, and $A = \alpha$ in the gamma case. If we set $a_n = n$, $b_n = [\log(n\pi/2) - \log\Gamma(1+A) - (AE + E - 1)]/n$ we get

$$\lim_{n\to\infty} E(e^{-sW_n}) = s^s. \tag{17}$$

This is the special case of (16) where $\alpha = 1$, $d = 0$, $c = 1$, $\beta = 1$, with $s = -it$.

An interesting aspect of the case $\lambda = \alpha$ is that the limiting distribution of W_n is unchanged if α is replaced by a $\sqrt{n}$-consistent estimate. Thus only tabulation of this specific distribution, whose generating function is (17), is required.

5   Reparameterization

As $\alpha \to 1$ the generating functions (14) and (15) tend to those corresponding to degenerate delta distributions located at $y = 1$ and at $y = -1$ respectively. Thus there is a discontinuity in the location of the limit distribution as α passes through unity, as well as a collapse into degeneracy. The phenomenon is not an intrinsic problem but is a curious quirk of the characterization (16) as far as our problem is concerned. It is cured by modifying the parameterization. Let

$$Z_n = \begin{cases} \frac{W_n}{|1-\alpha|} - \frac{sgn(1-\alpha)}{|1-\alpha|^{\alpha}}, & if\,\alpha \neq 1 \\ W_n, & if\,\alpha = 1 \end{cases} \tag{18}$$

α Pvalue	0.005	0.025	0.05	0.95	0.975	0.995
0.5	-1.287	-1.215	-1.154	253.2	1019.	25790.
1.0	-2.291	-1.800	-1.498	22.46	43.25	206.1
1.5	-2.883	-1.972	-1.476	8.896	12.769	32.111

TABLE 1. Selected Percentage Points of Z

where W_n is as defined in (5). Then if a_n and b_n are defined to give (14), (15) and (17) as the limiting generating functions for W_n, we get

$$\varphi_{Z,\alpha}(s) = \lim_{n \to \infty} E(e^{-sZ_n}) = \left\{ \begin{array}{ll} \exp[(1-\alpha)^{-\alpha}(s-s^\alpha)] & if \ \ \alpha < 1, \\ \exp[(\alpha-1)^{-\alpha}(s^\alpha-s)] & if \ \ \alpha > 1. \end{array} \right. \quad (19)$$

We then have

$$\lim_{\alpha \to 1} \varphi_{Z,\alpha}(s) = s^s. \quad (20)$$

The expression (19) is the special case of (16) where $d = |1-\alpha|^{-\alpha} \, sgn(1-\alpha)$, $c = |1-\alpha|^{-\alpha}$, and $\beta = sgn(1-\alpha)$, with $s = -it$. The percentage points of Z can then be tabulated to include $\alpha = 1$ without degeneracy or discontinuity.

6 Example

Table 1 gives the percentage points of Z as defined in (19) and(20) for selected α. The table is for illustration only; lack of space prevents a full tabulation.

As an illustration of an application we fitted the Weibull distribution of (3) to a sample of 47 observations of the time (in seconds) it took for light vans to pay at vehicle toll booths at the Severn Bridge River Crossing between England and Wales (Table 2). The maximum product of spacings (MPS) estimates, (see Cheng and Amin, 1983) of the parameters were $\tilde{\alpha} = 1.35$, $\tilde{\mu} = 2.91$, and $\tilde{\sigma} = 3.23$. The estimated value of α, indicates the problem is non-regular. We can thus use (8) to calculate a confidence interval for μ. Using $\lambda = 1$ this gave a 95% confidence interval for μ as (2.05, 3.02). Using $\lambda = \tilde{\alpha}$ gave a 95% confidence interval as (2.66, 3.08).

Simulation results not reported here indicate that the computed interval for the threshold tends to be too small when the shape and scale are estimated, but that a simple adjustment to take into account the variability in the shape and scale estimates can be used to compensate for this. We hope to report on these refinements elsewhere.

4.3	10.9	4.7	4.7	3.1	5.2	6.7	4.5	3.6	7.2	6.6	5.8
6.3	4.7	8.2	6.2	4.2	4.1	3.3	4.6	6.3	4.0	3.1	3.5
7.8	5.0	5.7	5.8	6.4	5.2	8.0	10.5	4.9	6.1	8.0	7.7
4.3	12.5	7.9	3.9	4.0	4.4	6.7	3.8	6.4	7.2	4.8	–

TABLE 2. Toll Payment Times (Seconds) of Light Vans at the Severn Bridge

References

Cheng, R.C.H and Amin, N.A.K. (1983). Estimating parameters in continuous univariate distributions with a shifted origin. *Journal of the Royal Statistical Society, B*, **45**, 394-403.

Cheng, R.C.H and Traylor, L. (1995). Non-regular maximum likelihood estimation problems. *Journal of the Royal Statistical Society, B*, **57**, 3-44.

Chow, T.L. and Teugels, J.L. (1978). The sum and maximum of i.i.d. random variables. In *Proceedings of 2nd Prague Symposium on Asymptotic Statistics.* (Eds Mandl, P. and Huskova, M.) 81-92, North Holland.

Lukacs, E. (1970). *Characteristic Functions.* Griffin, London.

Smith, R.L. (1985).Maximum-likelihood estimation in a class of non-regular cases. *Biometrika*, **72**, 67-92.

Zolotarev, V.M. (1964).On the representation of stable laws by integrals. *Trudi Mat. Inst. Steklova*, **71**, 46-50 English transl. in *Selected Translations Math. Stat. and Prob.*, **6**, 84-88. Amer. Math. Soc., Providence, R.I.

Optimal Design for Models Incorporating the Richards Function

G. P. Y. Clarke
L. M. Haines

ABSTRACT: Optimal designs for models incorporating the Richards function, which are based on criteria involving a second-order approximation to the mean square error of estimates of the parameters, are considered. Interest focuses on the shape parameter and the asymptote of the model, and it is shown that designs which, separately, minimize the mean square error for estimates of these parameters exhibit severe nonlinearity, but that designs which minimize compromise criteria, or designs formed as a combination of optimal designs for the individual parameters, have attractive properties.

KEYWORDS: Richards Function; Multiplicative Models; Optimal Design; Mean Square Error; Curvature Measures

1   Introduction

The Richards function is an asymmetric, sigmoidal curve used extensively as the deterministic component in nonlinear regression models for modelling growth in organisms and plants (Richards, 1959). Such models have been criticized for exhibiting severe nonlinearity, and indeed Ratkowsky (1990, p. 141) states:-

> The Richards model exhibits more undesirable nonlinear regression behaviour than almost any nonlinear regression model in common use. The continued use of this model is not recommended.

The broad aim of the present study is to explore the extent to which the undesirable nonlinear properties of models based on the Richards function can be alleviated by an appropriate, optimal choice of design points. More particularly, a specific example is introduced, and is used to illustrate the principles involved in the construction of such optimal designs.

2 Example

A set of growth data given in Ratkowsky (1983, p.88), and comprising measurements of the area of a cucumber cotyledon, y_i, taken at time, x_i, with $i = 1, \ldots, 9$, was selected for study, and modelled using a multiplicative nonlinear regression model based on the parametric form of the Richards function given by Nelder (1961), and formulated as

$$\ln y_i = \theta_4 - \theta_3 \ln \left\{ 1 + \exp \left[- \left(\frac{\theta_2 + \theta_1 x_i}{\theta_3} \right) \right] \right\} + \varepsilon_i, \quad i = 1, \ldots, 9, \quad (1)$$

where θ_3 is a shape parameter, θ_4 represents the $\ln y$ asymptote, and the terms, $\varepsilon_i, i = 1, \ldots, n$, represent independent errors with mean zero and constant variance, σ^2. The parameters, θ_1 and θ_2, relate to the coding of time, and are deemed to be unimportant. The least squares estimates of the parameters of model (1) for the chosen data set with x values, $0, 1, 2, 3, 4, 5, 6, 8, 10$, are given by $\hat{\theta} = [0.4664, -1.7435, 0.3608, 1.8713]$ and $\hat{\sigma}^2 = 0.009199$.

The root mean square intrinsic and parameter-effects curvatures for this model and design combination are given by $\gamma_{RMS}^N = 0.325$ and $\gamma_{RMS}^T = 1.644$ respectively, and indicate moderately severe nonlinearity. Curvature measures relating to the individual parameters, θ_3 and θ_4, include percentage biases of 21.35% and 0.37% respectively, and percentage contributions of second-order terms to the second-order approximation to the asymptotic variances of 57.63% and 12.04% respectively, and are most unsatisfactory, particularly for the shape parameter, θ_3.

3 Optimal Designs

Designs for model (1), which provide precise estimates of the two parameters of interest, θ_3 and θ_4, and at the same time exhibit mild nonlinearity, are now considered. The true parameter values are taken to be the least squares estimates, $\hat{\theta}$ and $\hat{\sigma}^2$, of the above example, exact designs comprising 9 points are considered, and the maximum separation of the x values, Δx, is retained as 10. A natural approach to precise parameter estimation is to minimize mean square error. This latter criterion cannot be calculated explicitly for nonlinear regression models however, and furthermore its first-order approximation, written V_I, does not accommodate nonlinearity. The criterion adopted here, therefore, is the second-order approximation to the mean square error developed by Clarke (1980), and expressed succinctly as

$$MSE = V_I + \sigma^2 V_{add} + \beta\beta', \quad (2)$$

where V_{add} is a matrix of second-order terms in the asymptotic variance of the parameter estimates, and β represents a vector of second-order approximations to the biases of these estimates. The latter two terms in (2) incorporate measures of nonlinearity indirectly through the second- and third-order derivatives of the expected responses with respect to the parameters.

The parameters of interest are θ_3 and θ_4, and some relevant properties of the designs which minimize the corresponding entries in the matrix, MSE, denoted $MSE(\theta_3)$ and $MSE(\theta_4)$ respectively, are presented in Table 1. Full details of these designs are given in Table A1 of the Appendix.

Table 1 : *Properties of designs minimizing $MSE(\theta_3)$ and $MSE(\theta_4)$*

Criterion	%Bias(θ_3)	%Bias(θ_4)	$MSE(\theta_3)$	$MSE(\theta_4)$
Example	21.35	0.37	0.2072	0.00433
Minimize $MSE(\theta_3)$	2.37	0.35	0.0314	0.00626
Minimize $MSE(\theta_4)$	142.62	0.03	3.923	0.00160

Clearly the design which minimizes $MSE(\theta_3)$ does not provide a precise estimate of θ_4, and, more particularly, the design minimizing $MSE(\theta_4)$ exhibits *disastrously* nonlinear properties for θ_3. It is therefore sensible to seek criteria which provide designs that perform well for the parameters, θ_3 and θ_4, simultaneously. A number of such compromise criteria, based primarily on the matrix, MSE, have been examined, and these are summarized in Table 2, together with measures of efficiency of parameter estimation for the associated optimal designs defined by

$$Eff(\theta_i) = \frac{MSE^\star(\theta_i)}{MSE(\theta_i)} \times 100,$$

where $MSE^\star(\theta_i)$ denotes the minimum attainable value of $MSE(\theta_i)$, and $i = 3$ and 4. The actual designs are presented in Table A1 of the Appendix. It is immediately apparent from Table 2 that the design minimizing the sum, $MSE(\theta_3) + MSE(\theta_4)$, favours the estimation of θ_3. This can be explained by the fact that, in general, the mean square error for θ_3 is much greater than that for θ_4, i.e. $MSE(\theta_3) >> MSE(\theta_4)$. It is further clear that this imbalance is, to a greater extent, redressed by minimizing the weighted sum, $\frac{MSE(\theta_3)}{MSE^\star(\theta_3)} + \frac{MSE(\theta_4)}{MSE^\star(\theta_4)}$. In fact the design corresponding to this latter criterion exhibits satisfactory nonlinear properties overall, and in particular, $\gamma_{RMS}^N = 0$, since the design has only four support points, $\gamma_{RMS}^T = 0.408$, and for the parameter estimates of θ_3 and θ_4, the percentage biases are 3.34% and 0.20% respectively, and the percentage contributions of second-order terms to the second-order approximation to the variances are 28.46% and 8.98% respectively. These measures compare favourably with those of the original design of the example.

Table 2 : % Efficiencies of selected 9-point optimal designs

Criterion	$Eff(\theta_3)$	$Eff(\theta_4)$
Example	15.17	37.03
Minimize $MSE(\theta_3)$	100	25.64
Minimize $MSE(\theta_4)$	0.80	100
Minimize $[\ MSE(\theta_3) + MSE(\theta_4)\ ]$	94.36	39.00
Minimize $[\ \frac{MSE(\theta_3)}{MSE^\star(\theta_3)} + \frac{MSE(\theta_4)}{MSE^\star(\theta_4)}\ ]$	71.13	58.48
Minimize γ_{RMS}^T	80.42	25.85
Minimize $\det(MSE)$	85.98	20.08

Designs which minimize the parameter-effects curvature, γ_{RMS}^T, and the determinant of the matrix, MSE, are included for completeness. These optimal designs are surprisingly similar, and clearly favour the precise estimation of θ_3 over that of θ_4.

It is tempting to consider an alternative, albeit simple, approach to the one described above for the construction of designs for the precise estimation of θ_3 and θ_4 simultaneously, namely to *combine* the two designs minimizing $MSE(\theta_3)$ and $MSE(\theta_4)$ to give the 18-point design,

$$\begin{pmatrix} -3.022 & 0.829 & 1.510 & 2.341 & 3.942 & 4.417 & 6.978 & 11.510 \\ 1 & 3 & 1 & 1 & 3 & 1 & 2 & 6 \end{pmatrix}. \quad (3)$$

This design is correctly appraised by comparing it with 18-point designs for which $\sigma^2 = 2\hat{\sigma}^2 = 0.018398$ and $\Delta x = 14.5315$, and the results of such a comparison, in terms of percentage efficiencies, with designs minimizing $MSE(\theta_3)$, $MSE(\theta_4)$, and an appropriately weighted sum thereof, are presented in Table 3. Full details of the optimal designs are given in Table A2 of the Appendix.

Table 3 : % Efficiencies for 18-point optimal designs with $\Delta x = 14.5315$

Criterion	$Eff(\theta_3)$	$Eff(\theta_4)$
Minimize $MSE(\theta_3)$	100	24.27
Minimize $MSE(\theta_4)$	0.58	100
Combined design	47.80	50.02
Minimize $[\ \frac{MSE(\theta_3)}{MSE^\star(\theta_3)} + \frac{MSE(\theta_4)}{MSE^\star(\theta_4)}\ ]$	74.69	57.05

It is surprising, and pleasing, to observe that the combined design performs well, particularly in estimating θ_4. In fact the nonlinearity associated with this design is moderate, and in particular $\gamma_{RMS}^N = 0.150$ and $\gamma_{RMS}^T = 0.451$, and for estimates of the parameters θ_3 and θ_4, the percentage biases are 2.75% and 0.08% respectively, and the percentage contributions of second-order terms to the second-order approximation to the variances are 21.24% and 2.39% respectively. The better estimation performance of the optimal design for the compromise criterion, $\dfrac{MSE(\theta_3)}{MSE^\star(\theta_3)} + \dfrac{MSE(\theta_4)}{MSE^\star(\theta_4)}$, should be offset against the fact that this latter design is based on 4 points of support and cannot accommodate tests of goodness-of-fit, whereas the combined design, (3), is based on 8 support points, and can therefore be used to perform such tests.

4   Conclusions

The second-order approximation to the mean square error of the estimators of the parameters of a nonlinear regression model is proposed as a basis for optimal design criteria. Its appeal is that it provides a measure of the precision in estimating the parameters of interest, and at the same time accommodates, at least indirectly, measures of nonlinearity associated with those parameters. The example introduced involves the Richards function, and demonstrates that a judicious choice of criteria based on functions of the approximate mean square error leads to optimal designs which exhibit mild nonlinearity, and that a direct combination of certain optimal designs is also satisfactory.

The results presented here essentially constitute a case study, and work is currently in progress to extend the study, and, in particular, to establish specific features of, and trends in, optimal designs for additive and multiplicative models based on the Richards function.

Acknowledgments: The authors wish to thank the University of Natal and the Foundation for Research Development, South Africa, for financial support.

References

Clarke, G.P.Y. (1980). Moments of the least squares estimators in a nonlinear regression model. *Journal of the Royal Statistical Society*, **42**, 227–237

Nelder, J.A. (1961). The fitting of a generalization of the logistic curve. *Biometrics*, **17**, 89–110

Ratkowsky, D.A. (1983). *Nonlinear Regression Modeling*, Marcel Dekker, New York

Ratkowsky, D.A. (1990). *Handbook of Nonlinear Regression Models*, Marcel Dekker, New York

Richards, F.J. (1959). A flexible growth function for empirical use. *Journal of Experimental Botany*, **10**, 290–300

Appendix

Details of the optimal designs discussed in the text are given in the following tables. Each design is specified by two rows, the first corresponding to the support points and the second to the associated numbers of observations.

Table A1 : *9–point optimal designs for* $\Delta x = 10.0$

Criterion	Design			
Minimize $MSE(\theta_3)$	−3.022 1	0.829 3	3.942 3	6.978 2
Minimize $MSE(\theta_4)$	1.510 1	2.341 1	4.417 1	11.510 6
Minimize [$MSE(\theta_3) + MSE(\theta_4)$]	−2.873 1	0.893 2	4.083 3	7.127 3
Minimize [$\frac{MSE(\theta_3)}{MSE^\star(\theta_3)} + \frac{MSE(\theta_4)}{MSE^\star(\theta_4)}$]	−2.123 1	0.967 2	4.126 2	7.877 4
Minimize γ_{RMS}^T	−2.895 2	0.009 3	4.001 2	7.105 2
Minimize det(MSE)	−3.597 2	0.808 2	3.808 3	6.403 2

Table A2: 18-*point optimal designs with* $\Delta x = 14.5315$

Criterion	Design			
Minimize $MSE(\theta_3)$	−6.549 2	0.324 5	3.900 7	7.983 4
Minimize $MSE(\theta_4)$	0.588 1	1.684 1	3.962 1	15.119 15
Minimize [$\frac{MSE(\theta_3)}{MSE^\star(\theta_3)} + \frac{MSE(\theta_4)}{MSE^\star(\theta_4)}$]	−5.611 1	0.601 3	4.128 5	8.921 9

Mixed Markov Renewal Models of Social Processes

R.B. Davies
G.R. Oskrochi

KEYWORDS: Logistic regression, Markov model, Random effects, Generalized linear models.

1   Introduction

Mixed Markov renewal models for movement between social states were proposed in the early 1980s (eg Flinn and Heckman, 1982). This is a promising class of model for analysing work and life history data because of its focus on categorical outcomes, its flexibility in representing both state dependence and duration effects, and its random effects specification. The random effects specification is particularly important given the mounting evidence that failure to allow for the inevitable omission of some relevant explanatory variables from any analysis risks serious inferential error not just on temporal dependencies but also on the effects of explanatory variables included in the model. A corollary of this problem is that the opportunity to provide some measure of control for omitted variables in observational studies is a major justification for collecting and analysing logitudinal data. However, mixed Markov renewal models are rarely used in social science research. Atleast in part this is because researchers have been deterred by the model specification and computational problems posed by the relatively simpler models used for event history analysis. Such methods are themselves only in routine use in a few areas of social science with established quantitative traditions.

The dimensionality of the error distribution is a critical issue; multivariate mixture models are notoriously difficult computationally and, in realistic formulations, it is rarely possible to obtain analytically tractable marginal probabilities by selecting conjugate distributions for the conditional and error probabilities. Those who have ventured to fit mixed Markov models to social science data have therefore tended to reduce the multivariate error distribution to univariate, following Flinn and Heckman (1982). For example, in a study of residential mobility, Pickles and Davies (1985) assume that the omitted variables may be summarised by a single unknown error term which appears, albeit with a different scale parameter, in each linear predictor.

In recent years, multivariate mixture models have been a focus of attention in other substantive areas (particularly biostatistics and epidemiology) and there have been relevant developments in a number of different areas of statistics (particularly multi-level modelling and Bayesian statistics). Moreover, with modern computing environments, the heavy computational demands of multivariate mixture models have become a less important consideration. It is therefore timely to reassess the specification and computational problems of using mixed Markov renewal models for social processes. In Section 2 of this paper we discuss the general issues which arise in fitting such models. In Section 3 we examine in detail the more promising approaches. Some empirical results and conclusions are presented in Section 4.

2   Model Fitting

A random effects model for longitudinal data has a conditional density function $g(S_i|\theta_i, X_i, \epsilon_i)$ for individual i where S_i is the sequence of states and durations, θ_i is a vector of structural parameters, X_i is a matrix of values of the explanatory variables (both exogenous and exogenous) at appropriate points of time, and ϵ_i is a vector of individual-specific error terms. Inference is based upon the integrated likelihood $g(S_i|\theta_i, X_i) = \int g(S_i|\theta_i, X_i, \epsilon_i)dF(\epsilon_i)$ where $F(\epsilon)$ is the multivariate distribution function of the random effects ϵ_i.

Three main approaches may be used to fit such a model:

1. Numerical integration, usually with Gauss-Hermatite quadrature to approximate the integral.

2. Adopting a nonparametric chacterisation of the mixing distribution, which effectively reduces the integration to a summation over an empirically identified set of mass points.

3. Monte Carlo simulation of the mixing distribution.

4. Simplifying the integrated likelihood by suitable approximations.

Additionally, Bayes approaches with Gibbs sampling are receiving considerable attention in epidemiological research. See, for example, Clayton (1994).

The nonparametric approach is particularly attractive because of evidence that longitudinal models are not always robust to alternative specifications of the error distribution (eg Heckman and Singer, 1984). The approach is very computer intensive but a number of authors have reported favourable experience with univariate error distributions (eg Laird, 1978; Wood and Hinde 1987; Davies, 1993), primarily because relatively few mass points are found to fully characterise the distributions. However, the limited evidence to date on multivariate error distributions is less encouraging with, for example, Davies and Pickles (1987) identifying 18 bivariate mass points in fitting a joint trip-timing store-choice model for grocery shopping. In these circumstances, the nonparametric approach cannot be recommended for routine use with multivariate integrals.

A variety of different methods have been proposed for approximating the integration, including Taylor expansions (Davies, 1984), and the use of posterior mode and curvature estimates within an EM algorithm (Stiratelli et al, 1984). None have attracted general interest and, in the absence of any promising developments in the literature for multivariate error distributions, we do not pursue these methods in this paper.

There have been few empirical comparisons of the remaining two approaches, numerical integration and Monte Carlo simulation. Moreover, no general advice is available on the number of points to be used in each case. There is a tendency to use as few as 8 or 10 points although Fahrmeir and Tutz (1994,7.4.3) and others recommend increasing the number of points until the results stabilise. This further increases the computational demands of these approaches. Comparison is further complicated by the fact that two quite different algorithmic methods of parameter estimation, direct maximisation of the loglikelihood and the EM algorithm, may be used within each approach. Nevertheless, it is clear from the evidence to date that there is little to chose computationally between numerical integration and Monte Carlo simulation for univariate error distributions but that the Monte Carlo approach is preferable in the multivariate case: the summation calculations increase exponentially with error dimensionality for numerical integration but linearly for Monte Carlo simulations. We therefore focus upon the Monte Carlo approach in the next Section.

3   A Model for Employment Status

Work and life history data is typically recorded at discrete times, usually monthly or annually. Let

$$Y_{it} = \begin{cases} 1. & 1 \text{ if individual i is employed at time t.} \\ 0 & \text{otherwise} \end{cases}$$

We adopt Logistic formulation for the conditional distribution of Y_{it} given Y_{it-1}:

$$P(Y_{it} = 1|Y_{it-1} = 1) = \frac{\exp(\eta_{1it})}{1 + \exp(\eta_{1it})}$$

and

$$P(Y_{it} = 1|Y_{it-1} = 0) = \frac{\exp(\eta_{2it})}{1 + \exp(\eta_{2it})},$$

where the parameter η_{it} depends on explanatory variables (note that duration enters as an explanatory variable) and unobserved individual specific effects:

$$\eta_{1it} = X'_{it}\beta_1 + \epsilon^*_{1i} \qquad \eta_{2it} = X'_{it}\beta_2 + \epsilon^*_{2i}$$

We assume that $\epsilon^*_{1i}, \epsilon^*_{2i}$ are two realization of a bivariate distribution function $(\epsilon^*_{1i}, \epsilon^*_{2i}) \sim N(\mu, \Sigma)$
or

$$\eta_{it} = \begin{bmatrix} \eta_{1it} \\ \eta_{2it} \end{bmatrix} = \begin{bmatrix} X'_{it} & 0 \\ 0 & X'_{it} \end{bmatrix} \begin{bmatrix} \beta_1 \\ \beta_2 \end{bmatrix} + Z_i \begin{bmatrix} \epsilon^*_{1i} \\ \epsilon^*_{2i} \end{bmatrix}$$

where Z_i is the design matrix for the random effects. The log likelihood of this model has the form

$$l(\beta, \Sigma) = \sum_{i=1}^{n} \log \iint f(Y_i|\epsilon^*_i, \beta, Y_{i,t-1})p(\epsilon^*_i, \Sigma)d\epsilon^*_{1i}d\epsilon^*_{2i}$$

where

$$f(Y_i|\epsilon^*_i, \beta, Y_{i,t-1}) = \prod_{t=1}^{T_i}\prod_{j=1}^{2} \left[\frac{\exp(\eta_{itj})^{y_{itj}}}{1 + \exp(\eta_{itj})}\right]^{y_{it-1j}}$$

and T_i is the number of time points for the ith individual.
This log-likelihood does not have a closed form. However a Cholesky decomposition for Σ allows us to simplifies the multivariate integration to an integration over several independent univariate distributions.
Writing $\epsilon^*_i = Qa_i$ where $Q = \begin{bmatrix} q_{11} & q_{12} \\ 0 & q_{22} \end{bmatrix}$ is the left Cholesky factor such that $\Sigma = QQ^H$, we have $a_{i \sim MVN}(0, I)$.
Note that the mean is absorbed into the intercept of the model. The model may now be written as $\eta_{it} = X'_{it}\beta + Z_iQa_i$ where $Z_i = (z_{i1}, z_{i2})$ and $z_{ij} = 0, 1$. Or $\eta_{it} = [X_{it}, a'_i \otimes Z_i]\begin{bmatrix} \beta \\ \theta \end{bmatrix}$ where $\theta = (\ q_{11} \quad 0 \quad q_{12} \quad q_{22}\ )'$ and $\otimes$ is a Kronoker product. Our Markov logistic model becomes

$$\eta_{it} = X'_{it}\beta + a_{i1}z_{i1}q_{11} + a_{i1}z_{i2}q_{21} + a_{i2}z_{i1}q_{12} + a_{i2}z_{i2}q_{22}$$

$$z_{i1} = \begin{cases} 1 & \text{if } Y_{it-1} = 1 \\ 0 & \text{otherwise} \end{cases} \qquad z_{i2} = \begin{cases} 1 & \text{if } Y_{it-1} = 0 \\ 0 & \text{otherwise} \end{cases}$$

note that $q_{21} = 0$.

Hence we convert the integration over a bivariate distribution to two integrals over univariate normal distributions. This is the multivariate generalization of the method suggested by Anderson and Hinde (1988). The main advantages is that we have passed the unknown variance Σ from $p(\epsilon_i, \Sigma)$ to $f(Y_i | \epsilon_i, \beta, Y_{i,t-1})$.

3.1   Direct Maximization

The score function for $\alpha = (\beta, \theta)$ has the form

$$S(\beta, \theta) = \left(\frac{\partial l(\beta, \theta)}{\partial (\beta, \theta)} \right) = \sum_{i=1}^{n} \left(\frac{\partial L_i(\alpha)/\partial \alpha}{L_i(\alpha)} \right) = 0$$

where $L_i = \iint f(Y_{it} | a_i, \alpha, Y_{i,t-1}) g(a_i) da_i$.
Using the Monte Carlo method to evaluate the above integral we obtain

$$S(\beta, \theta) = \sum_{i=1}^{n} \sum_{j=1}^{m} C_{ij}(\alpha) \left(\frac{\partial \log f(Y_{it} | d_{ij}, \alpha, Y_{i,t-1})}{\partial \alpha} \right)$$

where

$$C_{ij}(\alpha) = \frac{f(Y_{it} | d_{ij}, \alpha, Y_{i,t-1})}{\sum_{k=1}^{m} f(Y_{it} | d_{ij}, \alpha, Y_{i,t-1})} \qquad with \sum_{j=1}^{m} C_{ij}(\alpha) = 1$$

are weights, d_{ij} are simulated values drawn from the distribution of a_i as the Monte Carlo points with weights $C_{ij}(\alpha)$. See Fahrmeir and Tutz (1994, Sec. 7.4.3).
We used the Maxlik module from the Gauss application (this uses both quasi-Newton and Newton-Raphson methods) to maximize the likelihood for the result represented in Section 4.

3.2   The EM Algorithm

The EM-algorithm is based on the complete data log-density

$$\log f(Y, A, \alpha) = \sum_{i=1}^{n} \log f(Y_i \mid a_i, \alpha, Y_{i,t-1}) + \sum_{i=1}^{n} \log g(a_i)$$

where $Y = (Y_1, ..., Y_n)$ are the incomplete data and $A = (a_1, ..., a_n)$ are the unobserved values.
The EM-algorithm works via a function of α at (P+1)th E-step given by

$$M(\alpha \mid \alpha^{(p)}) = E \left\{ \log f(Y, A, \alpha \mid \alpha^{(p)}, Y) \right\}$$

This is the conditional expectation of $\log f(Y, A, \alpha)$, given the incomplete data Y and the last estimate of $\alpha^{(p)}$. This is a function of $\alpha^{(p)}$ and a_i but not α. Expectation is over the random effect distribution.

Using the Monte Carlo simulation to estimate this expectation gives

$$M(\alpha \mid \alpha^{(p)}) = \sum_{i=1}^{n} \sum_{k=1}^{m} C_{ik} \left[\log f(Y_i \mid d_{ik}, \alpha, Y_{i,t-1}) + \log g(d_{ik}) \right]$$

where again

$$C_{ik} = \frac{f(Y_{ii} \mid dik, \alpha, Y_{i,t-1})}{\sum_{k=1}^{m} f(Y_{it} \mid d_{ijk}, \alpha, Y_{i,t-1})} \qquad with \sum_{k=1}^{m} C_{ik} = 1$$

but, in contrast to direct maximization, C_{ik} is independent of α and it is therefore possible to calculate the second derivatives. Moreover, the derivative $\partial \log f / \partial \alpha$ can be considered as the score function of a GLM $E(Y_{itk}) = h(\eta_{itk})$ with linear predictor

$$\eta_{itk} = X'_{it}\beta + a_{ik1}z_{i1}q_{11} + a_{ik1}z_{i2}q_{21} + a_{ik2}z_{i1}q_{12} + a_{ik2}z_{i2}q_{22}$$

where $q_{21} = 0$. This requires the m times expansion of the data vector Y_{it} as $Y_{it1}, Y_{it2}, \ldots\ldots, Y_{itm}$ where m is the number of points chosen for the Monte Carlo simulation with weight factor C_{ik}.

The Markov model therefore reduces to two independent weighted GLMs with dependent prior weights. These weights have to be calculated at the (P+1 th) iteration based on previous parameter values $(\alpha^{(p)})$. These two independent GLM models are defined as

$$E(Y_{itk} = 1 \mid Y_{it-1k} = 1) = h(\eta_{1itk}), \quad \eta_{1itk} = X'_{it}\beta_1 + a_{ik1}z_{i1}q_{11} + a_{ik2}z_{i1}q_{12}$$

and

$$E(Y_{itk} = 1 \mid Y_{it-1k} = 0) = h(\eta_{2itk}) \qquad \eta_{2itk} = X'_{it}\beta_2 + a_{ik2}z_{i2}q_{22}$$

with dependent weight C_{ik}. This is a very attractive generalization of the methods proposed for univariate random effects by Hinde (1982) and Anderson and Hinde (1988).

Alternatively it is straight forward to calculate the expected conditional information matrix and then construct the M-step of the (usual) EM-algorithm by iteration scheme

$$\alpha_{k+1} = \alpha_k + I^{-1}(\alpha_k \mid \alpha^{(p)}) \frac{\partial M(\alpha \mid \alpha^{(p)})}{\partial \alpha}$$

where I^{-1} is the expected conditional information matrix given $\alpha^{(p)}$ (see Fahrmeir and Tutz, 1994).

4   Empirical Results

The methods in the previous Section were applied to monthly employment history data since leaving school for 239 men and 301 women in Rochdale , England. A typical set of results is shown in Table 1 with, for investigative purposes, just three explanatory variables: Age, duration in current state, and total duration in employment (HCAP). As may be anticipated, control for omitted variables results in substantial improvements in the loglikelihoods and large changes to the parameter estimates for the endogenous explanatory variables.

Our overall conclusions from an extensive programme of model fitting using the Rochdale and simulated data are listed below:

1. It is prudent to use atleast 20 points in the Monte Carlo simulations.

2. Surprisingly direct Newton-Raphson maximisation is circa 20-40 times slower than the EM algorithm approach which leads to (several) weighted independent GLMs with dependent weights.

3. The basic EM algorithm is circa 3 times slower than the EM algorithm with a GLM specification.

4. The speed of convergence for the EM algorithms is very sensitive to the starting values used.

5. The EM algorithms have a tendency to converge at suboptimum solutions; it is vital to check solutions by refitting with different starting values.

Current State	Variable	females Conventional Model p.e.	s.e	females Mixed Model p.e.	males Conventional Model p.e.	s.e	males Mixed Model p.e.
Not Employed	Constant	-1.41	2.63	-3.01	4.010	1.38	6.010
	Age	0.02	0.90	0.70	-2.00	0.47	-2.230
	LDUR	0.25	0.05	0.39	0.010	0.02	0.032
	HCAP	-0.03	0.03	-0.07	-0.045	0.02	0.062
Employed	Constant	4.66	2.03	-4.89	4.89	0.970	4.990
	Age	-0.30	0.70	3.39	-0.65	0.330	-0.600
	LDUR	0.14	0.02	0.11	0.046	0.010	0.036
	HCAP	-0.07	0.03	-0.26	0.019	0.016	0.026
Random effect	q_{11}			-2.18			-0.058
	q_{12}			0.33			0.190
	q_{22}			1.14			0.799
Log-likelihood		-1256.69		-1052.99	-3138.28		-3088.39

TABLE 1. Mixed Markov model

Acknowledgments: We are grateful to the Economic and Social Research Council for funding this research (research grant R000233850)

References

Anderson D A and Hinde J (1988). Random effects in generalised linear models and the EM algorithm, Communications in Statistics, 17, 3847-3856.

Clayton D. (1994). Generalised linear mixed models. In W Gilks, S Richardson, and D Spiegelhalter (Eds), Markov Chain Monte Carlo in Practice. (Chapman and Hall: London).

Davies R B. (1984). A generalised beta-logistic model for longitudinal data with an application to residential mobility, Environment and Planning, 16, 1375-1386.

Davies R B. (1993). Nonparametric control for residual heterogeneity in modelling recurrent behaviour, Computational Statistics and Data Analysis, 16, 143-160

Davies R B and Pickles A R (1987). A joint trip timing, store choice model including feedback effects and nonparametric control for omitted variables, Transportation Research A, 21, 345-361.

Fahrmeir L and Tutz G (1994). Multivariate Statistical Modelling Based on Generalised Linear Models. (Springer-Verlag: New York)

Flinn C and Heckman J J (1982). New methods for analysing individual event histories. In S Leinhardt (Ed), Sociological Methodology 1982, (Jossey Bass: San Francisco), pp99-140.

Heckman J J and Singer B (1984). A method for minimising the impact of distributional assumptions in econometric models for duration data, Econometrica, 52, 271-320.

Hinde J (1982). Compound Poisson regression models. In R Gilchrist (Ed), GLIM 82, Proceedings of the International Conference on Generalised Linear Models. (Springer-Verlag: Berlin), pp109-121.

Laird N (1978). Nonparametric maximum likelihood estimation of a mixing distribution, Journal of the American Statistical Association, 73, 805-811.

Pickles A R and Davies R B (1985). The longitudinal analysis of housing careers. Journal of Regional Science, 25, 85-101.

Stiratelli R, Laird N, and Ware J H (1984). Random-effects models for serial observation with binary response, Biometrics, 40, 961-971.

Wood A and Hinde J (1987). Binomial variance component models with nonparametric mixing distributions: a GLIM approach. In R Crouchley (Ed), Longitudinal Data Analysis, (Avebury: Aldershot, UK), pp110-128.

Statistical Inference Based on a General Model of Unobserved Heterogeneity

Ekkehart Dietz
Dankmar Böhning

ABSTRACT: In this paper, a family of Finite Mixed Generalized Linear Models is considered. A straightforward general EM-algorithm for estimating any model from this family by standard GLM-software is given. After discussing the particular problems of statistical inference arising when FMGLMs are used, three estimators of standard errors of the parameter estimates are compared by means of example data and some simulations.

KEYWORDS: Unobserved Heterogeneity; Overdispersion; Mixed Generalized Linear Models

1 Introduction

The most widely applied methods of statistical analysis are based on Generalized Linear Models. If no model from this broad family seems to fit any particular data set sufficiently, one usually takes generalizations of this family into consideration. One kind of generalization of the GLMs are the **Finite-Mixed GLMs**, which have attracted some interest in recent years. FMGLMs are not GLMs in the usual sense but they are strongly related to this model family as follows: A FMGLM is true for particular observational data if a GLM holds for this data and one additional unobserved categorical explanatory variable. Such a situation is often called *unobserved heterogeneity* .

There are three advantages of such models as compared with the most prominent generalizations of the GLMs, the **Quasi-Likelihood Models**. Firstly, such models do not only allow for a better fit of data if GLMs fail, but they can also give a simple and natural explanation for lack of fit of GLMs like overdispersion. It is attributed to the fact that the population, from which the data are drawn from, consists of several subpopulations and the unobserved categorical explanatory variable is just the indicator of the subpopulation membership. Secondly, the FMGLMs constitute a very broad family and models from that family are appropriate for a lot of different situations in practical data analysis. Thirdly, there is a simple general procedure to obtain maximum likelihood estimates of parameters for all models of this family by standard GLM software, as shown in Section 3. However, there seems to be some lack of computational and theoretical tools to provide statistical inference based on FMGLMs. This problem is discussed in Section 4. After that, three particular estimators of standard errors of parameter estimates will be considered in more detail.

2   The model

In this section we give a formal definition of the FMGLMs. Let y be a response variable, X a vector of explanatory variables, and

$$y \mid X \sim \sum_{j=1}^{c} p_j(X, \Lambda) * f(y, \vartheta_j(X)) \tag{1}$$

a finite mixture model, where $p_j(.)$ is a real function, $0 \leq p_j(X, \Lambda) \leq 1$, $\sum_j p_j(X, \Lambda) = 1 \, \forall X$, Λ is a real parameter(vector), f(.) is a density function from the exponential family, $\vartheta_j(X) = g(\mu_j) = a_j + X B_j$ is the parameter vector of the j-th *mixture component*, g(.) is the common link function of all mixture components, and (a_j, B_j) are the (mostly unknown) coefficients of component specific linear predictors.

We will refer to the *complete data representation* of this model throughout this paper. For this, a c-vector Z with components z_j, having value 1 if (y,X) is an observation from the k-th component of the mixture and 0 else, is introduced. It follows from 1 that, for the *complete* data, the usual GLM

$$y \mid Z, X \sim f(y, \vartheta(Z, X)) \tag{2}$$

where

$$\vartheta(Z, X) = g(\mu) = LP$$

and
$LP = a_1 z_1 + \cdots + a_c z_c + z_1 * (X^{(\mu)} B_1) + \cdots + z_c * (X^{(\mu)} B_c)$
or more parsimonious
$LP = a_1 * z_1 + \cdots + a_c z_c + X^{(\mu)} B = ZA + X^{(\mu)} B$
holds. $X^{(\mu)}$ denotes a subvector of X. Furthermore, 1 assumes implicitly a multinomial model MN for the indicator variables,

$$p_j(X) = P(z_j = 1 \mid X) = MN(Q(X^{(p)}, \Lambda)), \tag{3}$$

where Q(.) denotes the parameter vector of MN, which may depend on an-other subvector $X^{(p)}$ of X and on unknown parameters Λ.

Many kinds of unobserved heterogeneity can be modeled by 1 and 2, respectively. In the most simple case, $B_1 = B_2 = \cdots = B_c = 0$ and $p_j(X) = P(z_j = 1 \mid X) = p_j \; \forall j$, 1 reduces to *mixture distributions*. If only the second condition holds, and only the parameters of the mixture components depend on covariables, one obtains *mixed regression models*. Furthermore, the FMGLM family contains models for missing or misclassified categorical covariables, toxicological threshold models, and outlier models.

The scope of model 1 can be enlarged by further generalizations. One possibility is to allow for different error structures in each mixture component. One example is the **Z**ero **I**nflated **P**oisson model considered by Lambert (1992) , where a Poisson model and a one-point distribution are mixed.

Also different link function for each mixture components could be assumed Another useful generalization is to allow for multivariate response variables y. We have modeled unobserved heterogeneity in longitudinal data in such way, for example.

3 Ml-Estimation

Variants of the EM-algorithm can be used to obtain ml-estimates of the unknown parameters of FGLMs. In this section, we represent a general EM-procedure, which can be applied for each model of this family and which is straightforward to implement in GLIM programs.

In order to obtain ml-estimates $\hat{\Psi}$ of the unknown parameters of 1 for sample data (X_i, y_i), $i = 1, \cdots, n$, one has to maximize the log likelihood function

$$LL(\Psi) = \sum_{i=1}^{n} log(\sum_{j=1}^{c} p_j(X_i, \Lambda) * f(y_i, \vartheta_j(X_i, a_j, B_j))) \qquad (4)$$

where $\Psi = (a_1, \cdots, a_c, B_1, \cdots, B_c, \Lambda)$.

As starting point of the EM-algorithm one uses the log likelihood of 2

$$LL(\Psi) = \sum_{i=1}^{n} \sum_{j=1}^{c} z_{ij}(log(p_j(X_i, \Lambda)) + log(f(y_i, \vartheta_j(X_i, a_j, B_j)))), \qquad (5)$$

which is often called the *complete data log likelihood*. Within the EM-algorithm, **E**xpectation and **M**aximation steps are alternated repeatedly, in order to increase the conditional expectation of 5 and, by that, the value of 4. Using an initial value of Ψ, say $\Psi^{(0)}$, the

E-step :

requires the calculation of the conditional expectation of 4 given the observed data (X_i, y_i), $: i = 1, \cdots, n$ and the current estimate of Ψ, $\Psi^{(m)}$ say, obtained in the previous step. This can be done by replacing each indicator variable z_{ij} by its expectation conditional on X_i and $\Psi^{(m)}$, given by

$$
\begin{aligned}
E(z_{ij} \mid \Psi, y_i, X_i) &= w_{ij}^{(m)} = P(z_j = 1 \mid y_i, X_i, \Psi^{(m)}) \qquad (6) \\
&= \frac{p_j(X_i, \Lambda^{(m)}) f(y_i, \vartheta_j(X_i, a_j^{(m)}, B_j^{(m)}))}{\sum_{r=1}^{c} p_r(X_i, \Lambda^{(m)}) f(y_i, \vartheta_r(X_i, a_r^{(m)}, B_r^{(m)}))}.
\end{aligned}
$$

In order to improve the parameter estimate in the
M-Step :
this parameter value $\Psi^{(m+1)}$ is calculated, which maximizes the conditional expectation of (1.5)

$$
\sum_{i=1}^{n} \sum_{j=1}^{c} w_{ij}^{(m)} log(p_j(X_i, \Lambda)) + \sum_{i=1}^{n} \sum_{j=1}^{c} w_{ij}^{(m)} log(f(y_i, \vartheta_j(X_i, a_j, B_j)))
$$

obtained in the E-step. Both sums may be maximized separately. They have the form of a log likelihood function of a multinomial model and a GLM respectively, each for w_{ij}-weighted observations. That is, the M-step can be executed by standard GLM software.

4   Statistical Inference

When data are analysed by GLIM on base of FMGLMs instead of GLMs applying the algorithm described in the previous section, the additional input of the program is the number of components, c, and a model for the unobserved indicator variables. As additional output, one obtains alternative (possibly several component-specific) effect estimates of each covariable and estimates of the (conditional) distribution of the unobserved categorical covariable.

If the number of components, c, is known, as e.g. in the situation of missing or misclassified categorical covariables mentioned above, the **Likelihood Ratio Test** can be used for statistical inference on the effects of covariables, in the usual way.

Mostly, however, the number of components is unknown and has to be estimated from the data. For the estimation of mixture distributions, this can be done by an algorithm for nonparametric estimation of mixing distributions suggested by Böhning et al. (1992) and implemented in the program package C.A. MAN. It is difficult, however, to generalize this algorithm for the more general models considered in this paper.

A method to estimate c being applicable to FMGLMs generally is suggested in Celeux and Diebolt (1985) . Unfortunately, this method does not provide any significance levels or power estimates. It is also not possible to use the LRT in the usual way to decide if the true number of components is c', say, or c'' because its asymptotic distribution is usually unknown in that case (Böhning et al.(1994)).

One way out is to estimate the critical value of the LRT by simulation of the null model. We have implemented this method in the program DISMAP, which generates disease maps based on FMGLMs, where the number of mixture components is one of the parameters of main interest.

We will not pursue the problem of estimating c further and, instead of this, consider the problem of estimating standard errors of the parameter estimates assuming that c is a priori known or estimated in a first step of the analysis.

One way could be to leave the EM approach and to use the Newton-Raphson algorithm, which has shown to be feasible for some kinds of FMGLMs, or to use Gibbs-Sampling methods, which, in principle, seems to be feasible for all FMGLMs. Both methods provide the asymptotical covariance matrix of parameter estimates as a by product.

To stay within the EM framework, it has been suggested to compute the information matrix by numerical differentiation of the log likelihood function after the EM-algorithm has converged.

Meng and Rubin (1991) suggested an algorithm, which executes the numerical differentiation within the EM-iterates. Alternatively, we will focus on three estimators of standard errors of the parameter estimates, which have proven to be numerically stable and straightforward to program in GLIM.

5   Three Estimators of Standard Errors

Let ψ denote a single parameter of a FMGLM, $\hat{\psi}$ its ml-estimator, and $LL_{\psi=w}$ the supreme of the log likelihood function given $\psi = w$. ψ may be either an intercept a_j or a component specific effect b_{j_s} or a common effect b_s or a mixing weight p_j, where j and s are the indices for the mixture component and the covariable, respectively.

Now we define three estimators of the standard errors of $\hat{\psi}$

PL-estimator:

The **P**rofile **L**ikelihood estimator is primarily an estimator of confidence intervals. Those values ψ^+ of ψ are defined as being inside the 95 percent confidence interval, say, which fulfill the condition $2(LL_{\psi=\hat{\psi}} - LL_{\psi=\psi^+}) < crit_{95}$, whereas all the other values are defined as lying outside. $crit_{95}$ is equal to 3.84 for intercept and effect parameters and has to be estimated for the weights. The bounds of the confidence interval can be estimated by computing the profile likelihood for an appropriate grid of ψ values. If the PL confidence interval is compact and approximately symmetric with respect to $\hat{\psi}$, which is usually the case, if the sample size is sufficiently large, a standard error of $\hat{\psi}$ can be estimated by

$$\sigma_{LP}(\hat{\psi}) = \frac{\hat{\psi}_u - \hat{\psi}_l}{2 * 1.96},$$

where $\hat{\psi}_u$ and $\hat{\psi}_l$ denote the upper and the lower bound, of the interval, respectively.

LR-estimator:
This estimator is defined by

$$\sigma_{LR}^2(\hat{\psi}) = \frac{\hat{\psi}^2}{2 * (LL_{\psi=\hat{\psi}} - LL_{\psi=0})},$$

The justification of this estimator is given by the fact, that the Wald-test and the LRT are equivalent asymptotically.

MI-estimator:
A Multiple-Imputation sample for given data (y_i, X_i), $i = 1, \cdots, n$, is defined as an augmentation of this data by simulated values of the unobserved indicator variables Z_i, which are drawn from a c-point distribution with parameters w_{ij} defined in 6 at convergence, for each data point (y_i, X_i), respectively. In order to compute the MI-estimator, a number, m say, of such MI-samples has to be generated and for each sample the ml-estimate of ψ and its standard error has to be estimated. Let σ_q^* denote the estimats of the standard error of the parameter estimate by the q-th MI-sample. The MI-estimator of the standard error of $\hat{\psi}$ is then defined by

$$\sigma_{MI}^2(\hat{\psi}) = \frac{2}{m} \sum_i^m \sigma_i^{*2}$$

The justification of this estimator is given in Dietz (1995).

6   Example

The fabric faults data considered by Aitkin (1994) and others are used to compare the three estimators. The response variable is the number of faults in a bolt of fabric of length l. The logarithm of l is considered as the only covariable.

Fitting a poisson model gives a deviance of 64.5 on 30 df. The estimate of the effect parameter in this model is b=0.997 (0.176), so that proportionality (b=1) is obviously well supported but the deviance shows substantial overdispersion.

Fitting mixed poisson models with c=2 and c=3 and a common effect parameter gives a deviance of 49.38 in both cases, so that the two-component mixture might be appropriate. Its parameter estimates are given in the second column of Table 1. The profile likelihood function proves to be

Table 1. Estimates of standard errors of
parameter estimates of a 2-component
Poisson mixture model of the fabric data

parameter		s.e. estimator			
	m.l.e.	PL	LR	MI	LE
a_1	-3.13	1.52	1.36	1.52	1.52
a_2	-2.36	1.50	1.03	1.56	1.59
b	0.80	0.23	0.20	0.24	0.23
p	0.80	0.13	-	0.13	0.12

smooth and symmetric, approximately, so that a compact 95 percent PL confidence interval has been obtained. The respective standard errors are given in column 3 of Table 1. To compute the confidence interval of the weight parameter p, the MC-estimated critical value of the LRT is used instead of 3.84. The Columns 4 and 5 contain the respective results of the other two estimators. Note, that the LR-estimator cannot be applied to the weight parameter. Additionally, in column 6 are the estimates obtained by the program LE of BMDP. The LR-estimator lead to the smallest values, which are probably too small, as we will see from the next section.

7 Some Simulations

To obtain more detailed information on the small sample properties of the MI-estimators and the LR-estimators, simulation experiments have been carried out. For each of the sample sizes 50, 100, 200, and 200, 1000 Monte Carlo samples have been generated for two models. The first model was a two-component mixed poisson model having two intercept parameters and one common effect parameter, quite similar to the model used for the example data. The other one was a similar two-component mixed normal regression. The weights of the mixture components were chosen as 0.4 and 0.6, respectively, in both cases. The LR- and the MI-estimators of each sample were used to compute the respective 95 and 99 percent confidence intervals and the probabilities, that the true value of the common effect parameter, 1, is within the estimated intervals, were estimated from the respective 1000 replications. It turned out that in all but one case, the nominal coverage probabilities were fulfilled. The exception was the LR-estimator at the Poisson model and sample size 50. Both estimators turned out to be more conservative for the normal model than for the Poisson model, where the LR-estimator was a little bit more conservative than the MI-estimator. Further investigations, which also include component specific parameters and mixture models with more than 2 components have to be done to get a more complete picture.

8 Conclusions

The reservations against the use of FMGLMs in data analysis coming from a deficit of reliable tools for statistical inference seem to be surmountable. This is important because FMGLMs are often more appropriate than other kinds of generalizations of the GLM. Even if the unobserved heterogeneity is coming from an unobserved continuous covariable, the finite mixture approach is often at least as good in its approximation of the truth as models assuming a parametric random effects distribution. Indeed, the nonparametric ml-estimator of the mixing distribution is always a finite one. Because both the MI-estimator and the LR-estimator are straightforward to program, they might be very useful in developing general software for FMGLMs.

Acknowledgments: This research is supported by the German Research Foundation

References

Aitkin (1994) An EM algorithm for overdispersion in generalized linear models. *Proceedings of the 9th International Workshop on Statistical Modelling Exeter 1994*

Böhning, D., Schlattmann, P., and B. Lindsay (1992) Computer Assisted Analysis of Mixtures (C.A.MAN): Statistical Algorithms. *Biometrics*, **48**, 283-303

Böhning, D., Dietz, E., Schlattmann, P. and Lindsay, B.G. (1994) The distribution of the likelihood ratio for mixtures of densities from the one-parameter exponential family. *Annals of the Institute of Statistical Mathematics.*, **46**, 373-388

Celeux, G. and J.Diebolt (1985) The SEM algorithm: a probabilistic teacher algorithm derived from the EM algorithm for mixture problems. *Computational Statistics Quarterly*, **2** 1

Dietz, E. (1995) Estimating standard errors of finite mixtures models by MCEM-steps and multiple imputations. *The Annals of Statistics* to appear

Jansen, R.C. (1993) Maximum likelihood in a generalized linear finite mixture model by using the EM algorithm. *Biometrics* **49**, 227-232

Lambert, D. (1992) Zero-inflated poisson regression, with an application to defects in manufacturing. *Technometrics*, **34** , 1-14.

An Extended Model for Paired Comparisons

Regina Dittrich
Reinhold Hatzinger
Walter Katzenbeisser

ABSTRACT: The aim of this paper is to present a log-linear formulation of an extended Bradley-Terry model for paired comparisons that allows for simultaneous modelling of ties, order effects, categorical subject-specific covariates as well as object-specific covariates.

KEYWORDS: Bradley-Terry model, preference scales, covariates, subject-specific effects

1 The Bradley-Terry Model

The method of paired comparisons addresses the problem of determining the scale values of a set of objects on a preference continuum that is not directly observable. Paired comparisons are judgemental tasks that typically involve repeatedly exposing an individual to a pair of J objects (e.g., brands, attributes, ...) one at a time and asking for a judgement about which element in the pair is more preferred.

One of the most prominent models that covers such situations is due to Bradley and Terry (1952) and may be written as

$$\Pi_{jk} = \frac{\pi_j}{\pi_j + \pi_k},$$

where Π_{jk} is the probability that object j is preferred to object k; π_j and π_k are parameters describing the location of the objects on the preference scale; since the right hand side is invariant under change of scale, identifiability is obtained by the requirement that $\sum_i \pi_i = 1$. Letting $\pi_i = \exp\{\theta_i\}$, one obtains for the log-odds

$$\ln \frac{\Pi_{jk}}{1 - \Pi_{jk}} = \theta_j - \theta_k , \quad j < k.$$

For each pair $j < k$ let n_{jk} be the number of comparisons and suppose that j is preferred X_{jk} times and k is preferred $X_{kj} = n_{jk} - X_{jk}$ times.

Assuming the n_{jk} comparisons to be independent having the same probability Π_{jk}, the X_{jk} are binomially distributed with parameters n_{jk} and Π_{jk}. If additionally comparisons for different pairs of objects are also independent, the Bradley-Terry model may be fitted using ordinary methods for logit models.

The Bradley-Terry model has been extensively discussed in the literature (for surveys see Bradley, 1984, or David, 1988) and various extensions have been proposed. These extensions concern e.g., the presence of ties (Rao and Kupper, 1967; Davidson, 1970; Kousgaard,1976): ties occur when each judge is allowed a third option, that of no preference. Order effects, (i.e., the possibility that the order in which the objects are presented to a judge) may also influence the preference (Davidson and Beaver, 1977). Individual judge effects (Kousgaard, 1976; Tutz, 1989) and the incorporation of concomitant variables for the objects (Kousgaard, 1984; Tutz, 1989) are further extensions to the BT-model.

The aim of this paper is to present a log-linear formulation of the Bradley-Terry model which allows for simultaneous modelling of all these effects: ties, order effects, categorical covariates describing the judges, and object-specific covariates describing properties of the objects.

2   Extensions of the Basic Model

In order to incorporate all the above effects let us first rewrite the basic Bradley-Terry model as a log-linear model (Sinclair, 1982). In comparing object j with object k, the number of preferences for j, X_{jk}, and the number of preferences for k, X_{kj}, respectively, can be regarded as forming a contingency table, from which the diagonal entries are missing. The random variables X_{jk} are assumed to follow a Poisson distribution. Conditional on fixed $\sum_j \sum_k X_{jk} = m$, the X_{jk}'s are multinomially distributed; their expected values can be formulated as a multiplicative model. Let

$$\Pi_{jk} = \frac{\pi_j}{\pi_j + \pi_k} = \frac{\sqrt{\pi_j/\pi_k}}{\sqrt{\pi_j/\pi_k} + \sqrt{\pi_k/\pi_j}} ,$$

$$\Pi_{kj} = \frac{\pi_k}{\pi_j + \pi_k} = \frac{\sqrt{\pi_k/\pi_j}}{\sqrt{\pi_j/\pi_k} + \sqrt{\pi_k/\pi_j}} ,$$

and denote by $m_{j|jk}$ the expected number of comparisons in which object j is preferred to object k, $m_{j|jk} = m\Pi_{jk}$, has log-linear representation

$$\ln m_{j|jk} = \mu_{jk} + \frac{1}{2}\theta_j - \frac{1}{2}\theta_k ,$$

$$\ln m_{k|jk} = \mu_{jk} - \frac{1}{2}\theta_j + \frac{1}{2}\theta_k ,$$

where $\mu_{jk} = \mu_{kj} = \ln m - \ln \left\{ \sqrt{\pi_j/\pi_k} + \sqrt{\pi_k/\pi_j} \right\}$ and $\theta_j = \ln \pi_j$. Since there are $J \cdot (J-1)$ cells in the contingency table and $\binom{J}{2}$ μ-parameters and $J-1$ θ's to be estimated, the residual degrees of freedom are $\frac{(J-1)(J-2)}{2}$. Ties as well as order effects are easily incorporated via the Davidson and Beaver parameterisation (Sinclair, 1982). In general, let $\Pi_{jk \cdot j}$ be the probability that object j is preferred over object k when j is presented first and additionally let $\Pi_{0 \cdot j}$ denote the probability of no decision. The probabilities are specified by:

$$\Pi_{jk \cdot j} = \frac{\gamma \pi_j}{\gamma \pi_j + \pi_k + \nu \sqrt{\pi_j \pi_k}},$$

$$\Pi_{kj \cdot j} = \frac{\pi_k}{\gamma \pi_j + \pi_k + \nu \sqrt{\pi_j \pi_k}},$$

$$\Pi_{0 \cdot j} = \frac{\nu \sqrt{\pi_j \pi_k}}{\gamma \pi_j + \pi_k + \nu \sqrt{\pi_j \pi_k}};$$

the probabilities $\Pi_{jk \cdot k}$ that object j is preferred to object k, when k is presented first, are specified analogously.

The log-odds $\ln \Pi_{jk \cdot j}/\Pi_{kj \cdot j}$ are given by $\theta_j - \theta_k + \lambda$, where $\lambda = \ln \gamma$. Therefore, for two evenly-matched objects j and k, there is an advantage for preferring j over k if $\lambda > 0$, i.e. if $\gamma > 1$. The order effect parameter γ can therefore be interpreted as a log-odds ratio, i.e.:

$$\ln \frac{\Pi_{jk \cdot j}/\Pi_{kj \cdot j}}{\Pi_{jk \cdot k}/\Pi_{kj \cdot k}} = \ln \gamma^2 = 2\lambda.$$

The parameter ν may be interpreted as a discrimination constant: for evenly matched objects j and k, $\pi_j = \pi_k$, $\gamma = 1$ and so there is an advantage in favour of a decision if $\nu < 1$ because for the odds we have: $\Pi_{jk \cdot j}/\Pi_{0 \cdot j} = \Pi_{kj \cdot j}/\Pi_{0 \cdot j} = 1/\nu > 1$.

The model can be described in log-linear form by means of the following equations:

$$\ln m_{jk \cdot j} = \mu_{jk} + \frac{1}{2}\theta_j - \frac{1}{2}\theta_k + \lambda,$$

$$\ln m_{kj \cdot j} = \mu_{jk} - \frac{1}{2}\theta_j + \frac{1}{2}\theta_k,$$

$$\ln m_{0 \cdot j} = \mu_{jk} + \delta,$$

where

$$\mu_{jk} = \ln m - \ln \left\{ \gamma \pi_j + \pi_k + \nu \sqrt{\pi_j \pi_k} \right\} + \frac{1}{2} \ln \pi_j + \frac{1}{2} \ln \pi_k,$$

$\lambda = \ln \gamma$, $\delta = \ln \nu$, and, of course, $\theta_j = \ln \pi_j$. $\ln m_{jk \cdot k}$ is given in an analogous way by symmetry considerations. There are $3 \cdot J \cdot (J-1)$ cells in this contingency table and $J^2 + 1$ parameters to be estimated, the residual degrees of freedom are $J \cdot (2J - 3) - 1$.

As well as considering ties and order effects, it might be interesting to consider the effect of covariates describing the judges on the probabilities Π_{jk}. The log-linear formulation can easily be extended to incorporate categorical covariates. To illustrate the approach, assume that the judges are additionally classified according to one categorical covariate, such as sex. Let $\Pi_{jk \cdot j|l}$ be the probability that object j is preferred over object k when object j is presented first and the judges are classified into covariate category l, $l = 1, \ldots, L$; the probabilities $\Pi_{jk \cdot j|l}$ etc. can then be specified as:

$$\Pi_{jk \cdot j|l} = \frac{\exp\{\lambda_{jl}^1 + \lambda\}\pi_j}{\exp\{\lambda_{jl}^1 + \lambda\}\pi_j + \pi_k + \nu\sqrt{\pi_j \pi_k}},$$

$$\Pi_{kj \cdot j|l} = \frac{\pi_k}{\exp\{\lambda_{jl}^1 + \lambda\}\pi_j + \pi_k + \nu\sqrt{\pi_j \pi_k}},$$

$$\Pi_{0 \cdot j|l} = \frac{\nu\sqrt{\pi_j \pi_k}}{\exp\{\lambda_{jl}^1 + \lambda\}\pi_j + \pi_k + \nu\sqrt{\pi_j \pi_k}},$$

where the λ_{jl}^1 parameters represent the effect of the subject-specific covariate, measured on the l-th level on the joice of object j, and $\gamma = \exp\{\lambda\}$. The probabilities $\Pi_{jk \cdot k|l}$ are again defined in a symmetric manner. The log-odds $\ln\{\Pi_{jk \cdot j|l}/\Pi_{kj \cdot j|l}\}$ are given by

$$\ln \frac{\Pi_{jk \cdot j|l}}{\Pi_{kj \cdot j|l}} = \theta_j - \theta_k + \lambda + \lambda_{jl}^1;$$

for two evenly-matched objects j and k and when there is no order effect, i.e. $\lambda = 0$, there is an advantage for preferring j over k given the judges are classified in the l-th covariate category, if λ_{jl}^1 is positive. Moreover, the λ^1 parameters can be interpreted as a log-odds ratio:

$$\ln \frac{\Pi_{jk \cdot j|l+1}/\Pi_{kj \cdot j|l+1}}{\Pi_{jk \cdot j|l}/\Pi_{kj \cdot j|l}} = \lambda_{j,l+1}^1 - \lambda_{jl}^1.$$

If $m_{jk \cdot j|l}$ is the expected number of preferences in covariate class l for object j when compared to object k, with object j being presented first, then the log-linear representation of this model is given by

$$\ln m_{jk \cdot j|l} = \mu_{jkl} + \frac{1}{2}\theta_j - \frac{1}{2}\theta_k + \lambda_{jl}^1 + \lambda,$$

$$\ln m_{kj \cdot j|l} = \mu_{jkl} - \frac{1}{2}\theta_j + \frac{1}{2}\theta_k,$$

$$\ln m_{0 \cdot j|l} = \mu_{jkl} \qquad\qquad\qquad + \delta,$$

where

$$\mu_{jkl} = \ln m - \ln\{\exp\{\lambda_{jl}^1 + \lambda\}\pi_j + \pi_k + \nu\sqrt{\pi_j\pi_k}\} + \frac{1}{2}\ln\pi_j + \frac{1}{2}\ln\pi_k \,,$$

$m_{jk\cdot k|l}$ is defined analogously. The contingency table consists of $3 \cdot J \cdot (J-1) \cdot L$ cells and there are $J \cdot (J-1) \cdot L + (J-1) + (J-1)(L-1) + 2$ parameters to be estimated; the degrees of freedom are therefore $L(J-1)(2J-1) - 2$. In order to evaluate the effect of concomitant variables for the object on the probabilities Π, let us reparameterize the object parameters as a linear combination of P covariates, which represent P properties of the objects. Let us replace $\frac{1}{2}\theta_j = \alpha_j$ by the linear predictor

$$\alpha_j = \sum_{\nu=1}^{P} x_{j\nu}\beta_\nu \,,$$

where the $x_{j\nu}$'s denote the fixed covariates describing the ν'th property of object j und the β's are regression coefficients. Because of $\alpha_j - \alpha_k = \sum_\nu y_{jk\nu}\beta_\nu$ and analogously $\alpha_k - \alpha_j = -\sum_\nu y_{jk\nu}\beta_\nu$, where $y_{jk\nu} = x_{j\nu} - x_{k\nu}$, the log-linear formulation of the Bradley-Terry model with ties, order effects, P object-specific covariates – as well as one categorical subject-specific covariate is given by

$$\ln m_{jk\cdot j|l} = \mu_{jkl} + \sum_\nu y_{jk\nu}\beta_\nu + \lambda_{jl}^1 + \lambda \,,$$

$$\ln m_{kj\cdot j|l} = \mu_{jkl} - \sum_\nu y_{jk\nu}\beta_\nu \,,$$

$$\ln m_{0\ \cdot j|l} = \mu_{jkl} \qquad\qquad + \delta,$$

where the parameters μ_{jkl}, λ^1, λ, and δ are defined as above. The parameters π_j are reparameterized by $\pi_j = \exp\{2\sum_\nu x_{j\nu}\beta_\nu\}$.

3 Concluding Remarks

The approach of modelling the effect of one categorical subject-specific covariate and the effect of concomitant variables for the objects on the probabilities Π_{jk} can be generalized in an obvious way.
(i) The covariate in the example above is considered as nominal. If, however, the covariate is ordinal, we can replace the parameters λ_{jl}^1 by a linear structure. One way to achieve this is to assign fixed scores $\{u_l\}$, which reflect the ordering of the categories, for example $u_l = l$. This leads to the following specification:

$$\ln m_{jk\cdot j|l} = \mu_{jkl} + \frac{1}{2}\theta_j - \frac{1}{2}\theta_k + u_l\beta_j + \lambda \,;$$

88   Regina Dittrich, Reinhold Hatzinger, Walter Katzenbeisser

for unit spaced scores, $u_{l+1} - u_l = 1$, the parameters β_j can be interpreted as log-odds ratios:

$$\ln \frac{\Pi_{jk \cdot j|l+1}/\Pi_{kj \cdot j|l+1}}{\Pi_{jk \cdot j|l}/\Pi_{kj \cdot j|l}} = \beta_j .$$

(ii) Obviously, one can consider more than one subject-specific covariate.
(iii) Relationships between subject-specific covariates may be incorporated by means of interaction effects, for example

$$\ln m_{jk \cdot j|lr} = \mu_{jklr} + \frac{1}{2}\theta_j - \frac{1}{2}\theta_k + \lambda_{jl}^1 + \lambda_{jr}^2 + \lambda_{jlr}^{12} + \lambda .$$

For ordinal factors, the relationship between two factors may be modelled via a linear-by-linear association (Agresti, 1990): let $\{u_l\}$ and $\{v_r\}$ denote again fixed scores, representing the ordering of the covariate categories, we obtain the model

$$\ln m_{jk \cdot j|lr} = \mu_{jkl} + \frac{1}{2}\theta_j - \frac{1}{2}\theta_k + \lambda_l^1 + \lambda_r^2 + u_l v_r \beta_j + \lambda ,$$

where the parameters β_j can be interpreted as log-odds ratios: for unit-spaced scores, we have

$$\ln \frac{\Pi_{jk \cdot j|lr}/\Pi_{kj \cdot j|l(r+1)}}{\Pi_{jk \cdot j|(l+1)r}/\Pi_{kj \cdot j|(l+1)(r+1)}} = \beta_j .$$

(iv) Analogously to Bradley and El-Helbawy (1976), the α's may have a factorial structure; the objects are then considered only as factor combinations (cf. also Tutz (1989), p.88).
(v) Possible interaction effects between object- and subject-specific covariates can easily be studied.
An advantage of the log-linear formulation of the extended Bradley-Terry model is that all the μ-, λ-, δ-, and β- parameters can be estimated by GLIM using using poisson error and log-link. The design matrix consists of column vectors with suitable entries under μ, λ, etc.. Hypothesis tests or goodness of fit tests can be done in the usual way within the GLM framework.

4   An Example

The study aims to analyse students choice of foreign universities for studying one semester abroad. Paired comparisons between different universities in London (LO), Paris (PA), Milan (MI), Barcelona (BA), St. Gall (SG) and Stockholm (ST) were carried out with 303, mainly first year, students. Personal information about students included sex, main discipline of study (STUD), knowledge of English (ENG), French (FRA), Spanish (SPA), Italian (ITA) and Swedish. Additionally they were asked if they were considering spending one semester abroad, if they intended to take an international

degree (AB), if they were working (ARB), if they were living at home and
if they could afford a semester abroad without funding (STIP). All covariates were measured on two levels. The universities were characterised by
specialisation, by size and accommodation facilities.

A log linear model was fitted including the above subject covariates. The
substitution of the objects by their properties resulted in a far poorer fit and
was not investigated further. After elimination of unimportant parameters,
the final model is given in the following table (nuisance parameters μ are
not shown):

Parameter	Estimate	s.e.	Parameter	Estimate	s.e.
LO	1.59	0.081	ITA(2)	1.67	0.042
PA	1.03	0.100	ARB(2).PA	0.95	0.288
MI	1.51	0.151	ARB(2).MI	0.67	0.292
SG	0.29	0.100	ARB(2).BA	0.70	0.286
BA	0.88	0.160	STUD(2).PA	0.57	0.120
U	-1.28	0.049	STUD(2).SG	-0.41	0.117
ARB(2)	-3.12	0.076	AB(2).SG	0.30	0.114
STUD(2)	-0.56	0.032	ENG(2).SG	0.44	0.124
AB(2)	-0.55	0.031	FRA(2).PA	-1.29	0.123
ENG(2)	-1.01	0.034	SPA(2).BA	-0.87	0.163
FRA(2)	-0.55	0.032	ITA(2).MI	-1.60	0.158
SPA(2)	1.87	0.044			

TABLE 1. Parameter Estimates for the University Preference Data.

Considering first the influence of universities (=objects), there is a highly
increased tendency to prefer London and Paris. The highly significant negative parameter U ($=\nu$) indicates that there is a strong tendency for a
decision.

All object comparisons must be seen relative to Stockholm and subject
comparisons relative to a basic group of students not having the characteristics discussed below.

Students working as well as studying have an increased chance for prefering
Paris, Milan and Barcelona. Those who can afford studying abroad without
funding have a higher chance for prefering Paris and have a reduced inclina-
tion to go to St. Gall whereas those who are willing to take an international
degree have a increased chance to prefer St.Gall.

Those students specialising in trade business have a increased chance to
prefer Paris and a decreased chance to prefer St.Gall .

A consistent but not surprising result was found for the knowledge of languages. Students not speaking French (Spanish, Italian) are less likely to
prefer Paris (Barcelona, Milan). Those students judging their English as

not being good have a higher chance for prefering St. Gall which might be explained by the fact that it is the only German speaking university in the study.

References

Agresti, A. (1990). *Categorical Data Analysis*. J.Wiley, New York.

Bradley, R.A. (1984). Paired Comparisons: Some Basic Procedures and Examples. In: Krishnaiah, P.R., Sen, P.K. (Eds.): *Handbook of Statistics*, Vol. 4, North Holland, Amsterdam.

Bradley, R.A., El-Helbawy, A.T. (1976). Treatment contrasts in paired comparisons: Basic procedures with applications to factorials. *Biometrika*, **63**, 255-262.

Bradley, R.A., Terry, M.E. (1952). Rank Analysis of Incomplete Block Designs. I. The Method of Paired Comparisons. *Biometrika*, **39**, 324-345

David, H.A. (1988). *The Method of Paired Comparisons*. Griffin, Oxford University Press.

Davidson, R.R. (1970). Extending the Bradley-Terry model to accommodate ties in paired comparison experiments. *JASA*, **65**, 317-328.

Davidson, R.R., Beaver, R.J. (1977). On extending the Bradley-Terry model to incorporate within-pair order effects. *Biometrics*, **33**, 693-702.

Kousgaard, N. (1984). Analysis of a sound field experiment by a model of paired comparisons with explanatory variables. *Scand.J.Statist*, **11**, 51-57.

Kousgaard, N. (1976). Models for paired comparisons with ties. *Scand.J.-Statist*, **3**, 1-14.

Rao, P.V., Kupper, L.L. (1967). Ties in paired-comparison experiments: a generalisation of the Bradley-Terry model. *JASA*, **62**, 194-204.

Sinclair, C.D. (1982). GLIM for preference. In: Gilchrist, R. (ed.): *GLIM 82. Proceedings of the International Conference on Generalised Linear Models*, Springer Lecture Notes in Statistics 14.

Tutz, G. (1989). *Latent Trait-Modelle für ordinale Beobachtungen*, Lehr- und Forschungstexte Psychologie, Springer-Verlag.

Indirect Observations, Composite Link Models and Penalized Likelihood

Paul H. C. Eilers

ABSTRACT: The composite link model (CLM) is applied to the estimation of indirectly observed distributions. Coarsely grouped histograms and mixed discrete distributions are examples. The CLM can be very ill-conditioned, but a discrete penalty solves this problem. The optimal smoothing parameter is found efficiently by minimizing AIC. Three applications are presented.

KEYWORDS: composite link function, generalized linear model, mixing distribution, penalized likelihood.

1 Introduction

Some types of observations might be called indirect. There is some unobservable variable, u, of which we like to estimate its distribution. We do observe another variable, v, which has a certain relationship to u, as given by a matrix Q of conditional probabilities. Both u and v are discrete (approximations to continuous) variables. For each value of v, many values of u are possible.

Some examples may illustrate thisp problem. Let u be the time of infection of an individual by HIV, and let v be the time at which AIDS is diagnosed. The columns of Q are the distributions of incubation times following infection at the corresponding time units. Medical records give the distribution of v over time, from which we like to estimate the time course of HIV infections.

A second example concerns v's that are a histogram of coarsely discretized values of u. We like to know the distribution of u on a fine grid, detailed enough to be a good representation of the unknown continuous distribution. The rows of Q now describe how the cells on the fine grid are grouped, to give the coarse bins in which v is reported.

A third example is the modelling of discrete observations with an overdispersed distribution. Instead of assuming that each event has the same Poisson distribution with one fixed parameter λ, we assume that λ varies according to some unknown distribution γ, which we like to estimate. The columns of Q now are Poisson distributions with different λ's on some well-chosen grid.

In the following we will see how this type of problem can be treated as a Composite Link (Function) Model, to be abbreviated as CLM, as presented by Thompson and Baker (1981). However, a direct application of that model will not work, because the estimating equations are singular: the number of unknowns is (much) larger than the number of observations. Because we are to estimate a distribution, it is reasonable to expect a relatively smooth result. We can enforce smoothness with an appropriate penalty and remove the singularity that way. The necessary modifications to the CLM algorithm are minor.

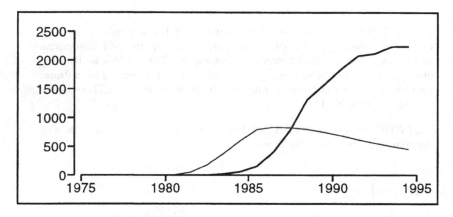

FIGURE 1. Yearly counts of reported AIDS cases (thick line). The thin line shows a model distribution of incubation times (multiplied by 10^4), starting in 1979.

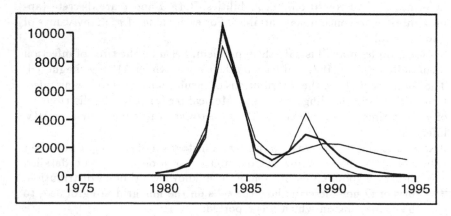

FIGURE 2. Estimated distribution of HIV infections. Thick line: $\alpha = 0.2$; thin line with short right tail: $\alpha = 0.05$; thin line with long right tail: $\alpha = 1$.

2   Theory

Let y_i, $i = 1 \ldots m$, give the observations, as a time series of counts or a histogram. Let the distribution to be estimated be γ_j, $j = 1 \ldots n$. Then we have that $E(y) = \mu y_\bullet = Q\gamma y_\bullet = C\gamma$, where $y_\bullet = \sum_{i=1}^m y_i$ and $C = Qy_\bullet$. For the computation of the likelihood we can work with the multinomial distribution with probabilities $p_i = \mu_i / \sum_{i=1}^m \mu_i$ and sample size $y_\bullet$, but we get an equivalent result (see chapter 13 of Bishop et al., 1975) by assuming that each y_i comes from a Poisson distribution with expectation μ_i. Then there is no need for the constraint $\sum_{i=1}^m p_i = 1$.

All elements of γ will be positive if we set $\gamma_j = e^{\eta_j}$. With this choice we arrive at the CLM. Thompson and Baker (1981) formulate it as follows:

$$E(y) = \mu = C\gamma; \quad \gamma = h(\eta); \quad \eta = X\beta. \tag{1}$$

Here $h(.)$ is the link function that transforms the linear predictor η into an "intermediate predictor" γ. A linear combination of the elements of γ gives the expectations of the observations.

Thompson and Baker give an algorithm for the estimation of β: let

$$\check{X} = CHX, \quad z = y - \mu + \check{X}\beta, \tag{2}$$

where H is a diagonal matrix with elements $h_{jj} = \partial\gamma_j/\partial\eta_j$. Regress z on $\check{X}$ with weights (in the case of Poisson distributed y_i) $w_i = 1/\mu_i$.

In our applications, $X = I$ and thus $\eta = \beta$. A somewhat simplified algorithm follows: the iterative solution of

$$\check{X}^T W \check{X} \eta = \check{X}^T W z. \tag{3}$$

As mentioned in the introduction, $m < n$, which makes this system of equations singular. A way out is to subtract a smoothness penalty from the log likelihood. The penalty (Whittaker, 1923) is:

$$P = \frac{\alpha}{2} \sum_{j=d+1}^n (\Delta^d \eta_j)^2, \tag{4}$$

where $\Delta\eta_j = \eta_j - \eta_{j-1}$, $\Delta^2\eta_j = \Delta(\Delta\eta_j)$, and so on; $0 \leq d \leq 3$. With $d = 0$ we have a ridge penalty, while $d = 2$ gives a discrete version of the integral of the squared second derivative, a penalty that is well known from smoothing splines. Only a small modification of (3) is needed:

$$(\check{X}^T W \check{X} + \alpha D^T D)\eta = \check{X}^T W z, \tag{5}$$

where D is a matrix with $n - d$ rows and n columns, such that $D\eta = \Delta^d\eta$. We recover (3) when $\alpha = 0$. The matrix $\alpha D^T D$ can be constructed beforehand; it incurs hardly extra computational work. Experience shows that a uniform distribution ($\eta = -\ln n$) works well as a starting point.

To find a good value for α, we minimize Akaike's Information Criterion (AIC), which is equivalent to

$$AIC = Dev(y|\mu) + 2Dim = 2\sum_{i=1}^{m} y_i \ln(y_i/\mu_i) + 2Dim, \qquad (6)$$

where $Dev(y|\mu)$ is the deviance and Dim is the effective dimension of the model. For the latter we follow the suggestion of Hastie and Tibshirani (1990) to take the trace of the smoother or "hat" matrix S that is implicit in the linearized smoothing problem (7):

$$\hat{z} = \check{X}\hat{\eta} = \check{X}(\check{X}^T W \check{X} + \alpha D^T D)^{-1}(\check{X}^T W)z = Sz. \qquad (7)$$

3   Applications

In this section, we look at applications of the penalized CLM. At first sight, three rather different problems are analyzed. But they have identical structures, with appropriate Q matrices to describe the relationships between the unknow distribution γ and the observed distribution y.

Figure 1 shows yearly counts of reported AIDS-cases among homosexual and bisexual men in France (data courtesy of Siem Heisterkamp). AIDS is a manifestation of infection by HIV. Column j of Q gives the distribution of the time of diagnosis, following infection in year j. For the purpose of illustrating the CLM, a simplified Q is used here. The columns are shifted replicas of one distribution, also shown in figure 1. Note the large spread: the average incubation time is about 10 years. In a more realistic study one would use quarterly data and a Q matrix that reflects changes in medical care, as well as seasonal effects in the reporting of AIDS cases.

We want to estimate the time series of HIV infections, $\gamma y_\bullet$. This is the "backcalculation problem". Heisterkamp (1995) proposes a difference penalty for stability. He uses a general optimization routine in S-plus and finds long computation times. The penalized CLM algorithm converges readily in about 10 to 20 iterations and takes a few seconds to find a solution. To find an optimal value for α, it was varied on a nice grid (down from 1000, 500, 200, 100, and so on) and the value that gave the minimum of AIC was found to be 0.2 ($AIC = 26.6$, $d = 2$). The corresponding time series $\gamma y_\bullet$ is drawn with a thick line in figure 2. Rather large changes of α give small changes in AIC near its minimum. For $\alpha = 0.05$ ($AIC = 26.8$) and $\alpha = 1$ ($AIC = 27.0$), the corresponding series are plotted too. We see that the sharp peak around 1983 does not change much, but that the most recent estimates of infections are strongly influenced by the amount of smoothing. This seems logical: because of the long average incubation time, we have very weak information about recent infections.

In the second example we look at a histogram y with coarse bins. The unknown distribution γ is based on much smaller bins, say 5 or 10 times narrower than those of y. In Eilers (1991) this problem was analyzed in some detail, but without reference to the CLM and AIC. The columns of Q have a very simple structure: they consist of $m-1$ zeros and one 1. If $q_{ij} = 1$, this means that bin j of γ is part of bin i of y. Stated otherwise: the rows of Q show which b ins of γ form each bin of y.

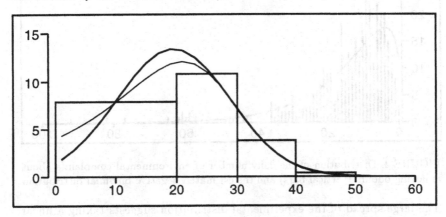

FIGURE 3. Histogram of lead concentrations in blood and estimated distributions; thin line: $d=2$, $\alpha = 100$, $AIC = 7.6$; thick line: $d=3$, $\alpha = 10^6$, $AIC = 6.2$.

Figure 3 shows a distribution from a paper by Hasselblad et al. (1980), giving many data on lead concentrations in the blood of New York City inhabitants (the raw numbers are 79, 54, 19, 2, 0; in the figure they are divided by the bin width). In that paper a technique is developed to estimate a lognormal distribution from grouped data. Here we find a nonparametric distribution with the penalized CLM, using a bin width of 2 for γ and $d=2$ in the penalty. The minimum of AIC (7.6) occurs for $\alpha = 2$; the corresponding detailed distribution is shown in figure 3 with a thick line. Note that the leftmost group of the histogram is wider than the other ones. With $d=3$, we find that AIC decreases monotonically with increasing α. For $\alpha = 10^6$ ($AIC = 6.2$) the fitted distribution is drawn with a thin line in figure 3.

The last example concerns an overdispersed discrete distribution. Our agency registers environmental complaints about annoying odours, dust, noise, and other blessings of modern life, about 15000 each year. The number of complaints per day is very variable, as can be seen from figure 4, showing data for the year 1988. The deviance of a Poisson distribution is 2382. A negative binomial distribution performs much better with a deviance of 121.8, but, as can be seen from figure 4, it shows systematic departures, especially at the left tail. We try to model the data as a mixture of Poisson distributions. Take a grid of values λ_j, $j = 1 \ldots n$, and construct the matrix Q in such a way that $c_{ij} = \lambda_j^{i-1} e^{-\lambda_j}/(i-1)!$. Thus column j of Q is a Poisson distribution with expectation λ_j. Let γ be the mixing distribution of the λ's. We are back again at the penalized CLM.

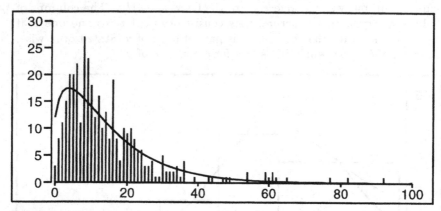

FIGURE 4. Distribution of the daily number of environmental complaints about annoying odours (vertical bars) and the estimated negative binomial distribution

The large spread of the experimental distribution suggests taking a linear grid for $\ln \lambda$, instead of for λ itself. As before, we seek the minimum of AIC by varying α on a nice grid. With $d = 2$, we find that $\alpha = 50$ is best, with $AIC = 101.4$. For the negative binomial distribution in figure 4, $AIC = 2*2 + 121.8 = 125.8$. The mixed model gives a marked improvement. With $d = 3$, AIC shows a monotone decrease with increasing α, approaching the value 98.2. This means that effectively the mixing distribution of $\ln \lambda$ is a (discrete approximation to) a normal distribution. Figure 5 shows the data distribution, the estimated mixed distribution ($d = 3$, $\alpha = 10^6$, $AIC = 98.2$), and the estimated distribution of λ.

Space does not allow a detailed comparison to rival techniques for modelling mixed discrete distributions. See Böhning et al., (1993) for a recent publication. The program they describe computes the mixing distribution as a set of (unequally spaced) discrete probability masses. Even though this may result in essentially the same fit to the data, one can argue that a smooth distribution makes more sense, both from a theoretical and a practical point of view. It is unlikely to envision λ as drawn from a set of point probablity masses; also a smooth γ gives more sensible results in pseudo-Bayes estimates of λ for individual days.

4   Discussion

The penalized CLM works well in many applications. One might not like the discretization of the distribution γ. A solution is to use a B-spline basis for the matrix X, as for log-spline density estimation (Kooperberg and Stone, 1991), with a penalty on the B-spline coefficients (Eilers and Marx,

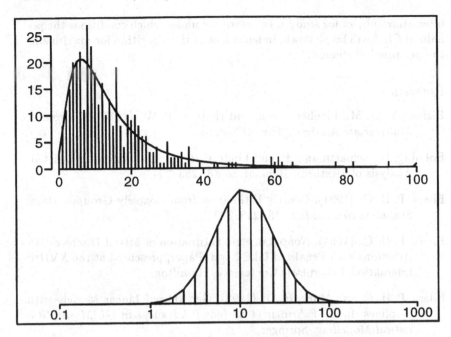

FIGURE 5. Upper graph: distribution of the daily number of environmental complaints about annoying odours (vertical bars) and fitted mixture of Poisson distributions. Lower graph: the estimated distribution of λ.

1992). We cannot avoid some form of discretization however, because of the matrix H in $\check{X} = CHX$ (equation 2): in each iteration it depends on the current estimate of γ. Each element of $\check{X}$ then is a (numerical) integral of the product of a row of C with a column of X (a B-spline), weighted by γ. Instead of a simple sum, a more sophisticated formula like the trapezoidal or Simpson's rule should be an improvement.

In this paper the penalized CLM was used to the analysis of counts. But many more applications can be envisioned: overdispersed binomial data (Eilers, 1994), binary data with grouped explanatory data and deconvolution of normal observations are good candidates. It also possible to extend the approach to the estimation of mixture distributions in mixed models. These results will be reported elsewhere.

Several details of the computations deserve further investigation. In the applications presented here, AIC was minimized for values of α for which convergence was obtained. But experience has shown that for relatively small values of α, the iterations in (5) will not converge. This is especially the case for higher values of d, the order of the differences in the penalty. In some sense, this is to be expected: the weaker the penalty, the weaker the stabilizing effect it will have. But it was also found that with damping schemes convergence is slower, but smaller values of α can be used. An interesting subject for study is to determine under which conditions the penalized CLM will break down, independent of the algorithm for maximizing the penalized likelihood.

References

Bishop, Y. M. M., Fienberg, S. E. and Holland, P. W. (1975). *Discrete Multivariate Analysis*. The MIT Press.

Böhning, D., Schlattmann, P. and Lindsay, B. (1992). Computer-assisted Analysis of Mixtures. *Biometrics*, **48**, 283–304.

Eilers, P. H. C. (1991). Density Estimation from Coarsely Grouped Data. *Statistica Neerlandica*, **45**, 255–270.

Eilers, P. H. C. (1994). Nonparametric Estimation of Mixed Discrete Distributions with Penalized Likelihood. Paper, presented at the XVIIth International Biometric Conference, Hamilton.

Eilers, P. H. C. and Marx, B. D. (1992). Generalized Linear Models with P-splines. In: L. Fahrmeir *et al.* (eds.) *Advances in GLIM and Statistical Modelling*. Springer.

Hasselblad, V., Stead, A. G. and Galke, W. (1980). Analysis of Coarsely Grouped Data from the Lognormal Distribution. *Journal of the American Statistical Association*, **75**, 771–778.

Hastie, T. J. and Tibshirani, R. J. (1990). *Generalized Additive Models*. Chapman and Hall.

Heisterkamp S. (1995). *Quantitative Analysis of AIDS/HIV: Development of Methods to Support Policy Making for Infectious Disease Control*. Ph. D. Dissertation, Leiden.

Kooperberg, C. and Stone. C. J. (1991). A Study of Logspline Density Estimation. *Computational Statistics and Data Analysis*, **12**, 327–348.

Thompson, R. and Baker, R. J. (1981). Composite Link Functions in Generalized Linear Models. *Applied Statistics*, **30**, 125–131.

Whittaker, E. T. (1923). On a New Method of Graduation. sl Proceedings of the Edinburgh Mathematical Society, **41**, 63–75.

Model Estimation in Nonlinear Regression

Joachim Engel
Alois Kneip

ABSTRACT: Given data from a sample of noisy curves in a nonlinear parametric regression model we consider nonparametric estimation of the model function and the parameters under certain structural assumptions. An algorithm for a consistent estimator is proposed and examples given.

KEYWORDS: Regression Models; Smoothing; Self–Modeling Regression; Analysis of Samples of Curves

1 Introduction

The observations in many experiments of biomedicine and the physical sciences represent noisy data of a smooth curve not just for one but for many experimental units. At consecutive times t (doses, locations or, generally speaking, levels of the covariate) observations Y are collected for a whole sample of units. For example, in botanical growth analysis the size $Y_{i,j}$ of each of, say, m plants is measured at n different time points $t_j, j = 1, \ldots, n$ leading to the model

$$Y_{ij} = f_i(t_j) + \epsilon_{ij}, j = 1, \ldots, n, i = 1, \ldots, m \tag{1}$$

with iid residuals ϵ_{ij} and $E(\epsilon_{ij}) = 0, \text{Var}(\epsilon_{ij}) = \sigma_i^2$.
While under some general smoothness assumption the regression curves f_i of the individual experimental units can be estimated without parametric specifications, this approach neglects any relationship that these curves may enjoy among themselves. After all, they all are realizations of one and the same (physical, chemical, biological etc.) process.
Models for statistical analysis of such multivariate data range from prespecified parametric models and the classical principle component approach (Rao, 1958) to completely nonparametric methods (Kneip and Gasser, 1992). Parametric specifications of the regression model such as

$$f_i(t) = f(t, \theta_i) \quad \text{for some unknown parameter } \theta_i \text{ and known } f \tag{2}$$

suffer from the fact, that a specific parametric model is rarely available in practice. It commonly is obtained through a subjective choice and lacks henceforth an objective foundation.

Instead of specifying the model function f a priori, we propose a semi-parametric approach which allows to solve the problem of model selection through model estimation.

The idea of estimating model function and individual parameters in non-linear regression has been introduced as SElf-MOdeling Regression (SE-MOR) by Lawton, Sylvestre and Maggio (1972) and by Kneip and Gasser (1988). One situation that allows SEMOR methods to work, is the shape invariant model which assumes that all individual curves f_i are obtained from a general shape function μ through parametric transformations of the axes, i.e. $f_i(t) = \theta_i^1 \mu((t - \theta_i^3)/\theta_i^2) - \theta_i^4, \theta_i = (\theta_i^1, \ldots, \theta_i^4) \in \mathbb{R}_+^2 \times \mathbb{R}^2$. Under some normalizing conditions for the parameters θ_i Kneip and Engel (1995) study the asymptotics of an algorithm that estimates at the same time the parameters θ_i, the model function μ and hence the individual regression curves f_i. A slightly different algorithm under shape invariance has been proposed by Eilers (1993).

Section 2 introduces the notation, discusses the problem of identifiability and defines a large class of problems for which SEMOR works. Crucial is the development of a concept that makes the simultaneous estimation of parameters and model function a well-defined problem. We specify requirements for the structural relationship among the individual regression curves f_i. While identifiability is discussed in a determinstic framework, estimation in SEMOR is considered in Section 3. A general algorithm is proposed and various examples given. Section 4 shows how SEMOR works in practice. In modeling human growth from begin of puberty to adulthood, we apply the proposed algorithm under the assumption of linear shape invariance.

2   Notation, Identifiability

Problem:
Given are data $(t_j, Y_{i,j}) \in J \times \mathbb{R}, J \subset \mathbb{R}$ representing the measurements of the ith experimental unit at the jth design point t_j, we consider regression models of the form (1). We assume that the regression functions f_i are unknown, but share some qualitative structure. Each regression function f_i is a parametric realization of an underlying model function of the form $f_i(t) = f(t, \theta_i)$ for θ_i belonging to some parameter space $\Theta \subset \mathbb{R}^d$. If the model function f were known, estimation of f_i would be a case of parametric nonlinear regression. Based on common structural properties to be specified below, our objective is to estimate the model function f as well as the individual parameters θ_i.

Goal:

Based on (t_j, Y_{ij}) estimate $f(\cdot, \cdot) \in C(J \times \Theta, \mathbb{R})$ and $\theta_i \in \Theta$.

As stated, this goal is not obtainable without further restrictions because of the problem of identifiability: defining $f^\star(., \theta) = f(., h^{-1}(\theta))$ for some one-to-one map h of the parameter space and denoting $\theta^\star = h(\theta)$ we have that $f^\star(., \theta^\star) = f(., \theta)$.

We solve the problem of identifiability by imposing restrictions on the parameter space and the set of model functions. Consider the following formalization: we define an operator S between function spaces as follows:

$$S : M \times \Theta \longrightarrow G \quad \text{where} M, G \subset C(J, \mathbb{R}^k), \Theta \subset \mathbb{R}^d, G = S(M \times \Theta)$$

Identifiability in our context becomes a matter of inverting the operator S. Therefore, we require from the operator S

1. For all $\theta_1, \theta_2 \in \Theta, \mu_1 \in M$ there exists $\mu_2 \in M$ with $S(\mu_1, \theta_1) = S(\mu_2, \theta_2)$.

2. (a) If $S(\mu, \theta_1) = S(\mu, \theta_2)$ then $\theta_1 = \theta_2$

   (b) If $S(\mu_1, \theta) = S(\mu_2, \theta)$ then $\mu_1 = \mu_2$.

In application we usually have $m \geq 1$ noisy curves $g_i \in G, i = 1, \ldots, m$ where M and G are subsets of function spaces of the same dimension. This implies that $M \times \Theta$ is usually larger than G. Hence the mapping S cannot be one-to- one. Given an element $g \in G$ there are no uniquely identifiable elements $\mu \in M$ and $\theta \in \Theta$ such that $S(\mu, \theta) = g$. In order to guarantee identifiability, we restrict function space and parameter space as follows:

Identifiability:

Let $M_0 \subset M, \Theta_0^{(m)} \subset \Theta^m = \Theta \times \ldots \Theta$. S then induces a mapping $S^{(m)}$: $M_0 \times \Theta_0^{(m)} \longrightarrow G^m$. In the following the 6-tuple $\Sigma = (M, M_0, \Theta, \Theta_0^{(m)}, G, S)$ with is called SEMOR model. The SEMOR model Σ is identifiable if $S^{(m)}$ is one-to-one. The smallest number m such that Σ is identifiable is called rank of Σ. $M \times \Theta^m$ is much larger than $M_0 \times \Theta_0^{(m)}$. Formally, we could say we introduce an equivalence relation on $M \times \Theta^m$:

$$\left(\mu, \begin{pmatrix} \theta_1 \\ \vdots \\ \theta_m \end{pmatrix} \right) \sim \left(\mu^\star, \begin{pmatrix} \theta_1^\star \\ \vdots \\ \theta_m^\star \end{pmatrix} \right) \Leftrightarrow S(\mu, \theta_i) = S(\mu^\star, \theta_i^\star), \quad i = 1, \ldots, m.$$

Defining a normalization operator

$$N : M \times \Theta^m \longrightarrow M_0 \times \Theta_0^{(m)}, \tag{3}$$

the equivalence classes $M \times \Theta^m / \sim \, \cong M_0 \times \Theta_0^{(m)}$ consist of the class of "normalized" elements.

Examples:

1. The bilinear shape invariant model (rank $= 1, d = 4$)
   $M = C(J, \mathbb{R}), \Theta = \mathbb{R}_+^2 \times \mathbb{R}^2, S(\mu, \theta) = \theta^1 \mu((t - \theta^3)/\theta^2) + \theta^4$
   We consider the following normalizing conditions:
   $M_0 = \{\mu \in M \mid \mu \text{ has 2 uniquely identifiable extrema}\}$
   $\Theta_0^{(m)} = \{(\theta_1, \ldots, \theta_m) \mid 1/m \sum_{i=1}^m (\theta_i^1, \theta_i^2, \theta_i^3, \theta_i^4) = (1, 1, 0, 0)\}$.
   $S^{(m)} : M_0 \times \Theta_0^{(m)} \to G$ is one-to-one: let $S^{(m)}(\mu, \theta) = S^{(m)}(\bar{\mu}, \bar{\theta})$, i.e.

   $$\theta_i^1 \mu \left(\frac{t - \theta_i^3}{\theta_i^2} \right) + \theta_i^4 = \bar{\theta}_i^1 \bar{\mu} \left(\frac{t - \bar{\theta}_i^3}{\bar{\theta}_i^2} \right) + \bar{\theta}_i^4 \text{ for all } i = 1, \ldots, m. \quad (4)$$

   Denote the location of the extrema of μ by x_1, x_2, of $\bar{\mu}$ by $\bar{x}_1, \bar{x}_2$. Then (4) implies that

   $$\theta_i^2 x_1 + \theta_i^3 = \bar{\theta}_i^2 \bar{x}_1 + \bar{\theta}_i^3, i = 1, \ldots, m \quad (5)$$
   $$\theta_i^2 x_2 + \theta_i^3 = \bar{\theta}_i^2 \bar{x}_2 + \bar{\theta}_i^3, i = 1, \ldots, m \quad (6)$$

   By averaging (5) and (6) we conclude that $x_1 = \bar{x}_1, x_2 = \bar{x}_2$ implying that $\theta_i^2 = \bar{\theta}_i^2$ and $\theta_i^3 = \bar{\theta}_i^3$. By averaging (4) it follows $\mu = \bar{\mu}$. Evaluating (4) at two points with different μ value yields $\theta_i^1 = \bar{\theta}_i^1, \theta_i^4 = \bar{\theta}_i^4$. Alternative normalizing conditions for the shape invariant model are possible. For example, in Härdle and Marron (1990) normalization is done by using the first curve as reference, i.e. by requiring that $(\theta_1^1, \theta_1^2, \theta_1^3, \theta_1^4) = (1, 1, 0, 0)$. For a third possibility, see Eilers (1993).

2. Smoothed Principal Component Analysis (rank$= k \in \mathbb{N}, d = k^2$):
   $S(\mu, \theta) = (\sum_{i=1}^k \theta_{ji} \mu_i(t), j = 1, \ldots, k)'$ where $\mu \in C(J, \mathbb{R}^k), \Theta = \mathbb{R}^{k^2}$. To guarantee identifiability we define $M_0 = \{(\mu_1, \ldots, \mu_k) \in M \mid \int_J \mu_i(t)\mu_j(t)dt = \delta_{ij}\}, \Theta_0 = \{(\theta_{i,j}) \in \Theta \mid \theta \text{ invertible and } \theta'\theta = \text{diag}(\vartheta_{11}, \ldots, \vartheta_{kk}) \text{with } \vartheta_1^2 \geq \ldots \geq \vartheta_{kk}^2\}$. Then $S : M_0 \times \Theta_0 \longrightarrow C([0, 1], \mathbb{R}^k)$ is one-to-one:
   Given $g = (g_1, \ldots, g_k)' \in G$, the symmetric matrix $(\int gg')^{1/2}$ allows the unique decomposition as $(\int gg')^{1/2} = PDP'$ where P is an orthogonal matrix and $D = \text{diag}(d_1, \ldots, d_k)$ a diagonal matrix with $d_1^2 \geq d_2^2 \geq \ldots \geq d_k^2$. Then $\theta = PD^{1/2}$ and $\mu = \theta^{-1}g = D^{-1/2}P'g$ are the unique elements in $M_0 \times \Theta_0^{(m)}$ with $g = \theta\mu$.

3. Principal components with shifts: (rank$=k \in \mathbb{N}, d = 4k^2$)
   $S(\mu, \theta) = \sum_{i=1}^k \theta_{ij}^1 \mu_i((t - \theta_{ij}^3)/\theta_{ij}^2) + \theta_{ij}^4$, where $M = C(J, \mathbb{R}^k), \Theta = (\mathbb{R}_+^2 \times \mathbb{R}^2)^k$. Let $A_j = \cup_{i=1}^n \text{supp}\mu_j ((. - \theta_{ij}^3)/\theta_{ij}^2)$. Define $M_0 = \{\mu \in M \mid A_i \cap A_j = \emptyset \text{ for all } i \neq j, \mu_j \text{ has 2 uniquely identifiable extrema}\}$ and let $\Theta_0^{(m)} = \{(\theta_{ij}^1, \theta_{ij}^2, \theta_{ij}^3, \theta_{ij}^4) \in \Theta \mid 1/n \sum_{i=1}^n (\theta_{i,j}^1, \theta_{ij}^2, \theta_{ij}^3, \theta_{ij}^4) = (1, 1, 0, 0)\}$ for all $j = 1, \ldots, k$.

3   Estimation in SEMOR

Assume an identifiable SEMOR model $\Sigma = (M, M_0, \Theta, \Theta_0^{(m)}, G, S)$. The operator $S : M \times \Theta \longrightarrow G$ induces for all $\theta \in \Theta$ a bijective operator

$$S_\theta : M \longrightarrow G, \quad S_\theta(\mu) = S(\theta, \mu)$$

S_θ is one-to-one and onto. This follows immediately from property (1) and (2) of Section 2.

Loss function:

Given functions $(g_1, \ldots, g_m) \in G^m$, generated from the SEMOR model Σ, i.e. there exist "true" normalized parameters $(\theta_1^\star, \ldots, \theta_m^\star)' \in \Theta_0^{(m)}$ and a "true" model $\mu^\star \in M_0$ such that $g_i = S(\mu^\star, \theta_i^\star)$ for all $i = 1, \ldots, m$. When estimating the unknown true parameters by $(\theta_1, \ldots, \theta_m) \in \Theta_0^{(m)}$ and the model function μ by $1/m \sum_{i=1}^n S_{\theta_i}^{-1}(g_i)$ we incur a loss given as follows:

$$L(\theta_1, \ldots, \theta_m) = \sum_{i=1}^m \int_J \left\| S\left(1/m \sum_{j=1}^m S_{\theta_j}^{-1}(g_j), \theta_i \right) - g_i \right\|_2^2.$$

Obviously, $L \geq 0$. Assume $L = 0$. Then $S\left(1/m \sum_{j=1}^m S_{\theta_j}^{-1}(g_j), \theta_i\right) = g_i = S(\mu^\star, \theta_i^\star)$ for all $i = 1, \ldots, m$. If $(\theta_1, \ldots, \theta_m) \in \Theta_0^{(m)}$, it follows by the identifiability of the model that $\theta_i = \theta_i^\star$, $1/m \sum_{j=1}^m S_{\theta_j}^{-1}(g_j) = \mu^\star$, i.e. the loss function assumes the value 0 only for the true parameters.

Estimation Procedure:

In applications the "true" individual curves g_i are not available. Therefore we replace the unknown functions g_i by nonparametric estimates, say, based on the kernel estimator or local polynomial regression estimators. Then the objective is to minimize:

$$\hat{L}(\theta_1, \ldots, \theta_m) = \sum_{i=1}^m \int_J \left\| S\left(1/m \sum_{j=1}^m S_{\theta_j}^{-1}(\hat{g}_j), \theta_i \right) - \hat{g}_i \right\|_2^2.$$

Minimizing $\hat{L}$ leads (depending on the structure of Σ) to a high $(m \times d)$ dimensional least squares problem, which is computationally quite impractical. Therefore, we propose using backfitting in the following

Algorithm:

Step 1: Start with initial estimates $\hat{\theta}_i^{(0)} \in \Theta_0^{(m)}$ and let

$$\hat{\mu}^0 = \frac{1}{m} \sum_{j=1}^m S_{\hat{\theta}_j^{(0)}}^{-1}(\hat{g}_j).$$

Step 2: For $u = 1, \ldots, u^\star$

(a) Solve the m separate d–dimensional least square problems

$$\bar{\theta}_i^{(u)} = \text{argmin} \int_J \|S(\hat{\mu}^{(u-1)}, \theta) - \hat{g}_i\|^2 \quad , \quad i = 1, \ldots, m.$$

(b) Set

$$\bar{\mu}^{(u)} \;=\; \frac{1}{m} \sum_{i=1}^m S_{\bar{\theta}_i^{(u)}}^{-1}(\hat{g}_i) \tag{7}$$

(c) Then "normalize" parameter and model estimates (see (3))

$$(\hat{\mu}^{(u)}, \hat{\theta}_i^{(u)}) = N(\bar{\mu}^{(u)}, \bar{\theta}_i^{(u)}).$$

Step 3: Finally, set

$$\hat{\theta}_i \;=\; \hat{\theta}_i^{(u^\star)}$$

$$\hat{\mu} \;=\; \frac{1}{m} \sum_{i=1}^m S_{\hat{\theta}_i}^{-1}(\hat{g}_i^\star). \tag{8}$$

Here $\hat{g}^\star$ denotes a nonparametric estimate of g based on an amount of smoothing that is allowed to be different from the smoothing applied to obtain $\hat{g}_i$ in step 1 and 2.

Remarks:

1. In many applications, with clever initial estimates a few iterations suffice in order to achieve the best obtainable convergence rate of $\sqrt{n}$ for the parameter estimates (see Kneip and Engel, 1995).

2. While the estimate for the model μ in (7) is used to derive appropriate parameter estimates, the goal in (8) is to obtain good estimates for the model function μ. This being a slightly different objective makes it advantageous to allow for a different amount of smoothing in (8) than in (7).

Assumptions:

1. The inital parameter estimates $\hat{\theta}_i^{(0)}$ are a.e. consistent for θ_i.

2. The individual curve estimates $\hat{g}_i$ and $\hat{g}_i^\star$ are a.e. consistent for g_i.

3. The one-to-one operator $S^{(m)} : M_0 \times \Theta_0^{(m)} \to G$ is continuous in the L_2 norm and has a continuous inverse.

4. The normalizing operator N is continuous.

Theorem 1 *Assume an identifiable SEMOR model Σ. Under above assumptions $\hat{\theta}_i$ and $\hat{\mu}$ are a.e. consistent estimators for θ and μ.*

Proof: From the assumptions it follows immediately that $\hat{\mu}^0$ is consistent.

$$\|S(\hat{\mu}^{(u-1)}, \bar{\theta}_i^{(u)}) - \hat{g}_i\| \leq \|S(\hat{\mu}^{(u-1)}, \hat{\theta}_i^{(u-1)}) - g_i\| + \|g_i - \hat{g}_i\|$$

$$\leq \|S(\hat{\mu}^{(u-1)}, \hat{\theta}_i^{(u-1)}) - S(\mu, \hat{\theta}_i^{(u-1)})\| + \|S(\mu, \hat{\theta}_i^{(u-1)}) - g_i\| + \|g_i - \hat{g}_i.\|$$

By assumptions all three terms on the right converge to 0. Consistency of $\hat{\theta}_i$ now follows from the continuity of the normalizing operator while consitency of $\hat{\mu}$ is a direct consequence of the continuity of the S^{-1}.□

4   Example: Modelling Human Growth

The data stem from the Zürich longitudinal growth study. The sample consists of measurements on body size of $N = 112$ girls and $N = 120$ boys, taken yearly or half–yearly (in puberty), from birth to adulthood. We are interested in estimating the velocity of human growth, separately for boys and girls . We assume that the linear shape invariant model (see Example 1 above) with $J = [\bar{\tau}, 20]$ holds, where $\bar{\tau}$ denotes average puberty entrance age. The puberty entrance age for each child is estimated as that time point where the nonparametrically estimated velocity curve has a local minimum before the pubertal spurt (PS). For details, see Gasser et al. (1985a, 1985b). A slight modification has to be done due to the fact that we are dealing with derivatives and not with the regression functions themselves. We apply kernel estimatore tailored for estimating derivatives (Gasser and Müller, 1984). To apply the above algorithm initial estimators are needed for the shift parameters along the time axis, i.e. for $\theta_i^{(2)}$ and $\theta_i^{(3)}$. These were obtained via the locations of two extremal points τ_{1i} and τ_{2i} as the age of maximal acceleration of the i–th child in the pubertal growth spurt and the age of maximum velocity in pubertal growth spurt of the ith child.

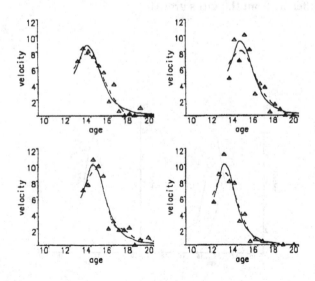

FIGURE 1. Comparison of two different estimates of the growth curve for four boys: a kernel estimate based on measurements for the ith child only (dashed line) and a SEMOR estimate (solid line). The triangles represent the divided differences of the data.

These functionals are determined from nonparametric estimates of growth acceleration curves (see Müller, 1985) . The initial parameter estimators are obtained as follows (compare (5) and (6)):

$$\hat{\theta}_{i2}^0 = \frac{\hat{\tau}_{i2} - \bar{\hat{\tau}}_{i2}}{1/m \sum_{i=1}^m (\hat{\tau}_{i2} - \bar{\hat{\tau}}_{i2})}, \hat{\theta}_{i3}^0 = \hat{\tau}_{i1} - \hat{\theta}_{i2}^0 \frac{1}{m} \sum_{i=1}^m \hat{\tau}_{i1}$$

As Kneip and Engel (1995) show, in above algorithm $u^* = 3$ iterations suffice to obtain $\sqrt{n}$ convergence for the parameter estimates $\hat{\theta}_i$. Further, in this particular application, all vertical shift parameters are set $\theta_i^4 = 0$ since growth ends after puberty. Provided shape invariance holds, above algorithm leads to estimators with much higher efficiency compared with nonparametric estimates of the individual curves f_i. Kneip and Engel (1995) show that when using kernel estimators the rate of MISE convergence for estimated individual curves increases from $n^{-2k/2k+1}$ to $(mn)^{-2k/2k+1}$ when using kernels of order k. Finally, Figure 2 compares the estimated model functions $\hat{\mu}_{\text{boys}}, \hat{\mu}_{\text{girls}}$ for boys and girls. The picture on the left shows the estimated model functions after 3 iteration steps representing average growth during PS for girls and for boys. The picture on the right shows $\hat{\mu}_{\text{boys}}$ and $\vartheta_1 \hat{\mu}_{\text{girls}}((t - \vartheta_3)/\vartheta_2)$, where the parameters ϑ are obtained by a least squares fit to the boys model. It suggests that the pubertal growth of boys is structurally not different from the girl's growth.

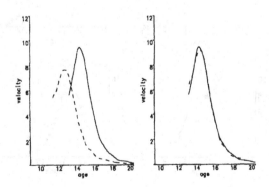

FIGURE 2. Estimate of model function for human growth between puberty entrance and age 20. Solid line: boys, dotted line: girls [left]. Model for boys and model for girls fitted to boys model[right].

References

Eilers, P. (1993). Estimating Shapes with Projected Curves. *8th International Workshop on Statistical Modelling. Louvain*

Härdle, W. and Marron, S. (1990). Semiparametric Comparison of Regression Curves. *Annals of Statistics*, **18**, 63 – 90

Gasser, T. and Müller, H. G. (1984). Estimating Regression Functions and their Derivatives by the Kernel Method. *Scandinavian Journal of Statistics*, **11**, 171 – 185

Gasser, T., Müller, H.G., Köhler, W., Prader, A., Largo, R., Molinari, L. (1985a). An Analysis of the Mid–growth and Adolescent Spurts of Height Based on Acceleration. *Annals of Human Biology* **12**, 129–148

Gasser, T. Köhler, W., Müller, H.G., Largo, R., Molinari, L., Prader, A. (1985b). Human Height Growth: Correlational and Multivariate Structure of Velocity and Acceleration. *Annals of Human Biology* **6**, 501–515

Kneip. A. and Engel, J. (1995). Model Estimation in Nonlinear Regression under Shape Invariance. *Annals of Statistics,***23**

Kneip, A. and Gasser, T. (1988). Convergence and Consistency Results for Self-modeling Nonlinear Regression. *Annals of Statistics*, **16**, 82 – 112

Kneip, A. and Gasser, T. (1992). Statistical Tools to Analyze a Sample of Curves. *Annals of Statistics*, **20**, 1266–1305

Lawton, W.H., Sylvestre, E.A., Maggio, M.S. (1972). Self-modeling Nonlinear Regression. *Technometrics*, **14**, 513 – 532

Müller, H.G (1985). Kernel Estimators of Zeros and Location and Size of Extrema of Regression Functions. *Scandivanian Journal of Statistics,***12**, 221 – 232

Rao, C.R. (1958). Some Statistical Methods for Comparison of Growth Curves. *Biometrics*, **14**, 1 – 17

Pearson statistics, goodness of fit, and overdispersion in generalised linear models

C.P. Farrington

1   Introduction

The Pearson statistic is commonly used for assessing goodness of fit in generalised linear models. However when data are sparse, asymptotic results based on the chi square distribution may not be valid. McCullagh (1985) recommended conditioning on the parameter estimates and obtained approximations to the first three conditional moments of Pearson's statistic for generalised linear models with canonical link functions. This paper presents a generalisation of these results to non-canonical models, derived in Farrington (1995). A first order linear correction term to the Pearson statistic is defined which induces local orthogonality with the regression parameters, and leads to substantial simplifications in the expressions for the first three conditional and unconditional moments. Expressions are given for Poisson, binomial, gamma and inverse Gaussian models. The power of the modified statistic to detect overdispersion is assessed, and the methods are applied to adjusting the bias of the dispersion parameter estimate in exponential family models.

2   A family of modified Pearson statistics

Let $Y_i, i = 1...n$, denote independent random variables sampled from an exponential family distribution with mean μ_i and variance function V_i, where V_i is a function of μ_i. Suppose that a generalised linear regression model has been defined with inverse link function h, $n \times p$ model matrix X and regression parameters β:

$$h^{-1}(\mu) = \eta = X\beta.$$

The maximum likelihood estimates of β may be obtained as solutions of the p estimating equations $g_r(\beta) = 0$, $r = 1...p$, where:

$$g_r(\beta) = \sum_{i=1}^{n} \frac{(y_i - \mu_i)}{V_i} \frac{\partial \mu_i}{\partial \beta_r}.$$

In order to assess goodness of fit, the model is embedded in a wider family with variance function ϕV_i. Departures from the null model with $\phi = 1$ are assessed by means of the additional unbiased estimating equation $g_{p+1}(\beta, \phi) = 0$ defined by:

$$g_{p+1}(\beta, \phi) = \sum_{i=1}^{n} \left[\frac{(y_i - \mu_i)^2}{V_i} - \phi \right] + \sum_{i=1}^{n} a_i(y_i - \mu_i)$$

where a_i is a function of μ_i. This gives rise to a family of goodness of fit statistics:

$$P_a = \sum_{i=1}^{n} \frac{(y_i - \hat{\mu}_i)^2}{\hat{V}_i} + \sum_{i=1}^{n} \hat{a}_i(y_i - \hat{\mu}_i).$$

The choice $a_i = 0$ yields the Pearson statistic $P = P_0$. McCullagh (1985) showed that, for models with canonical link functions, the Pearson statistic P and the regression parameter estimates are asymptotically uncorrelated, and derived approximations to the first three moments of the unconditional and conditional distribution of P. These results may be extended to the family of modified Pearson statistics P_a, with arbitrary link functions. Full details are given in Farrington (1995). In particular for generalised linear models the P_a and the regression parameters are asymptotically uncorrelated. A special case of particular importance arises when:

$$a_i = -V'_i/V_i$$

where the prime denotes differentiation with respect to μ_i. This choice of a_i minimizes the unconditional variance of P_a, and removes the dependence of its unconditional expectation on the first-order bias of the regression parameter estimates. In addition it induces substantial simplifications in the approximations to the first three unconditional and conditional moments of P_a. Let P^* be the modified Pearson statistic corresponding to this choice of a_i. Thus:

$$P^* = \sum_{i=1}^{n} \frac{(y_i - \hat{\mu}_i)}{\hat{V}_i} - \sum_{i=1}^{n} \frac{\hat{V}_i'}{\hat{V}_i}(y_i - \hat{\mu}_i).$$

The first three conditional moments of P^* are:

$$E(P^*|\hat{\beta}) = n - p - \frac{1}{2} \cdot \sum_{i=1}^{n} \frac{\hat{V}_i''}{\hat{V}_i} \hat{h}'^2 \hat{Q}_{ii} + O(n^{-1/2})$$

$$var(P^*|\hat{\beta}) = (n-p)(\hat{\rho}_4 - \hat{\rho}_3^2 + 2) + O(n^{1/2})$$

$$\kappa_3(P^*|\hat{\beta}) = (n-p)\left(\hat{\rho}_6 - 3\hat{\rho}_{35} + 3\hat{\rho}_{34}^2 + 12\hat{\rho}_4 - 8\hat{\rho}_3 + 8\right) + O(n^{1/2})$$

where $Q = X(X^T W X)^{-1} X^T$ with $W = \text{Diag}(h_i'^2)/V_i$ is the approximate covariance matrix of the estimated linear predictors, and:

$$\hat{\rho}_{rs}^t = n^{-1} \sum_i (\hat{\kappa}_{ri}/\hat{V}_i^{r/2})^t (\hat{\kappa}_{si}/\hat{V}_i^{s/2}).$$

These expressions are easily evaluated, for instance using GLIM in which the estimated Q_{ii} are available through the vector %vl.

3 Goodness of fit

The choice $a_i = -V_i'/V_i$ is shown in Farrington (1995) to induce local orthogonality between P^* and β so that the modified Pearson statistic P^* depends only weakly on β, given the maximum likelihood estimate. Thus the conditional distribution of P^* may be used to assess goodness of fit even if a sufficient reduction of the data is not available. Upper tail probabilities may be calculated, based on the approximate standardised variable:

$$Z = (P^* - E(P^*|\hat{\beta}))/var(P^*|\hat{\beta})^{1/2}$$

either directly by reference to the standard normal distribution, or, as suggested by McCullagh (1986), using an Edgeworth approximation.

3.1 Poisson models

For Poisson data $V_i = \mu_i$, $a_i = -1/\mu_i$ and hence:

$$E(P^*|\hat{\beta}) \doteq n - p$$
$$var(P^*|\hat{\beta}) \doteq 2(n-p)$$
$$\kappa_3(P^*|\hat{\beta}) \doteq 8(n-p)\left(1 + \frac{1}{2n} \cdot \sum_{i=1}^n \hat{\mu}_i'^{-1}\right).$$

3.2 Binomial models

In this case $V_i = m_i \pi_i (1 - \pi_i)$, $a_i = -(1 - 2\pi_i)[m_i \pi_i (1 - \pi_i)]^{-1}$ hence:

$$E(P^*|\hat{\beta}) \doteq n - p + \sum_{i=1}^n \frac{\hat{h}_i'^2}{m_i^2 \hat{\pi}_i (1 - \hat{\pi}_i)} \hat{Q}_{ii}$$
$$var(P^*|\hat{\beta}) \doteq 2(1 - p/n) \sum_{i=1}^n \frac{m_i - 1}{m_i}$$
$$\kappa_3(P^*|\hat{\beta}) \doteq 8(n-p)$$
$$\cdot \left(1 - \frac{1}{n} \cdot \sum_{i=1}^n \frac{5m_i - 4}{m_i^2} + \frac{1}{2n} \cdot \sum_{i=1}^n \frac{m_i - 1}{m_i^2}[\hat{\pi}_i(1 - \hat{\pi}_i)]^{-1}\right).$$

In the special case $m_i = m$ the approximate conditional expectation reduces to $n - p(1 - m^{-1})$.

3.3   3.3 Gamma models

For gamma data with known index ν, $V_i = \mu_i^2/\nu$, $a_i = -2/\mu_i$ and so:

$$
\begin{aligned}
E(P^*|\hat{\beta}) &\doteq n - (1 + 1/\nu)p \\
var(P^*|\hat{\beta}) &\doteq 2(n - p)(1 + 1/\nu) \\
\kappa_3(P^*|\hat{\beta}) &= 8(n - p)(1 + 1/\nu)(1 + 4/\nu).
\end{aligned}
$$

3.4   Inverse Gaussian models

For inverse Gaussian data with known index λ, $V_i = \mu_i^3/\lambda$, $a_i = -3/\mu_i$. Hence:

$$
\begin{aligned}
E(P^*|\hat{\beta}) &\doteq n - p - 3 \cdot \sum_{i=1}^{n} \frac{\hat{h}_i'^2}{\hat{\mu}_i^2}\hat{Q}_{ii} \\
var(P^*|\hat{\beta}) &\doteq 2(n - p)(1 + 3\bar{\mu}/\lambda) \\
\kappa_3(P^*|\hat{\beta}) &\doteq 8(n - p)\left(1 + \frac{27\bar{\mu}}{2\lambda} + \frac{81\bar{\mu}^2}{2\lambda^2}\right)
\end{aligned}
$$

where $\bar{\mu} = n^{-1}\sum \hat{\mu}_i$, $\bar{\mu}^2 = n^{-1}\sum \hat{\mu}_i^2$.

4   Overdispersion

Overdispersion with respect to a reference distribution is a common cause of lack of fit, and arises frequently in practice. In this section we concentrate on a simple form of overdispersion in which the variance function V_i is scaled by a constant dispersion factor ϕ.

4.1   Power to detect overdispersion

The power of the tests based on the approximations to the conditional distributions of P^* and P, with and without the Edgeworth correction, and of the nominal chi-square test were compared in a simulation study. For $n = 50$, binomial $B(m, p_i)$ observations with $m = 2$ were generated, resulting in very sparse data. Overdispersion was introduced by allowing the probabilities p_i to vary about their expected values π_i according to a beta distribution, thus producing beta-binomial data with dispersion parameter $1 \leq \phi < 2$. A strong univariate regression effect on the logistic scale was imposed on the π_i, and a logistic binomial model was fitted. For each of a selected set of values of ϕ 10,000 simulations were generated and the proportion with an upper tail probability less than 0.05 was calculated for each test statistic. The results are shown in Figure 1.

Figure 1: Power to detect overdispersion

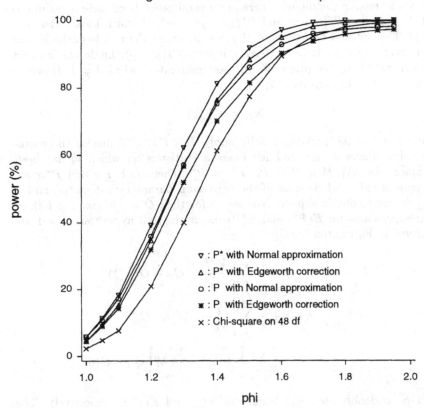

The test based on the conditional modified Pearson statistic P^* performs better than either the conditional P or the nominal chi-square test on 48 degrees of freedom. Perhaps surprisingly, for values of ϕ in excess of 1.75, the test based on the conditional distribution of P has slightly lower power than the chi-square. The Edgeworth correction for the test statistics based on P and P^* produced in type I errors marginally closer to the nominal value of 0.05: 0.0452 compared to 0.0578 without the Edgeworth correction for P^*, 0.0448 instead of 0.0579 for P. In contrast the type I error for the chi-square test was 0.0226. For high values of ϕ the Edgeworth correction had little effect, though it occasionally resulted in negative tail probabilities, which were counted as 'significant'.

4.2   Estimating the dispersion parameter

Once overdispersion has been established, and the true variance function may be assumed to be of the form ϕV_i, an estimate of ϕ is required with which to rescale the model. There are several possible consistent estimates of ϕ, including $P/(n-p)$ and $P^*/(n-p)$, both of which have order n^{-1} bias. The optimal properties of $P^*/(n-p)$ under the null hypothesis $\phi = 1$ provide some basis for preferring it over $P/(n-p)$. Little can be said in general about the properties of these estimators when $\phi \neq 1$. However when the third moments of Y_i satisfy:

$$\kappa_{3i} = \phi^2 V_i' V_i$$

then some of the previous results apply. Let $P\phi$, $P^*\phi$ denote the corresponding Pearson and modified Pearson statistics, in which V_i has been replaced by ϕV_i; thus $P = P_1$, $P^* = P_1^*$. Then both $P\phi$ and $P^*\phi$ are asymptotically independent of the regression parameter estimates, and it is possible to obtain explicit expressions for their $O(n^{-1})$ bias. The following expressions for $E(P^*)$ and $E(P)$ under the null hypothesis $\phi = 1$ are derived in Farrington (1995):

$$E(P^*) \;=\; n - p - \frac{1}{2} \cdot \sum_{i=1}^{n} \frac{V_i''^2}{V_i} h_i'^2 Q_{ii} + O(n^{-1})$$

$$E(P) \;=\; n - p + \frac{1}{2} \sum_{i,j=1}^{n} \frac{V_i'}{V_i} \frac{h_i''}{V_j} h_i' h_j' Q_{ij} Q_{jj}$$

$$- \frac{1}{2} \sum_{i=1}^{n} \left[\frac{V_i''}{V_i} h_i'^2 + \frac{V_i'}{V_i} h_i'' \right] Q_{ii} + O(n^{-1}).$$

Let α, β denote the $O(1)$ terms of $E(P)$ and $E(P^*)$, respectively. The corresponding terms for $E_\phi(P_\phi)$, $E_\phi(P_\phi^*)$ are then $\phi\alpha$ and $\phi\beta$, where E_ϕ denotes the expectation with respect to the true distribution. It follows that:

$$E_\phi(P/(n-p)) \;=\; \phi + n^{-1}\phi^2\alpha + O(n^{-2})$$
$$E_\phi(P^*/(n-p)) \;=\; \phi + n^{-1}\phi[\beta + (\phi-1)\alpha] + O(n^{-2})$$
$$E_\phi(P/E(P)) \;=\; E_\phi(P^*/E(P^*)) \;=\; \phi + n^{-1}\phi(\phi-1)\alpha + O(n^{-2})$$

The estimator $P^*/E(P^*)$ combines lesser bias with ease of estimation, and is unbiased when $\phi = 1$.

4.3   Bias reduction

These results may be applied to reducing the $O(n^{-1})$ bias in the dispersion parameter estimate in exponential family models, such as gamma and inverse Gaussian models. Consider for example the problem of estimating the dispersion parameter ϕ for gamma distributed data with index ν, where $\phi = 1/\nu$. McCullagh and Nelder (1989) propose the estimator $P/(n-p)$ where P is the Pearson statistic based on the exponential distribution, with $V_i = \mu_i^2$. This estimate has $O(n^{-1})$ bias $\phi^2 \alpha$. In this case $E(P^*) = n - 2p$ and so the estimator $P^*/E(P^*)$ is $P^*/(n-2p)$, with $O(n^{-1})$ bias $\phi(\phi-1)\alpha$. The value of α depends on the link function used. For the identity link function, $\alpha = -p$. With the log link,

$$\alpha = -2p + \sum_{i,j=1}^{n} \frac{\mu_i}{\mu_j} Q_{ij} Q_{jj}$$

while for the reciprocal link,

$$\alpha = -3p + 2 \sum_{i,j=1}^{n} \mu_i^4 Q_{ij} Q_{jj}$$

where in both cases Q is the approximate covariance matrix of the linear predictors under the exponential model.

5 Example

Table 1 shows the proportions of children in two groups with evidence of past infection by Human Herpes Virus type 6 (HHV6), by age measured in completed weeks.

Table 1: numbers tested and positive for HHV6 by age

Age (weeks)	Group A n	pos	Group B n	pos	Age (weeks)	Group A n	pos	Group B n	pos
25-29	9	3	6	5	70-74	10	10	8	6
30-34	10	5	6	6	75-79	9	8	4	4
35-39	14	7	8	5	80-84	11	9	2	1
40-44	10	8	12	12	85-89	11	8	4	3
45-49	9	4	12	10	90-94	3	3	1	1
50-54	9	4	15	14	95-99	10	8	0	0
55-59	16	13	8	8	100-104	8	6	1	1
60-64	22	9	7	5	105-109	8	7	0	0
65-69	7	7	4	4	110-114	9	9	1	1

The question of interest is whether the incidence in group B is higher than in group A. The simplest model assumes a constant age specific incidence λ and a multiplicative group effect. This leads to a generalised linear model with binomial error, number with past infection as response, and complementary log log link function:

$$\ln(-\ln(1 - p_i(a))) = \ln(a) + \alpha + \beta_i$$

where a = age, $p_i(a)$ = proportion in group i infected by age a, and $\alpha = \ln(\lambda)$. Fitting in GLIM with the age interval midpoints as a continuous variable gives a deviance of 47.32 and a Pearson statistic of 50.05 on 32 degrees of freedom. The upper tail probability for the Pearson statistic based on the chi-square distribution is 0.022. The binomial denominators however are rather small, calling into question the validity of small dispersion asymptotics. In contrast the modified Pearson statistic is 37.38 with approximate conditional mean 32.11, conditional variance 50.03 and conditional standardised third moment 0.673. The observed value of P^* is therefore well within the centre of the distribution, showing no evidence of a serious misfit. In conclusion, the incidence in group A is 0.0219 per week, 95 percent CI 0.0181 to 0.0266, and the incidence ratio in group B compared to group A is 1.73, 95 percent CI 1.26 to 2.39.

References

Farrington C.P. (1995). On assessing goodness of fit of generalised linear models to sparse data. Preprint.

McCullagh P. (1985). On the asymptotic distribution of Pearson's statistic in linear exponential family models. *International Statistical Review*, **53**, 61–67.

McCullagh P. (1986). The conditional distribution of goodness of fit statistics for discrete data. *Journal of the American Statistical Association*, **81**, 104–107.

McCullagh P. and Nelder J.A. (1989). Generalised Linear Models, 2nd edition. London: Chapman & Hall.

M-estimation: Some Remedies

Robert Gilchrist
George Portides

ABSTRACT: This paper is concerned with two problems facing M- estimators. M-estimators bound the influence of large residuals, or more generally, large deviations from the mean. In doing so, however, they can become inconsistent, in particular in the case of non-Normal Generalized Linear Models (GLMs), thus leading to biased estimates. We present a method for correcting such biased estimates without needing to alter the standard estimation procedures used in most statistical packages. Another problem facing M-estimators is that they do not take into account the (potentially high) leverage of a point. The cause of high leverage could be the mis-recording of a single explanatory variate value. We investigate the Mean Shift Outlier Model (MSOM) as discussed by Cook and Weisberg (1982), and show that this can lead to a method for reducing the leverage of a suspect point. However, we also note that in other cases, the method would increase the leverage (albeit removing the influence) and therefore some other approach is needed. Our discussion is based on the Normal case, although we have in mind extensions of these proposals in the GLM context.

KEYWORDS: Bias Correction; GLMs; Imputation; Influential Design Points; M-estimators; MSOM

1 A Brief Account on the Theory of GLMs

Following Francis et. al. (1993), we have units $i = 1, ... n$ of the response variable y_i, which we assume to follow the exponential family of distributions

$$l_i(\theta_i, \phi; y_i) = \log f(y_i; \theta_i, \phi) = \frac{y_i \theta_i - b(\theta_i)}{a(\phi)} + c(y_i, \phi), \qquad (1)$$

for some functions a, b and c. Here θ is the canonical parameter and $a(\phi)$ is the scale parameter, which we initially assume to be fixed or known. Furthermore, we have $E[Y] = \mu = b'(\theta)$ and $Var[Y] = a(\phi)b''(\theta)$ where primes denote differentiation w.r.t. θ.

In the presence of p explanatory variates, their effect is accounted for by the linear predictor $\eta_i = \sum_{j=0}^{p} x_{ij}\beta_j$ for $i = 1, ..., n$ and $x_{i0} = 1$, with the β_js being the unknown parameters to be estimated. The mean is functionally related to the linear predictor through $\eta_i = g(\mu_i)$ for some known link function g.

The MLE of $\beta' = (\beta_0, ..., \beta_p)$ can then be obtained by solving the following system of equations

$$\sum_{i=1}^{n} s_j(\underline{\beta}; y_i, x_{ij}) = 0, \quad j = 0, ..., p, \tag{2}$$

where $s_j(y; \mu) = \partial l_i / \partial \beta_j$ is the score function. Solution of (2) may be obtained through the iteratively reweighted least squares procedure (IRLS), see Francis et. al. (1993), pp 264–267.

2   M-estimators for GLMs

The score function defined above is unbounded to large deviations from the mean, hence the presence of outliers can produce unreliable estimates. One possible way of tackling this problem is through the use of M-estimators, see for example Huber (1981), which possess the property of bounding the influence function or some other variant of it, see for example Hampel et. al. (1986) section 2.1. In general, a M-estimator of the mean parameter is defined by

$$\sum_{i=1}^{n} (\psi(\mu_i; y_i) - B(\mu_i)) = 0, \tag{3}$$

where

$$B(\mu) = E_\mu [\psi(\mu; Y)] \tag{4}$$

is introduced to ensure Fisher consistency of the estimator.

The ψ-function usually takes a piecewise form in such a way that observations far way from the mean or the central part of the distribution have a bounded influence on the estimates. Alternatively, because M-estimators are commonly obtained through the IRLS procedure, the ψ-function can be expressed as $\psi(\mu; y) = s(\mu; y)w(\mu; y)$ where $s(\mu; y) = \partial l / \partial \mu = (y - \mu)/\phi v(\mu)$ and $w(\mu; y)$ is a piecewice weight function.

Example 1: Field and Smith (1994)

Define the set $A_i(\mu_i, p) = \{y : q \leq F(y; \mu_i) \leq 1 - q\}$ and $k(\mu_i) = \sup_{y \in A_i} |s(\mu_i; y)|$, where F is the underlying distribution function. The proposed weight function is then given by

$$w(\mu_i; y_i) = \min_{y_i} \left[\frac{k(\mu_i)}{|s(\mu_i; y_i)|}, 1 \right].$$

This weight function bounds the scores greater than the supremum of scores attained in the center of the (estimated) underlying distribution. In an iterative estimation scheme the bound $k(\mu)$ will change according to the current estimate of μ.

Example 2: Field and Smith (1994)

$$w(\mu_i; y_i) = \begin{cases} F(y_i; \mu_i)/q & \text{, if } F(y_i; \mu_i) < q \\ 1 & \text{, if } q \leq F(y_i; \mu_i) \leq 1 - q \\ (1 - F(y_i; \mu_i))/q & \text{, if } F(y_i; \mu_i) > 1 - q. \end{cases}$$

In this example, points lying in the $100(1 - 2q)\%$ central part of the distribution are unaffected, while points outside this region are downweighted according to their distribution function.

2.1 Computation

Attempting to solve (3) directly can prove to be extremely difficult, especially as most standard statistical packages cannot tackle problems of this kind. In addition the analytic form of $B(\mu)$ needs to be known in order to avoid repetitive numerical integration for each unit in an iterative scheme. Alternatively, we may obtain an unbiased estimate of the mean based on an indirect approach. Using the IRLS procedure, we may obtain a biased estimate, $\tilde{\mu}$, which satisfies the inconsistent estimating equation

$$\sum_{i=1}^{n} \psi(\tilde{\mu}_i; y_i) = \sum_{i=1}^{n} w(\tilde{\mu}_i; y_i) s(\tilde{\mu}_i; y_i) = 0, \tag{5}$$

and then adjust $\tilde{\mu}$ to $\hat{\mu}$ by solving

$$E_{\hat{\mu}}[\psi(\tilde{\mu}_i; Y_i) = 0. \tag{6}$$

Solution of (6) can be given iteratively using the Newton-Raphson method, i.e. solving

$$\hat{\mu}^{(r+1)} = \hat{\mu}^{(r)} - \frac{E_{\hat{\mu}^{(r)}}[\psi(\tilde{\mu}; Y)]}{E_{\hat{\mu}^{(r)}}[\psi'(\tilde{\mu}; Y)]}, \tag{7}$$

where $\psi' = \partial\psi/\partial\mu$ and superscripts indicate the current iteration with $\hat{\mu}^{(0)} = \tilde{\mu}$.

This method is computationally intensive when expectation of the ψ- function and its derivative cannot be evaluated analytically. This is due to the fact that expectations will have to be evaluated numerically for each individual point given its mean (Example 2). For continuous variables the Gaussian Quadrature can be used to perform numerical integration, while for discrete ones summation is required over all possible response values. If however, analytic expressions can be obtained (Example 1), then bias correction can be applied on the whole vector of points and their corresponding means.

3   Reducing Influence of a Single Covariate Value

We have discussed correcting M-estimators for bias; however, M-estimators tackle extreme response values without considering the leverage at a point, i.e. the influence of covariates on parameter estimates. For example, suppose we have a design matrix $X = \{x_{ij}\}$ such that one value of a single covariate, say x_{kl}, is poorly recorded. The resulting fit is likely to be affected, with an increased influence due to the increased leverage of the point. An M-estimator would simply consider a measure of deviation of the response value from its mean and downweight the unit accordingly, whilst a Bounded Influence (BI) estimator, see e.g. Hampel et. al. (1986) section 6.3., would consider both such a measure and any excessive leverage and downweight the unit. Although the latter method tackles such points, it does not take into account the reliable information of the point, i.e. the values y_k and $\left(X_{(l)}\right)_k$ where $\left(X_{(l)}\right)_k$ is the kth row of the design matrix excluding the $\underline{x}_l$ covariate. We here present two proposals for tackling this problem. Our discussion is based on the case of a single outlier in a multiple regression setting involving only continuous explanatory variates.

Cook and Weisberg (1982) discuss the *Mean Shift Outlier Model* (MSOM) as a means of identifuing a unit with misrecorded explanatory variates. The MSOM is written as

$$\underline{Y} = X\underline{\beta} + \underline{d}\lambda + \underline{\varepsilon}, \tag{8}$$

where $\underline{d}$ is a vector of zeros except of the kth element being equal to one. The model follows the standard linear regression assumptions, i.e. $E[\underline{\varepsilon}] = \underline{0}$ and $Var[\underline{\varepsilon}] = \sigma^2 I$. We may initially consider that the kth unit has misrecorded values in any of the explanatory variates $\underline{x}_j$, $j = 0, ...p$. The MSOM may then be written as

$$\underline{Y} = \left[X + \begin{pmatrix} \underline{0}' \\ \underline{m}'_k \\ \underline{0}' \end{pmatrix} \right] \underline{\beta} + \underline{\varepsilon}, \tag{9}$$

where $\underline{m}'_k \underline{\beta} = \lambda$ and $\underline{m}'_k = (m_{k0}, m_{k1}, ..., m_{kp})$. Based on the LSE of β, $\hat{\underline{\beta}}$, obtained through a regression of y on X, the LSE of λ is given by

$$\hat{\lambda} = \frac{\underline{d}'(1 - H)y_k}{\underline{d}'(1 - H)\underline{d}} = \frac{r_k}{1 - h_k}, \tag{10}$$

where $H = X(X'X)X'$ is the hat matrix, h_k is the kth diagonal element of H and r_k the raw residual from the regression of y on X. The estimate of λ is equivalent to the deletion residual $r_{(k)} = y_k - \hat{y}_{(k)}$.

Assuming Normality, we can test $\lambda = 0$ using the test statistic

$$T_i = \frac{r_k}{\hat{\sigma}\sqrt{1 - h_k}} \tag{11}$$

which follows a t-distribution with $N = n - p - 2$ degrees of freedom. This is the usual test statistic for testing the parameter λ for the additional variate $\underline{d}$. Cook and Weisberg (1982) suggest that an estimate of σ^2 could be calculated by the usual estimate for scale for the model with the extra parameter (which is equivalent to the scale as estimated by a model with the kth case deleted), i.e.

$$\hat{\sigma}^2 = s_{(k)}^2 = \frac{1}{N} \left\{ \sum_{i=1}^n e_i^2 - \frac{e_k^2}{1 - h_k} \right\}.$$

We may note that a problem with this procedure is that the acceptance region for the test (11), which can be written as $h_k > 1 - \left(r_k / \sigma t_{N, 1-a/2} \right)^2$, will tend not to reject $\lambda = 0$ for large h_k.

3.1   Proposal 1: A Variation of the MSOM

The MSOM technique does not provide us with information to decide which of the $m_{k0}, ..., m_{kp}$ is non-zero. We can arbitrarily choose one, say $m_{kl} \neq 0$, the rest being set to zero. So, we here initially assume that we have an influential point $(y_k, x_{k1}, ..., x_{kp})$ with a high leverage caused by x_{kl}. In a similar manner as in the MSOM, the LSE of $\lambda = m_{kl}\beta_l$ will adjust the value of x_{kl} such that $\hat{y}_{(k)} = y_k$. In other words, the technique will eliminate the influence of the unit as a whole, which is equivalent to deleting it. Hence, based on the estimated model with the kth case deleted we can obtain an estimate of m_{kl} using

$$\hat{m}_{kl} = \frac{\hat{\lambda}_k}{\hat{\beta}_{(k)l}} = \frac{e_k}{(1 - h_k)\hat{\beta}_{(k)l}}. \tag{12}$$

We can use $\hat{m}_{kl}$ to adjust the poorly recorded x_{kl} by $\hat{x}_{kl} = x_{kl} + \hat{m}_{kl}$, i.e. $\hat{x}_{kl}$ places y_k on the line $\underline{\hat{y}}_{(k)}$. At first sight this seems attractive. However, it should be noted that this approach can, in fact, increase the leverage of the point, by changing x_{kl} to a value $\hat{x}_{kl}$ which is further away from the body of $(x_{k1}, ... x_{kp})$. This might well be the opposite of what we would like to do! But even if the new $\hat{x}_{kl}$ reduces leverage, we would still need to incorporate an extra variability term to allow for the uncertainty in x_{kl}. We here present two alternative ways for performing this:
(i) We view the MSOM as a heteroscedastic model where

$$Var[Y_i] = \begin{cases} \sigma^2 & , \text{if } i \neq k \\ \sigma^2 + Var[\lambda] & , \text{if } i = k. \end{cases} \tag{13}$$

If $\lambda = 0$ is rejected based upon the t-statistic (11), then T_k follows a non-central t-distribution with N degrees of freedom and non-centrality parameter $\delta = T_k$. Hence, we have, see Pearson and Hartley (1972) p 64,

$$Var[\lambda] = \frac{\sigma^2}{1 - h_k} Var[t_N] \tag{14}$$

$$= \frac{\sigma^2}{1 - h_k} \left[\frac{N}{N-2}(1 + \delta^2) - E[t_N]^2 \right] = \sigma^2 C(N, \delta, h_k),$$

where $E[t_N] = (N/2)^{1/2} \dfrac{\Gamma((N-1)/2)}{\Gamma(N/2)} \delta$. We may note that $C(N, \delta, h_k)$ can be evaluated entirely from the regression of y on X. Therefore, instead of fitting the MSOM we can use a weight function of the form

$$w_i = \begin{cases} 1 & , \text{ if } \lambda_i = 0 \\ (1 + C(N, \delta, h_k))^{-1/2} & , \text{ otherwise} \end{cases} \tag{15}$$

in the regression of y on $\hat{X}$, where $\hat{X}$ uses $\hat{x}_{kl}$. Such a weighting scheme will, in principle, achieve homoscedasticity, even though, it will have no effect on the resulting parameter estimates as the influence of point k has already been eliminated. It will, however, have an effect on the variance-covariance matrix of the parameter estimates. In particular, the standard errors of the parameter estimates will increase and in this way incorporate the overdispersion of case k.

(ii) In the previous approach the variability of x_{kl} was reflected through an extra the variability in y. Here instead, we propose to incorporate the variability of x_{kl} directly in the estimation procedure. An estimate of the variability of x_{kl} is

$$\widehat{Var}[m_{kl}] = \frac{\widehat{Var}[\lambda]}{\hat{\beta}^2_{(k)l}}, \tag{16}$$

where $\widehat{Var}[\lambda]$ is based upon (14). This term may then be loaded onto the $(ll)^{\text{th}}$ diagonal element of $\hat{X}'\hat{X}$. We note that this is a form of ridge regression, and contrary to the previous approach the estimates will now be affected. See also Goldstein (1995).

3.2   Proposal 2: Imputing an Influential Covariate Value

We have suggested a means of adjusting an explanatory variate which can reduce leverage and hence reduce influence. However, we have noted that the procedure based on the MSOM can increase leverage (although removing influence). This generally, would seem undesirable; hence other approaches need to be considered. A possible approach might be to impute a value for the covariate value causing the influence. This may be done using standard imputation techniques, for example through a regression of (i) $\underline{x}_l$ on $X_{(l)}$, or (ii) $\underline{x}_l$ on $X_{(l)}$ and $\underline{y}$. However, the main disadvantage of such an approach is that it does not take into account the observed, although, influential value x_{kl}. It does however, retain information provided by both the response value and the remaining covariate values. Furthermore, this approach requires a distributional assumption for $\underline{x}_l$ which contradicts the L.S. assumption of fixed (non-stochastic) explanatory variates.

3.3 Choosing Which Covariate to Adjust

Up to now, we have discussed how we may adjust the value of a covariate which produces the influence in the corresponding case. A problem we might face is on choosing the covariate whose value is to be adjusted. A simple approach is to adjust each covariate value at a time and choose the one which produces the greatest reduction in leverage (provided that a reduction is achieved at all).

3.4 Example

The data set considered here was taken from Francis et. al. (1993) p343. The analysis investigates the relative effects of temperature (TEMP) and calcium concetrations (PCACHL) in the early salt face on the final calcium concentrations (PCALC) of the metal phase in a certain chemical process. We assume a linear relationship between response and covariates. In an initial investigation using Cook's distances, see Cook and Weisberg (1982), and leverage points we see that observation 28 distinctively possesses the largest influence. The outlying nature of this point is verified through the MSOM where the point produces the largest λ, namely $\hat{\lambda} = -.9847$. However, the estimated test statistic T=-1.737 based on $s_{(i)}^2 = .1362$, is not as significant as we would have expected it to be. In applying our proposal initially to adjust TEMP we obtain $m = -91.2$, which produces a reduced leverage of $h = .188$, while in adjusting PCACHL, we have $m = -3.6$ which gives an increased leverage, namely $h = .5819$. Thus we conclude that the high leverage, and hence influence, of this unit is most likely caused by the high TEMP value. In the table below we provide the estimates obtained (a) from the original data, (b) from the adjusted TEMP using proposal (3.1i), and (c) from the adjusted TEMP using proposal (3.1ii). Values given in brackets correspond to the standard errors of the estimates.

Estimates	(a)	(b)	(c)
CONSTANT	-16.63	-20.42	-17.20
	(4.607)	(4.806)	(4.190)
TEMP	.005478	.01080	.007401
	(.003325)	(.004429)	(.003666)
PCACHL	.2625	.2738	.2539
	(.05022)	(.04569)	(.04332)

4   Conclusion

The method of M-estimation may be used in the GLM context to obtain estimates resistant to outlying response values. In their use, however, we face the problem of biased estimates and the presence of outlying covariate values. We have discussed how bias can be corrected without altering the standard estimation techniques, i.e. the IRLS procedure. Furthermore, we have presented a proposal for treating an outlying covariate value in the Normal model, without losing the reliable information provided by the remaining variates at the suspected point. Our investigation leads us to suggest an approach along the following lines: Firstly, we could search for the most influential point using the usual Cook's distance. The MSOM is then fitted and the hypothesis $\lambda = 0$ tested. If λ appears to be insignificant we may consider imputing a new value for the covariate. On the other hand, if $\lambda = 0$ is rejected, we apply our proposal (section 3.1) and check if the adjusted $\hat{x}_{kl}$ reduces the leverage of the point. If a reduction is achieved, we can incorporate a measure of variation due to $\hat{x}_{kl}$. Otherwise, we consider imputing a value for x_{kl}. Further investigation is required regarding extension of this method to the non-Normal GLM context. In this respect, we note that Williams (1987) presents how the MSOM technique can be applied in the GLM context.

References

Cook, R.D. and Weisberg, S. (1982). *Residuals and Influence in Regression*. New York: Chapman and Hall.

Field C. and Smith B. (1994). Robust estimation - A weighted maximum likelihood approach. *Int. Statist. Rev.* **62**, 405–424.

Francis B., Green M. and Payne C. (1993). *The GLIM System Release 4 Manual*. Oxford: Oxford University Press.

Goldstein, H. (1995). *Multilevel Statistical Models, Second Edition*. London: Edward Arnold.

Hampel, F.R., Ronchetti, E.M., Rousseeuw, P.J. and Stahel, W.J. (1986). *Robust Statistics: The Approach Based on Influence Functions*. New York: John Wiley.

Huber P.J. (1981). *Robust Statistics*. New York: Wiley.

Pearson, E.S. and Hartley, H.O. (1969) *Biometrika Tables for Statisticians, Vol.2*. Cambridge–CUP.

Williams, D.A. (1987). Generalized Linear Model Diagnostics Using the Deviance and Single Case Deletion. *Appl. Statist.* **36**, 181–191.

Subject-Specific and Population-Averaged Questions for Log-Linear Regression Data

Ulrike Grömping

ABSTRACT: This paper illustrates the relation between subject-specific and population-averaged models (Zeger et al., 1988), especially for log-linear regression data with normal random effects. The emphasis is on simple special cases. The practical implications are discussed using an example from the literature.

KEYWORDS: GLMM; Log-Linear Regression; Subject-specific Effects; Population-Averaged Effects

1 Subject-specific and population-averaged models

Since Zeger et al. (1988), it has become more and more popular to discuss subject-specific (SS) models as opposed to population-averaged (PA) models. We will consider the difference between these models with respect to the expectation of a response variable especially in log-linear regression.

As a subject-specific log-linear model, consider the generalised linear mixed model (GLMM) with m clusters, i=1,...,m, a fixed parameter vector β (p×1) and individual parameters d_i (q×1), which are realisations of iid random vectors δ_i. The n_i observations of the i-th cluster are summarised in the $n_i \times 1$ response vector $y_i = (y_{i1},...,y_{in_i})^T$ and the two predictor matrices $X_i = (x_{i1}^T,...,x_{in_i}^T)^T$ and $Z_i = (z_{i1}^T,...,z_{in_i}^T)^T$ with (1×p) vectors x_{ij} and (1×q) vectors z_{ij} (j=1,...,n_i). The predictor matrix X_i is associated with the fixed parameter vector β, while Z_i affects the response via the random effects δ_i. The δ_i are assumed to be independent of the predictor matrices. Here, the important aspect of the GLMM is the model for the conditional expectation of the response, given the random effects, which is given by

$$
\begin{aligned}
E(y_{ij} \mid X_i, Z_i, \delta_i{=}d_i) \\
= E(y_{ij} \mid x_{ij}, z_{ij}, \delta_i{=}d_i) = \exp(x_{ij}\beta + z_{ij}d_i), j{=}1,...,n_i.
\end{aligned}
\tag{1}
$$

This model conditions on the realisations of the individual effects δ_i. Often the matrix $\mathbf{Z}_i$ consists of a subset of the columns of $\mathbf{X}_i$. This reflects the idea that some of the explanatory variables in $\mathbf{X}_i$ have individual effects which vary randomly over subjects. The parameter β describes the effect of the explanatory variables in $\mathbf{X}_i$ on the expected response of an average subject, i.e. a subject with $\delta_i=0$. Therefore, Zeger et al. (1988) called the effects collected in β SS effects. Note that a particular subject may be quite different from the average subject, especially if variances in $\mathbf{D}$ are relatively large. If the random effects δ_i vary substantially, model (1) may become useless, and one may want to find explanations in terms of characteristics of the subjects in order to explain part of the variation. This issue is not pursued here.

There are research questions which rather refer to the average response in the population, i.e. to a marginal or PA model that describes the marginal expectation $E_{\delta_i}(E(\mathbf{y}_{ij} \mid \mathbf{X}_i, \mathbf{Z}_i, \delta_i))$ obtained by "averaging out" the random effects. As an example consider data from a health survey being analysed with the aim of estimating the effect of a certain risk factor on the distribution of response in the population. In such situations, the SS parameter vector β from (1) does not give the appropriate answer.

This well-known fact is illustrated with the help of a plot of individual, SS and PA expectation curves for a simple special case: Suppose the GLMM

$$E(\mathbf{y}_{ij} \mid \mathbf{x}_{ij}, \delta_i = d_i) = \exp(\mathbf{x}_{ij}\beta + \mathbf{x}_{ij}d_i)$$

with a scalar $\mathbf{x}_{ij}$, $\mathbf{z}_{ij}=\mathbf{x}_{ij}$, $\beta=1$, and δ_i taking on the values $-1, -\frac{1}{2}, 0, \frac{1}{2}, 1$ with probability $\frac{1}{5}$ each. Then, for the fifth of the population with $\delta_i=-1$, $\mathbf{x}_{ij}$ has no influence at all; the maximum influence of $\mathbf{x}_{ij}$ on the expected response can be found in the fifth of the population with $\delta_i=1$. Figure 1 shows the individual expectation curves in dependence of $\mathbf{x}_{ij}=x$ for the five sub populations with different realisations of δ_i (dashed lines). The curve for the average subjects, i.e. the curve with $\delta_i=E(\delta_i)=0$, is marked by dots; this is the curve described by the SS parameter β. Obviously, for this simple special case the PA expectation of response, i.e. the average with respect to the distribution of δ_i (conditional on $\mathbf{x}_{ij}=x$) is obtained by calculating the pointwise arithmetic mean of the five individual expectation curves (solid line in figure 1).

2   Relation between SS and PA models in log-linear regression with normal random effects

From now on, the GLMM (1) will be considered for $\delta_i \sim \mathcal{N}(0, \mathbf{D})$ with a positive definite covariance matrix $\mathbf{D}$. If the data are assumed to arise from such a GLMM, but research questions pertain to the population rather than to an average subject, the PA response curve or, more generally,

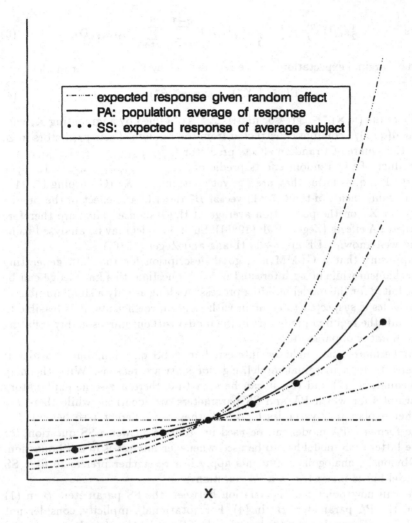

FIGURE 1. Curves of expected response for the special model described in the text with $x \in [-2,2]$

the marginal expectation $E(y_{ij} \mid X_i^*)$ of the response (with an $(n_i \times p^*)$-regressor matrix X_i^*) has to be modelled. Under model (1) with normal random effects, it is well-known that the marginal expectation can be explicitly calculated as the mean of a log-Normal distribution and is given by

$$E(y_{ij} \mid X_i, Z_i) = \exp(x_{ij}\beta + \tfrac{1}{2}z_{ij}Dz_{ij}^T). \qquad (2)$$

As $\quad \frac{1}{2}z_{ij}Dz_{ij}^T = \frac{1}{2} \sum_{t=1}^{p} z_{ijt}^2 D_{tt} + \sum_{t=1}^{q-1} \sum_{s=t+1}^{q} z_{ijt}z_{ijs}D_{ts},$ (3)

the marginal expectation can be represented by the log-linear model

$$E(y_{ij} \mid X_i^*) = \exp(x_{ij}^* \beta^*),$$ (4)

where the $(n_i \times p^*)$-regressor matrix X_i^* is obtained by augmenting X_i with the $q(q+1)/2$ columns constructed from the random effects predictors in Z_i as the square of a random effects predictor $(z_{ijr}^2)_{j=1,\dots,n_i}$, $r=1,\dots,q$, or the product of two random effects predictors $(z_{ijt}z_{ijs})_{j=1,\dots,n_i}$, $t=1,\dots,q-1$, $s=t+1,\dots,q$, as far as they are not yet contained in X_i (Grömping (1994)). The components of the $(p^* \times 1)$-vector β^* describe the effect of the predictors in X_i^* on the population average of the response. They are therefore called PA effects (Zeger et al. (1988)). Such a model may be analysed using the well-known GEE approach (Liang and Zeger (1986)).

Supposing that a GLMM is a good description for the data generating mechanism while being interested in a PA question, this knowledge can be exploited for the model building process: As long as only a small number of variables is suspected to come in with random coefficients, it is feasible to include the required predictors in the model without unreasonably inflating the number of parameters.

Furthermore, one might be interested in a SS question, but consider it easier to fit a marginal model, e.g. for software reasons. With the help of equations (2) and (3), it can be seen that there are some explanatory variables for which PA and SS parameters are identical, while there are other explanatory variables for which they are different (see below). For the former a PA model can be used in order to answer a SS question, for the latter a SS model has to be used whenever interest is in a SS question. (Obviously, analogous arguments apply if a researcher prefers to fit a SS model but is interested in a PA question.)

We will now point out the relation between the SS parameters β in (1) and the PA parameters β^* in (4). For notational simplicity, consider not only the marginal model (4) but also the SS model (1) with the augmented regressor matrix X_i^*, and think of the parameter vector β of the SS model (1) as being accordingly augmented with zeroes.

The true parameters for the marginal model (4) are obtained by adding the appropriate components of D to the true parameters of the SS model. Four cases can be distinguished:

(i) Parameters referring to explanatory variables that cannot be written as a square of a random effects predictor or a product of two random effects predictors are not affected, i.e. SS and PA parameters are identical for these explanatory variables.

(ii) For an explanatory variable that can be represented as the square of the r-th random effects predictor, β_k^* is obtained by adding $\frac{1}{2} D_{rr}$ to β_k.

As an example, consider the intercept parameter, if $\mathbf{Z}$ contains a column of ones.

(iii) If an explanatory variable can be represented as a product of the r-th and the s-th random effects predictors, $\beta_k^* = \beta_k + \mathbf{D}_{rs}$. This is the case for all random effects predictors, as long as a column of ones for the intercept parameter does also belong to $\mathbf{Z}$. (For binary predictors, cf. (iv).)

(iv) Finally, if a predictor has a representation as the square of the r-th random effects predictor as well as the product of s-th and t-th random effects predictors, $\beta_k^* = \beta_k + \frac{1}{2}\mathbf{D}_{rr} + \mathbf{D}_{st}$ (usually s=r). This may occur, if a column of ones for the intercept parameter and a binary variable (0/1) belong to $\mathbf{Z}$, so that the binary variable can be represented as its own square and as its product with the column of ones.

3   Example

Let us now consider a data set published by Thall and Vail (1990) that is used in Breslow and Clayton (1993) as an example for a SS model: 59 epileptic patients from a clinical trial are considered, who have been treated by placebo or an active drug respectively. The response to be modelled is the number of epileptic seizures in four subsequent two-week periods after treatment, i.e. there are four observations for each patient. As explanatory variables Breslow and Clayton (1993) incorporate a column of ones for the intercept parameter (ONE), the natural logarithm of a baseline number of epileptic seizures in a two-week period before treatment (ln(BASE)), the type of treatment (TRT, 0=placebo vs 1=active drug), interaction between baseline number of seizures and treatment (ln(BASE)×TRT), the natural logarithm of age (ln(AGE)) and the time after treatment, coded as a single variable with values -0.3, -0.1, 0.1, 0.3 (VISIT). Thus we obtain a model with 4×1 response vectors $\mathbf{y}_i$ and 4×6 regressor matrices $\mathbf{X}_i$. Breslow and Clayton (1993) fit a SS log-linear regression model with normal random effects for ONE and VISIT (their model IV), i.e. model (1) with 4×2 $\mathbf{Z}_i$-matrices and two normally distributed random effects with variances $\mathbf{D}_{11}$ and $\mathbf{D}_{22}$ respectively and covariance $\mathbf{D}_{12}$. We will not discuss whether or not the model is adequate and applicable in this situation. Instead we will consider the corresponding PA model and discuss the interpretation of the parameters.

According to the considerations in the previous section, for obtaining the correct marginal model we have to consider the columns of the matrices $\mathbf{Z}_i$. The first column of each $\mathbf{Z}_i$ is a four-dimensional 1; its square is also a four-dimensional 1 and is already contained in $\mathbf{X}_i$ (ONE). The second column of $\mathbf{Z}_i$ consists of the values of VISIT, its square will be denoted as VISIT^2 and has to be additionally included in the model as it is not yet present in $\mathbf{X}_i$. Finally, the product of the columns of $\mathbf{Z}_i$ is the variable VISIT which is already present in $\mathbf{X}_i$. Hence VISIT^2 is the only variable to be additionally included into the model in order to obtain a correct PA model (4).

Let us now deal with interpretations of the parameters. Obviously the explanatory variables ln(BASE), TRT, ln(BASE)×TRT and ln(AGE) can be represented neither as a square of a column of Z nor as a product of two columns of Z, so that, according to case (i) in the previous section, SS and PA parameters are identical for these variables. Therefore we can answer both questions with both models. Consider e.g. the treatment effect: In this example, assuming correctness of the model by Breslow and Clayton (1993), the effect of treatment on the response for an average subject is identical to the effect of treatment on the population average of the response.

Turning to the remaining explanatory variables, we see that they have either a representation as a square of a column of Z_i (ONE, VISIT^2) or as a product of two columns of Z_i (VISIT). Thus the PA parameters are different from the SS parameters for these variables (cf. cases (ii) and (iii) in the previous section), so that SS and PA models give answers to different questions. Consider the development of the number of seizures for the average subject over time after treatment: The estimated SS parameter (Breslow and Clayton (1993)) for the variable VISIT in the SS model described above is −0.26, and the coefficient for VISIT^2 in this model is 0. On the other hand, consider the development of the population average of the number of seizures over time after treatment: Pretending again that the estimated parameters of Breslow and Clayton (1993) are the true parameters, we now have −0.26 + (−0.01) = −0.27 as the parameter for the variable VISIT (D_{12}=−0.01), and we additionally have the non-zero parameter 0.2738 for VISIT^2 (as D_{22}=0.5476).

Figure 2 illustrates the difference between these models. The development over time is depicted as the ratio of the actual (SS or PA) expectation to the expectation at the first point in time. The SS development over time (dashed line) looks approximately linear, while the decrease of the PA expectation with time (solid line) shows a different shape. This illustrates nicely that one has to decide whether interest is in a PA or in a SS question, if one investigates the development over time. Here one would probably be interested in the development of the response of an average subject, i.e. in the SS question rather than in the PA question.

For this example note that omitting the variable VISIT^2 from the PA model should affect the intercept parameter but none of the other parameters, as VISIT^2 is uncorrelated to the other explanatory variables, so that its omission in a log-linear regression model should be (approximately) irrelevant according to Neuhaus and Jewell (1993). If one would fit the PA model without VISIT^2, one would even get an approximately correct answer for the SS question concerning development over time, as the PA parameter for VISIT is only slightly different from the SS parameter due to the small covariance between the two random effects.

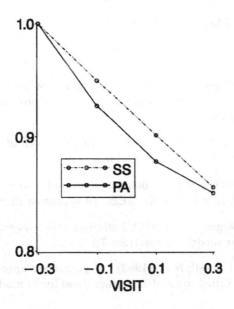

FIGURE 2. Development over time of the expected number of seizures

4   Discussion

It has been demonstrated, how the idea of an underlying GLMM (SS model) can be used for choosing an appropriate marginal model that is suitable for answering PA research questions. Often it is sufficient to include one or a few additional predictors into the model. Furthermore, for some regressors PA and SS parameters are identical, while they are different for other regressors. Hence there are some predictors for which the researcher has to decide if he or she is interested in a SS or a PA research question and has to choose the model accordingly. For other explanatory variables both questions will be answered by the same parameters so that the model choice can be governed by pure convenience considerations, if interest is in such variables only.

SS and PA effects describe different quantities, and it is well-known that they may be different. However, for the special case of log-linear models with normal random effects, three well-known papers touching upon the comparison between SS and PA models create the false impression that the only difference between SS and PA models concerns the intercept (Zeger et al. (1988), Breslow and Clayton (1993), Neuhaus and Jewell (1993)). For models with random intercept but no further random parameters, the intercept is indeed the only parameter that is affected (cf. also Diggle et al. (1994), p. 142). For more general mixed effects log-linear models, the actual relation between SS and PA effects has been shown by Grömping

(1994) and has also been given here.

References

Breslow, N.E. and Clayton, D. (1993). Approximate inference in generalised linear mixed models. *Journal of the American Statistical Association* **88**, 9-25.

Diggle, P., Liang, K.-Y., and Zeger, S.L. (1994). *Longitudinal data analysis*. Oxford University Press, Oxford.

Grömping, U. (1994). A note on fitting a marginal model to mixed effects log-linear regression data via GEE. *To appear in Biometrics*.

Liang, K.-Y. and Zeger, S.L. (1986). Longitudinal data analysis using generalised linear models. *Biometrika* **73**, 13-22.

Neuhaus, J.M. and Jewell, N.P. (1993). A geometric approach to assess bias due to omitted covariates in generalised linear models. *Biometrika* **80**, 807-815.

Thall, P.F. and Vail, S.C. (1990). Some covariance models for longitudinal count data with overdispersion. *Biometrics* **46**, 657- 671.

Zeger, S.L., Liang, K.-Y. and Albert, P.S. (1988). Models for longitudinal data: a generalised estimating equation approach. *Biometrics* **44**, 1049-1060.

Radon and Lung Cancer Mortality: An Example of a Bayesian Ecological Analysis

Ulrich Helfenstein
Christoph E. Minder

ABSTRACT: The purpose of this ecological study is the investigation of the relation of lung cancer mortality and radon in Switzerland, considering confounding risk factors and taking into account the particular spatial structure of the data. We make use of a Bayesian model pioneered by Clayton and Kaldor (1987) and by Besag et al. (1991): The Poisson model for observed mortality rates is extended to include two sources of extra-Poisson variation; a spatially 'unstructured' white noise process and a spatially 'structured' intrinsic conditional autoregressive process. Under the full model and the restricted models the radon effect is 'significant'. However, effect estimates and corresponding SE's vary considerably between models. We discuss and interpret these effects and conclude that there is no single best model explaining the data. Rather, we find that in such complex situations it is preferable to present the sequence of all relevant models to epidemiologists. Only an analysis based on epidemiological knowledge and this range of models allows one to reach an informed conclusion about the risk factor of interest.

KEYWORDS: Radon; Lung cancer; ecological regression; extra-Poisson variation; Bayesian modelling; spatial autocorrelation; BUGS; Gibbs sampling

1 Introduction

In this paper we present an example of 'complex Bayesian modelling' in a 'macroepidemiological' study. While in 'microepidemiological' studies individuals are the observational units, macroepidemiological studies have units consisting of groups of individuals. The basic questions of macroepidemiology concern the modelling and thus understanding of variability of diseases in *space* and *time*. Examples of studies and models concerned with *time* only are age–cohort models (e.g. Clayton and Schifflers, 1987) and intervention studies (e.g. Helfenstein et al. 1991).

In our study we investigate whether *spatial* variation of lung cancer mortality is related to radon concentration, taking into account the particular nature of spatial data.

Death rates or age–standardized mortality ratios are usual measures of disease impact in area–related or ecological regressions. Since areas (towns, districts etc.) may vary in population size, it appears natural to make use of 'weighted regression' where the weights are chosen to be a measure of population size. However, as explained by Pocock et al. (1981) and Breslow (1983), this approach may give too much weight to large areas: Not all factors influencing mortality may be known and included in the model. Thus, it is likely that only a part of the residual variation is due to the 'within area' or sampling variation. In studies based on large units such as countries the residual variance may be dominated by the unexplained part, suggesting that equal weighting is appropriate (e.g. in Helfenstein and Minder, 1990).

An additional particularity of spatial disease data is that areas close together may exhibit similarity of disease impact leading to spatial autocorrelation. Cook and Pocock (1983) presented a study with towns as observational units. They suggested to characterize the correlation between towns as a function of the distance between them.

Later Clayton and Caldor (1987) proposed an empirical Bayes model suited for data based on contiguous areas (e.g. districts). This model has been further developed by Besag et al. (1991) in the context of Bayesian image analysis, a field appearing at first sight remote from epidemiology. Our ecological analysis is based on this model whose characteristics are outlined below.

We ask whether age–standardized mortality from lung cancer is higher in districts with high radon concentration (above 200 Bq/m^3) than in the other districts. An important confounding risk factor for lung cancer is smoking. Unfortunately, Swiss mortality records do not contain any information on smoking, nor is information on smoking prevalence or intensity available on the district level. Therefore, we include in our study only persons where this factor is least pronounced, i.e. 'older women' (older than 60 years). As an additional known risk factor for lung cancer we include 'degree of urbanisation' (Clayton et al. 1993). We use Gibbs sampling for obtaining inferences about the model parameters, an approach which is becoming increasingly popular (Casella and George 1992, Gilks et al. 1994).

2   Statistical Analysis

Let O_i represent the observed number of deaths from lung cancer in district i ($i = 1, \ldots, 46$). The O_i are assumed to be Poisson distributed with $E(O_i) = \mu_i$. Then the full model may be written:

$$\log \mu_i = \log E_i + \alpha_1 \, \text{urban}_i + \alpha_2 \, \text{radon}_i + b_i + c_i$$

- E_i are the age–adjusted frequencies expected from the reference population.

- urban_i is the confounding effect of urbanization measured at three levels: -1: rural, 0: semi–rural, +1: urban.

- radon_i is the risk factor of interest: radon = 1 if radon concentration > 200 Bq/m^3, radon = 0 otherwise.

- b_i is an 'exchangeable' white noise process $N(0, \tau_{EX})$ where τ_{EX} is the inverse of the variance of b_i .

- c_i is an intrinsic conditional autoregressive Gaussian process with probability distribution

$$p(\mathbf{c}|\tau_{CAR}) \propto (\tau_{CAR})^{N/2} \, \exp[-\tau_{CAR}/2 \sum_{i \sim j} (c_i - c_j)^2]$$

where $i \sim j$ signifies that district i is adjacent to district j.

The hyperparameters τ_{EX} and τ_{CAR} controlling the random effects are given the noninformative priors described in Bernardinelli and Montomoli (1992). The above model containing fixed and random effects is an example of a 'generalized linear mixed model' (Breslow and Clayton 1993). The random effects b_i and c_i may be surrogates for unobservable or unobserved covariates. The b_i represent spatially unstructured ('exchangeable') effects. The c_i represent variables showing spatial structure or 'clustering' in the sense that contiguous districts are modeled to be more similar with regard to lung cancer mortality than districts further apart. Estimation of the hyperparameters of the random effects will provide a guide in the interpretation of effect estimates of radon and urbanisation.

The structure of the model is shown in Figure 1. Each node of the graph represents a variable. Circles represent unknown quantities and squares fixed quantities. Nodes O_i, E_i, and x_i are known quantities, while the other quantities need to be estimated. Under the 'conditional independence assumption' the joint posterior distribution may be represented as a product of simpler submodels associated with the 'stochastic' nodes:

$$[\mathbf{b}, \mathbf{c}, \alpha_1, \tau_{EX}, \tau_{CAR}|\mathbf{O}, \mathbf{E}, \mathbf{x}] \propto$$
$$[\mathbf{O}|\mathbf{b}, \mathbf{c}, \alpha_1, \mathbf{E}, \mathbf{x}] \times [\mathbf{b}|\tau_{EX}] \times [\mathbf{c}|\tau_{CAR}] \times [\alpha_1] \times [\tau_{EX}] \times [\tau_{CAR}].$$

Here, $[X, Y], [X|Y]$, and $[X]$ represent joint, conditional and marginal distributions respectively (covariate radon is omitted). The programming system BUGS (Gilks et al. 1994) with its built-in Gibbs sampler allows any one to approximate this complicated joint posterior distribution by sampling from all univariate full conditional distributions. The full conditional distribution for general node ν may be written as:

$$[\nu|\cdot] \propto$$

$$[\nu|\text{parents and brothers of}\,\nu] \prod [\text{children of}\,\nu|\text{parents of children of}\,\nu]$$

In Figure 1 'parent–children' relations are represented by directed edges and 'brother-relations' are represented by undirected edges. A more detailed presentation of these concepts may be found in Bernardinelli and Montomoli (1992). From the simulated joint posterior distribution we automatically obtain the marginal posterior distribution of individual parameters. Posterior means and SD's of parameters are then easily calculated from these marginal distributions.

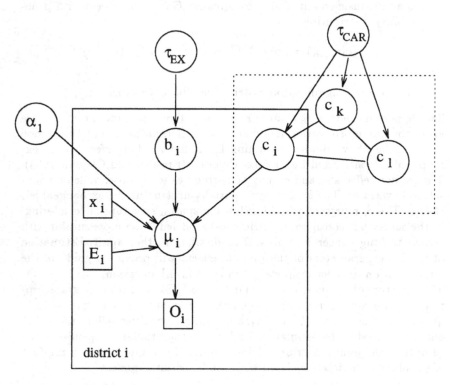

FIGURE 1. Graphical representation of the EX+CAR-model for radon and lung cancer mortality.

Table 1 shows a comparison of regression coefficients (posterior means) and corresponding SD's (posterior SD's) for four different models. The first block shows the simple Poisson-model (without any extra-Poisson variation). The z-ratios $\hat{\alpha}/\text{SD}(\hat{\alpha})$ are very large, in particular for the urbanization effect. However, the scale parameter was 5.0 indicating presence of overdispersion. The EX–model shown in the next block allows for non–autocorrelated extra–Poisson variation. The estimate of the radon effect remains the same as in the first model; however, its SD is markedly higher. The urbanization effect estimate is reduced by 20%, while its SD is doubled compared to the simple Poisson model. The CAR–model in block three allows for autocorrelated extra–Poisson variation. It shows a radon coefficient attenuated by 24% compared to the EX–model while its SD has hardly changed. The urbanisation effect is further reduced by 25%, while its SD remains almost the same. Finally, block four shows the EX+CAR–model incorporating both forms of extra–Poisson variation. It gives essentially the same results as the CAR–model.

In the EX+CAR–model we obtained the following estimates of hyperparameters $(\hat{\sigma} = \hat{\tau}^{-1/2})$: $\hat{\sigma}_{EX} = 0.0656$ with SD = 0.0323 and $\hat{\sigma}_{CAR} = 0.406$ with SD = 0.0557. The ratio $\hat{\sigma}/\text{SD}$ is 2.03 for the EX–effect and 7.29 for the CAR–effect, making the first barely significant.

TABLE 1. Estimated coefficients and SD's for radon level (radon) and urbanization (urban)

Model	$\hat{\alpha}$	$\text{SD}(\hat{\alpha})$	$\hat{\alpha}/\text{SD}(\hat{\alpha})$
Poisson–GLM			
radon	0.447	0.0884	5.06
urban	0.332	0.0260	12.8
EX–model			
radon	0.444	0.156	2.85
urban	0.265	0.0578	4.57
CAR–model			
radon	0.338	0.155	2.18
urban	0.199	0.0615	3.24
EX+CAR–model			
radon	0.344	0.162	2.13
urban	0.202	0.0637	3.17

3 Discussion

Both radon and urbanization effects on lung cancer mortality are 'significant' in all models. However, estimated coefficients and corresponding SD's vary strongly between models. In comparison, the simple Poisson–GLM not allowing for extra–Poisson variation gives high estimates of effects but small SD's. This is likely so since the model ignores any random influences arising from nonobserved covariates. Hence, Poisson effect estimates may be too large, and their SD's are likely too small. In the full EX+CAR–model the CAR–effect is much more pronounced than the EX–effect. In addition, no differences are found in estimates obtained from this full model and the simpler CAR–model. Thus, we are lead to choose the CAR–model. In this CAR–model the regression coefficient of radon is smaller than in the EX–model. This is an indication that the covariate effect of radon is confounded with location. From case–control studies performed on uranium–miners it is very likely that radon is a risk factor for lung cancer (see e.g. Samet 1989). Thus, if radon concentrations are spatially autocorrelated then the lung cancer mortality is expected to be autocorrelated too. Inclusion of the CAR–effect will consume part of this 'causal' autocorrelation. Thus, it is likely that in such a model the 'true' radon effect is underestimated. Clayton et al. (1993) suggest entertaining the CAR–model anyway because the danger of unmeasured confounders outweighs that of obtaining conservative estimates. Unknown risk factors which have a spatial structure are taken adequately into account by the CAR–effect. One may also argue in such cases that it is reasonable to present both models, the CAR–model representing a conservative attitude and the EX–model a more liberal one. The 'truth' lies then probably between the two, and it cannot be decided on statistical grounds which of these models is more adequate.

A different approach for deciding which of the two models is more adequate is based on a priori epidemiological considerations. Bernardinelli and Montomoli (1992) decide a priori for a CAR–model and against an EX–model in a study of breast cancer in Sardinia based on epidemiological arguments. However, Clayton and Bernardinelli (1992) study very similar breast cancer data with the EX+CAR–model and find that both the EX– and the CAR–components contribute about equally to the extra–Poisson variation. Thus, it appears that the EX+CAR–model is a very valuable aid in choosing between models, in particular, if no reliable prior knowledge is available.

The present study illustrates that in ecological analysis, even more than in other statistical endeavours, there may be no single 'best' answer to an estimation problem. The series of models presented in Table 1 covers several possibilities, from the statistically optimistical simple Poisson model to the maybe overly conservative CAR and EX+CAR–models. Only this range of analyses will allow an informed conclusion on the effect of radon on the lung cancer mortality of older women in Switzerland.

The approach illustrated here has other fields of application in medical statistics; hierarchical models with individual observations showing varying variances play a basic role in meta–analysis, a topic of prime interest in current methodological research in medicine. The main purpose of meta–analyses of clinical trials is to summarize quantitatively a set of studies and to obtain overall estimates of the treatment effects and corresponding confidence intervals. Individual studies are usually based on different sample sizes and other differences of study protocols. Inhomogeneity between studies is naturally characterized by a between study random effect (Der-Simonian and Laird 1986, Whitehead and Whitehead 1991). Confidence intervals for overall treatment effects were found to be markedly wider when allowing for random between study variation than when neglecting between study inhomogeneity and using a 'fixed effects model' (see e.g. Helfenstein and Steiner 1994). When it was of interest to obtain improved estimates of individual study effects, Bayesian hierarchical models similar to the EX–model described above have been used (Carlin 1992). We are convinced that similar methodology will be used increasingly in other areas of application, too.

References

Bernardinelli, L., Montomoli C. (1992). Empirical Bayes Versus Fully Bayesian Analysis of Geographical Variation in Disease Risk. *Statistics in Medicine*, **11**, 983–1007

Besag, J., York, J., Mollie, A. (1991). Bayesian Image Restoration, With Applications in Spatial Statistics (with Discussion). *Annals of the Institute of Statistical Mathematics*, **43**, 1–59

Breslow, N.E. (1984). Extra–Poisson Variation in Log–linear Models. *Applied Statistics*, **33**, 38–44

Breslow, N.E., Clayton, D.G. (1993). Approximate Inference in Generalized Linear Mixed Models. *Journal of the American Statistical Association*, **88**, 9–25

Carlin, J.B. (1992). Meta–analysis for 2×2 Tables: a Bayesian Approach. *Statistics in Medicine*, **11**, 141–159

Casella, G., George E.I. (1992). Explaining the Gibbs Sampler. *The American Statistician*, **46**, 167–174

Clayton, D.G., Bernardinelli, L., Montomoli, C. (1993). Spatial Correlation in Ecological Analysis. *International Journal of Epidemiology*, **22**, 1193–1202

Clayton, D.G., Kaldor, J. (1987). Empirical Bayes Estimates of Age–standardized Relative Risks for Use in Disease Mapping. *Biometrics*, **43**, 671–681

Clayton, D.G., Bernardinelli, L. (1992). Bayesian Methods for Mapping Disease Risk. In: Cuzick, J., Elliot, P.(eds). *Small Area Studies in Geographical and Environmental Epidemiology*. Oxford University Press, Oxford, 205–20

Clayton, D.G., Schifflers, E. (1987). Models for Temporal Variation in Cancer Rates. I: Age–period and Age–cohort Models. *Statistics in Medicine*, **6**, 449–467

Cook, D.G., Pocock, S.J. (1983). Multiple Regression in Geographical Mortality Studies, with Allowance for Spatially Correlated Errors. *Biometrics*, **39**, 361–71

DerSimonian, R., Laird, N. (1986). Meta-analysis in Clinical Trials. *Controlled Clinical Trials*, **7**, 177–188

Gilks, W.R., Thomas, A., Spiegelhalter, D.J. (1994). A Language and Program for Complex Bayesian Modelling. *The Statistician*, **43**, 169–78

Helfenstein, U., Minder, Ch. (1990). The Use of Measures of Influence in Epidemiology. *International Journal of Epidemiology*, **19**, 197–204

Helfenstein, U., Ackermann–Liebrich, U., Braun–Fahrlaender, Ch., Wanner, H.U. (1991). The Environmental Accident of 'Schweizerhalle' and Respiratory Diseases: a Time Series Analysis. *Statistics in Medicine*, **10**, 1481–1492

Helfenstein, U., Steiner, M. (1994). Fluoride Varnishes (Duraphat): A Meta-analysis. *Community Dentistry and Oral Epidemiology*, **22**, 1–5

Pocock, S.J., Cook, D.G., Beresford, S.A.A. (1981). Regression of Area Mortality Rates on Explanatory Variables: What Weighting is Appropriate? *Applied Statistics*, **30**, 286–95

Whitehead, A., Whitehead, J. (1991). A General Parametric Approach to the Meta-analysis of Randomized Clinical Trials. *Statistics in Medicine*, **10**, 1665–1677

Samet, J.M. (1989). Radon and Lung Cancer. *Journal of the National Cancer Institute*, **81**, 745–757

IRREML, a tool for fitting a versatile class of mixed models for ordinal data

A. Keen
B. Engel

ABSTRACT: Our aim is to develop models for ordered categorical data that are as general as for continuous data and allow for similar inferential procedures. The basic model is the common threshold or grouped continuous model, assuming a underlying continuous variable z which is observed imperfectly. Any family of continuous distributions is a candidate for approximating the distribution of z and a generalised linear mixed model may be specified for its parameters. The choice of distribution induces the link function that links the mean of the observed frequencies to one of the parameters of the distribution of z, usually the location. The remaining parameters of the distribution of z are parameters of this link function. The link parameters are estimated by local linearisation of the link function, which extends the model to an approximate generalised linear mixed model including linear contributions of the link parameters. All parameters of the model are estimated simultaneously by iterative reweighted REML. It is feasible to analyse fairly general models for the parameters of the distribution of z, in particular its location and scale parameter.

KEYWORDS: ordinal data, threshold model, variance components, generalised linear mixed model, parametric link

1   Introduction

In research it is common practice that at each observational unit different types of variables are observed, eg in an agricultural experiment not only the yield (amount or number), but also the quality class of the products. In such cases it is a great advantage for analysis and interpretation of results when for the different types of variables the relationship with influence factors can be modelled in the same way. The extension from linear models to generalised linear models (GLMs) has been a great step forward in this respect for regression type situations with one error stratum. In this way linear (regression) models for the mean of Poisson type variables (small numbers), binomial type variables (fractions, including 0-1 data) and more recently also ordered categorical data (in the sequel called ordinal data) can be specified, be it at possibly different scales. This also enables to compare different response variables with respect to the similarity of effects of influence factors. Software for fitting GLMs for Poisson and binomial data is widely available. For ordinal data, software to fit linear models for the mean of the underlying distribution also is available in some of the major packages, but is restricted to fitting linear models for the mean μ of z with fixed parameters only and a limited set of distributions for z.

In GLMs the variance of the response may depend on the mean, with constant or known dispersion factor. Less important, but sometimes not unimportant, is to model the dispersion as dependent on influence factors (Aitkin, 1987), McCullagh and Nelder (1989, chapter 10). A common approach is to alternate between weighted least squares for mean and dispersion, fitting the model for the mean conditional on the dispersion and the model for the dispersion conditional on the mean, maximising the likelihood or extended quasi likelihood. The alternating algorithm has the disadvantage that in both conditional analyses a partial criterion function is minimised, which in no way accounts for effects on the other parameter. This may lead to a unstable iteration process or even to local minima or unacceptable solutions, as 0 dispersion for some of the populations.

GLMs have been extended to GLMMs, extending the linear model at the link scale to a linear mixed model, i.e. extending the regression model to an analysis of variance model, a multi-level model, or a multi-strata model. With increasing complexity of models and data gathering this is a much needed extension of regression models, which also includes random parameters models (growth curves eg). For normal data see Laird and Ware (1982) and Ware (1985). The standard analysis for GLMs is Fisher scoring, which results in iterative reweighted least squares (IRLS), with the well-known link-transformed response variate and weights, see McCullagh and Nelder (1989). The link-transformed response variate accounts for the nonlinearity induced by the link function and weights result from the distribution for the observations as well as from the link transformation. Different approaches for analysing GLMMs have been proposed. We will adopt iterative reweighted REML (IRREML) as first published by Schall (1991), see also Breslow and Clayton (1993), Wolfinger (1993), Engel and Keen (1994), McGilchrist (1994). The estimation for GLMMs then is analogous to the IRLS algorithm for ordinary GLMs, but with REML-methodology replacing LS. Details may be found in Engel and Keen (1994). Engel, Buist and Visscher (1995) present a comparison between different related approaches and show eg that MAP from Gianola and Foulley, PQL from Breslow and Clayton and IRREML by Schall (1991) and Engel and Keen (1994) are identical with respect to the function that is minimised. Engel and Keen (1994) discussed also the use of weights by taking the expectation over the distribution of the random parameters. However, this can not be applied generally as the expectation does not always exist.

Standard software for REML, with facilities for introducing weights and restricting the residual variance to a fixed value (typically value 1) can be employed for GLMMs. In contrast to maximum likelihood estimation, which is quickly bogged down by the need for numerical integration, numerical restrictions are the same as for conventional mixed models for normal data.

Many papers on mixed models for binomial and Poisson data have been published, but for ordinal data the extension to mixed models has not been described in detail and standard software is not available. An early indication for the approach can already be found in Thompson (1979). A general approach has been discussed by Harville and Mee (1984). Particular cases have been discussed by Gilmour, Anderson and Rae (1985), Gianola and Foulley (1983).

With independent Poisson frequencies the model can be formulated as a generalised linear mixed model (GLMM) with parametric link function. Basic parameters of the link function are the cutpoints. With σ modelled as dependent on influence factors, the parameters of this model are extra parameters of the link function. Also if a shape parameter λ has to be estimated or is modelled as dependent on influence factors, the extra parameter(s) are added to the set of link parameters. For estimating the link parameters we use the procedure of Pregibon (1980), see also McCullagh and Nelder (1989, chapter 11).

The parameters of the link function act as nonlinear parameters, so that the model is not linear at the link scale, even if the model for the location μ of z is linear. This is even more so if the scale parameter σ of z is not constant, but depends on influence factors. The advantage of this approach is, that in this way the methodology developed for GLMMs can be used, including software developed for GLMMs. Because of the possibly large number of nonlinear parameters to be estimated, numerical problems can be expected. To minimise such problems a strategy is proposed to fit the complete model in a few successive steps, details of can be found in Keen and Engel (1994).

The algorithm has been implemented in procedure CLASS, written in Genstat 5 (1993). CLASS is available from the first author.

2   The model

Observations for unit i are class frequencies n_{ij}, for classes $j = 1...J$, with mean ν_{ij}. When several individuals or units correspond to the same combination of β, u and γ, they may be grouped together. $n_i(n_{i1}...n_{iJ})'$ is the vector of counts. The total over the classes is $N_i n_{i1} + ... + n_{iJ}$. For ungrouped data $n_{ij} 0$ or 1 and $N_i 1$ for all i. Both for grouped and ungrouped data, we will refer to $i1, 2, \ldots$ as the units. The underlying distribution z has location parameter μ, scale parameter σ and shape parameter(s) λ. Obviously ν_{ij} is a nonlinear function of μ_i, σ_i and λ_i:

$$\nu_{ij} F_{\mu_i, \sigma_i, \lambda}(\theta_j) - F_{\mu_i, \sigma_i, \lambda}(\theta_{j-1}) \tag{1}$$

with threshold θ_j such that class j corresponds to $\theta_{j-1} < z_i < \theta_j$. For location-scale families of distributions (1) can be written as:

$$\nu_{ij} F_{\lambda_i}((\theta_j - \mu_i)/\sigma_i) - F_{\lambda_i}((\theta_{j-1} - \mu_i)/\sigma_i) \tag{2}$$

with now F_{λ_i} in standard form. θ_0 is the lower bound of the support of z, usually $-\infty$, θ_J the upper bound, usually ∞. Let the pdf $f_\lambda 20 F_\lambda'$ be the corresponding derivative. Common choices are the normal or logistic distribution, without additional parameters λ. An example including a parameter λ is the family of Student's distributions with λ degrees of freedom. For use of the Student distribution in statistical inference for continuous data see Fraser (1979). Usually μ and σ^2 will be the mean and variance of z. However, mean and variance of z do not necessarily have to exist, e.g. for the Student distribution with one degree of freedom (Cauchy distribution) where $f(x)1/\{\pi(1 + x^2)\}$. The more general meaning of μ and σ are the location and the scale transformation of the standard form of the distribution respectively. Typically λ is assumed not to depend on the experimental units, while μ and σ will be modelled in terms of explanatory variables observed on these units:

$$g_\mu(\mu)\eta_\mu X_\mu \beta_\mu + Z_\mu u_\mu \quad \text{and} \quad g_\sigma(\sigma)\eta_\sigma X_\sigma \beta_\sigma + Z_\sigma u_\sigma.$$

Here g is the link function and η is the linear predictor. X and Z are design matrices for fixed effects β and random effects u, respectively. $E(u)0$, $\text{Var}(u_\mu)G_\mu$ and $\text{Var}(u_\sigma)G_\sigma$. Indices μ and σ refer to the parameter of z that is modelled. Obvious choices for link functions are those that remove constraints on the parameters, eg g_μ may be the identity link if z has support on the whole real axis and log link if z has support on the positive axis only. For g_σ the logarithmic link is a common choice, see e.g. Aitkin (1987), McCullagh and Nelder (1989, §5.2.2 and Ch. 10). Often G may be expressed as a diagonal matrix with elements $\sigma_{u1}^2 \ldots \sigma_{uc}^2$, the components of variance, along the diagonal. Observe that, when random effects are included, μ and σ^2 are conditional moments.

Conditional upon the random effects, n_i follows a multinomial distribution with total N_i and probabilities ν_{ij}. Location and scale of the distribution of z is arbitrary, but differences in location and scale between units are not, so one of the σ's is arbitrary as is one of the cutpoints or thresholds. We choose $\theta_1 0$ and σ from the standard form of the distribution (e.g. $\sigma \frac{\pi}{\sqrt{3}}$ for the logistic distribution).

3   Estimation

As we are most interested in modelling μ_i as a function of explanatory variates we write (1) and (2) as:

$$\nu_{ij} h_j(\mu_i; \theta_j, \theta_{j-1}, \sigma_i, \lambda) h_j(\mu_i; \xi_{ij})$$

with ξ_{ij} the collection of all parameters $(\theta_j, \theta_{j-1}, \sigma_i, \lambda)$. In the GLM formulation the function h_j is the inverse link function, relating the mean of the observed frequencies to the predictor scale μ. However, note, that the link could have been made with any other parameter of z, eg σ or λ. To keep the model as linear as possible the link should be made with the parameter that is considered to be affected most by other influencing factors (i.e. that contains most parameters). Estimation is discussed in two steps. First, we consider the parameters of the link function known and estimate the model in μ, then we discuss the estimation of ξ.

3.1   Estimation for known ξ

The linear predictor $\eta_{\nu_{ij}}\mu_i$ is related to the mean, say $\nu_{ij}N_i\pi_{ij}$, of n_{ij} by a function h_j, say, i.e. $h_j(\eta_{\nu_{ij}}; \xi)\nu_{ij}$. Parameters will be estimated iteratively. Suppose that $\pi_{ij}, \eta_{\nu_{ij}}, \ldots$ are evaluated for starting values for β, u and $\sigma_{u1}^2 \ldots \sigma_{uc}^2$. An adjusted dependent variable ζ is defined by:

$$\zeta_{ij}\eta_{\nu_{ij}} + (n_{ij} - \nu_{ij})/h_j{}'(\eta_{\nu_{ij}}; \xi_{ij}),$$

where $h_j{}'(\eta_{\nu_{ij}}; \xi_{ij})$ is the derivative with respect to η. Similar to Engel and Keen (1994) it is readily shown that $\zeta(\zeta_{11} \ldots \zeta_{1J}, \zeta_{21} \ldots \zeta_{2J}, \ldots)'$ has approximate first two moments:

$$E(\zeta)X\beta + Zu \quad \text{and} \quad Var(\zeta)ZGZ' + R$$

where $XX_\mu * 1_J$ and $ZZ_\mu * 1_J$, $*$ is the Kronecker product and 1_J a vector of length J with all elements equal to 1. It can be shown that the MMEs can be set up as if, conditional upon the random effects, the observations $n_{i1} \ldots n_{iJ}$ are independent Poisson counts with means $\nu_{i1} \ldots \nu_{iJ}$ (Keen and Engel, 1994). In that case R^{-1} will be a diagonal matrix of iterative weights $w_{ij}h_j{}'(\eta_{\nu_{ij}}; \xi)^2/\nu_{ij}$, similar to an ordinary GLM. Components of variance are updated by one step of the Fisher scoring algorithm for REML.

Function h is in general non-monotonic, which is at least a point of some theoretical inconvenience. The link function as an inverse of h exists only locally in the neighbourhood of points η_ν with $h_j{}'(\eta_\nu; \xi)$ β. Points where $h_j{}'(\eta_\nu; \xi)0$ are obviously problematic. When the link-trandformed dependent variate is modified by multiplication with h', the regressors, when modified in the same way, become equal to 0. Corresponding modified weights are 0 as well, so offending observations may be ignored in an iteration step.

3.2   Estimation for unknown ξ

What remains is to estimate the parameters in ξ. A method due to Pregibon (1980) (see also McCullagh and Nelder, 1989, §11.3) will be used, where η_ν is linearized with respect to ξ, in ξ^*

$$\eta_\nu \approx \eta_{\nu\cdot} - [-\partial\eta_\nu/\partial\xi]_{\xi\cdot}(\xi - \xi^*)\hat{\eta}_{\nu\cdot} - X_\xi\cdot\xi^* + X_\xi\cdot\xi.$$

Columns of the matrix of derivatives $X_\xi[-\partial\eta_\nu/\partial\xi]_\xi$, evaluated for the current estimates, are introduced as additional explanatory variables. This can be combined with the iterations described for known ξ, updating $X_{\xi\cdot}$ as a part of X after each iteration. If term $-X_\xi\cdot\xi^*$ is used as offset, ξ is estimated and not $\xi - \xi^*$, which would be the case when no offset was used. Use of the offset is necessary if ξ is a random parameter for which the variance has to be estimated from the data. Derivatives for parameters θ and γ (in the variance σ_i) have been derived in Keen and Engel (1994).

Analysis of the final approximate linear mixed model provides standard errors of estimates and test results as for an ordinary linear mixed model. Note, that, if σ has been modelled as to differ between treatments or to depend on covariates, these differences in σ are incorporated in the model and are therefore taken into account when establishing standard errors.

4   Examples

First a small example is included to illustrate the use of flexible models for gaining insight in how much information about effects is contained in the data. The second example is from our practice in statistical consultation for the Agricultural Research Department (DLO-NL) in The Netherlands and actually motivated our derivation of the model and analysis.

4.1   Example 1: therapeutic effect of nootropic agents

The example is from Uesaka (1993), who used it for illustrating a test for interaction between two treatments and a block factor based on logarithmic generalised odds ratios. The data is from a multicenter trial on the therapeutic effect of two nootropic agents on patients with multi-infarct disease. The effect was evaluated on an ordinal scale of five categories and patients were classified a posteriori into four groups according to the baseline severity of overall neurological symptoms. Fitting the fixed effects model with different underlying distributions resulted in the following test results (chisquare values) for treatment effect and interaction, along with the residual deviance as measure of goodness of fit.

Distribution	Main effect (1 df)	Interaction (3 df)	Mean deviance
normal	3.0	4.4	0.85
logistic	2.5	4.7	0.88
extreme value			
- left skew	2.7	4.4	1.19
- right skew		no convergence	
lognormal	2.3	4.0	0.85
Student		converges to the normal	
- 1 df		no convergence	
- 2 df	1.6	4.8	1.00
- 3 df	1.9	4.8	0.96
- 5 df	2.3	4.8	0.91
- 8 df	2.5	4.7	0.88
-20 df	2.8	4.5	0.86

This illustrates that there is little evidence for the existence of a main effect of treatment and an interaction between treatment and baseline severity. Use of the normal distribution may indicate a treatment effect, but this effect is not robust for the choice of underlying distribution. The results also illustrate that in such a small dataset many distributions fit well.

4.2   Example 2: potato damage

Data result from one of a series of experiments to reduce the damage of potato tubers due to a potato lifter, performed at the Institute of Agricultural Engineering in Wageningen, The Netherlands. One source of damage is the type of rod used in the machines and the data concern an experiment conducted to compare eight rods. As degree of damage is known to vary considerably between varieties and years, in order to guarantee information on differences between rods two energy levels and six varieties were involved in the experiment. Observations were obtained for almost all 188 combinations of rods, energy levels, varieties and also for three weight classes. The experiment involved 20 potato tubers per combination, i.e. 5500 tubers. Each tuber was dropped on the rod from a height determined by the energy level and the weight of the potato. To determine the damage each tuber was peeled and the degree of blue colouring was classified into one of four classes, from class 1 for undamaged to class 4 for severely damaged. The data has been presented in Keen and Engel (1994).

Variance components show conveniently how variation in the data is related to explanatory factors. In the first stage of the analysis therefore only main effects have been included as fixed effects, while all interactions of rod, energy level, weight class and variety were included as fixed effects. Normal underlying distributions were assumed with unit variance. Interactions of rod with all other factors seem to be important. These are the interactions that add to the variance of the estimates of rod effects. Estimated mean rod effects are therefore less precise than eg mean variety effects (standard error of difference 0.20).

It is of interest to check the assumption on equality of variances of the underlying distributions. Adding in the model dependence of the variance on the rods eg. indicates, that rods 4, 7 and 3 seem to have lower standard deviation for the underlying distribution than the other rods which may have equal standard deviations.

5 Discussion

Since the fixed effects β are introduced on the same underlying scale as the random effects u, the model is 'subject specific' in the terminology of Zeger, Liang and Albert (1988). Interpretation of effects β is in terms of conditional probabilities and in terms of marginal probabilities. The latter follow by integrating out the random effects. In general, expressions for marginal effects may be derived by Taylor series expansion, see Engel and Keen (1994). For more details see eg Engel, Buist and Visscher (1995).

A single population can be fitted by any distribution. The thresholds then are just transformations of the frequencies. The distribution determines the change in expected frequencies due to a shift in thresholds. So the relevancy of the choice of distribution starts when modelling shifts in the mean μ or differences in σ. Families of distributions involving shape parameters can be used for improving goodness of fit. The family of Student distributions may be useful as an alternative to the normal distribution, with degrees of freedom as scale parameter expressing kurtosis. An example where a change in distribution removes lack of fit is the (small) dataset on severity of nausea used by Farewell (1983) to illustrate his method of modelling overdispersion, see Keen and Engel (1994). A better fit is expected to reduce the bias of estimated effects and predicted class probabilities.

There is no need to restrict the underlying distribution to be continuous on the whole real axis. In some situations a gamma distribution or beta distribution may be appropriate. This presents only the difficulty that also μ must be constrained. This constraint can be removed by a proper link function for μ, as the log or logit. This extends the link function of ν with the link function of μ to a combined link function that relates ν to η. Use of a link function for the mean μ has not the same impact for ordinal data as for continuous data, due to arbitrariness in the thresholds. A variance of the form $\sigma^2 V(\mu)$, $g_\sigma(\sigma^2)\eta_\sigma$ will often be more appropriate. One might consider for example a distribution for positive z with constant coefficient of variation σ^2 and log link for the mean, while z is only observed, say, upto the nearest integer. A relevant link function for σ may be the log, which removes the non-negativity constraint for σ. As for σ it is possible to specify a mixed model for λ, but usually this will not be relevant.

For establishing goodness of fit or lack of fit, the mean deviance (based on the multinomial distribution) may be used if the expected frequencies ν_{ij} are not too low. A value for the mean deviance of about 1 should correspond with a good fit. An extra problem compared to a fixed effects model here is, that the expected frequences are conditional on random effects. An intuitive adaptation for randomness is to use the expected frequences in the usual way for the deviance, but to correct for degrees of freedom for estimating the random parameters when calculating the mean deviance. If the matrix N of frequencies is sparse, in fact no proper measure of goodness of fit is available. A general solution to the lack of fit problem then is to specify more general models. Lack of fit can occur due to two different sources: 1) in the rows of N due to a model in μ which is too restrictive and 2) in the cells of N, as deviations from the model in μ, due to a wrong choice of distribution (including a wrong model for σ or for λ). If no suitable larger model can be fitted that removes the lack of fit, the remaining lack of fit is often referred to as 'overdispersion'. Overdispersion in the rows can be remedied (or modelled) by a random component with different independent contributions for each row, i.e. modelling residual variation in μ. Overdispersion in the cells can be remedied by a random component with different contributions to each cell of N. In special situations more specific types of modelling may be applied, but this is outside the scope of this paper.

References

Aitkin, M.A. (1987). Modelling variance heterogeneity in normal regression using GLIM *Applied statistics*, **36**, 332-339

Breslow, N.E. and Clayton, D.G. (1993). Approximate inference in generalized linear mixed models. *Journal of the American Statistical Association*, **88**, 9-25

Engel, B., Buist, W. and Visscher, A. (1995). Inference for threshold models with variance components from the generalized linear mixed model perspective. *Genetics Selection Evolution*, **27**, 15-32

Engel, B. and Keen, A. (1994). A simple approach for the analysis of generalized linear mixed models. *Statistica neerlandica*, **48**, 1-22

Gianola, D. and Foulley, J.L. (1983). Sire eveluation for ordered categorical data with a threshold model. *Genetics Selection Evolution*, **15**, 201-223

Gilmour, A.R., Anderson, R.D. and Rae, A.L. (1985). The analysis of binomial data by a generalized linear mixed model. *Biometrika*, **72**, 593-599

Harville, D. and Mee, R.W. (1984). A mixed-model procedure for analyzing ordered categorical data. *Biometrics*, **40**, 393-408

Keen, A. and Engel, B. (1994). Analysis of a mixed model for ordinal data by iterative reweighted REML, GLW-DLO report LWA-94-21, May 1994. Submitted for publication.

Laird, N. M. and Ware, J.H. (1982). Random effects models for longitudinal data. *Biometrics*, **38**, 963-974

McCullagh, P. and Nelder, J.A. (1989). *Generalized linear models*. 2nd. edition, London: Chapman and Hall

McGilchrist, C.A. (1994). Estimation in generalized mixed models. *Journal of the Royal Statistical Society B*, **56**, 61-69

Pregibon, D. (1980). Goodness of link tests for generalized linear models. *Applied statistics*, **29**, 15-24.

Schall, R. (1991). Estimation in generalized linear models with random effects. *Biometrika*, **78**, 719-728.

Thompson (1979). Sire evaluation. *Biometrics*, **35**, 339-353

Uesaka, H. (1993). Test for interaction between treatment and stratum with ordinal responses. *Biometrics*, bf 49, 123-129

Wolfinger, R. (1993). Laplace's approximation for nonlinear mixed models. *Biometrika*, **80**, 791-795.

Ware, J.H. (1985). Linear models for the analysis of longitudinal studies. *The American Statistician*, **39**, 95-101

Towards a General Robust Estimation Approach for Generalised Regression Models

Jayalakshmi Krishnakumar

ABSTRACT: This paper presents a general methodology for robust estimation of the coefficients as well as the variance - covariance parameters of an econometric model, by combining the optimal B-robust estimation available for M-estimators and an original extension of the bounded-influence methods recently developed for variance components models. Special attention is given to the application of this procedure in the context of panel data models, particularly for the most well-known error component and random coefficient models.

KEYWORDS: M-estimator; Bounded Influence; Optimal B-robustness; Variance Components; Panel Data Models

1 Introduction

In this paper, we attempt to formulate a general methodology to obtain robust estimators for the coefficients as well as the (co)variance parameters in a generalised regression econometric model. The need for robust estimation is no longer to be recalled neither the basic concepts of robustness such as the influence-function , B-robustness , optimality and so on (cf. Huber (1981) and Hampel, Ronchetti, Rousseeuw and Stahel (1986)). In the context of classical linear regression models, several robust estimators, which aim to bound the influence function, are also well-known among econometricians who have started using them widely in their applications.

Turning to variance-components models, recent years have witnessed different robust proposals for effectively estimating the unknown parameters. Let us mention in particular the studies by Rocke (1983) , Fellner (1986) , Stahel and Welsh (1992) , Huggins (1993a, 1993b) and Richardson and Welsh (1993) . They can be essentially classified into two different approaches. The first one consists in assuming that the sample of observations is nested i.e. such that there always exists groups or sub-vectors of observations in-

dependent of each other enabling the resulting estimators to be regarded as
M-estimators and allowing for appropriate robustification (cf. Stahel and
Welsh (1992) and Huggins (1993a,1993b)). In the second approach, suitable
for non-nested models, pseudo-observations are substituted for the actual
observations whenever the difference between the actual value and the fitted value is large. Then estimation is carried out with the new set of values
(cf. Rocke (1983), Fellner (1986), and Richardson and Welsh (1993)).

We propose an (iterative) two-step procedure which combines the optimal
B-robust estimation (for the coefficients) and a *generalisation* of the robust
proposals of Rocke (1983), Fellner (1986) and Richardson and Welsh (1993)
for the variance-covariance parameters. In this way, we limit the influence
of any single observation on the estimators of both the coefficients and the
variance-covariance parameters thus yielding estimators associated with the
bulk of the data when contamination is present. This general methodology
is presented in the next section and in the subsequent sections we consider
its application to some frequently used panel-data models namely the two-
way Error-Component (EC) model and the Random-Coefficient (RC)
model. For a recent excellent survey of panel data models the reader is
referred to Matyas and Sevestre (eds.,1992).

2   Presentation of our Methodology in the General Framework

We start with the following standard representation of a generalised regression model

$$y = X\beta + \varepsilon \tag{1}$$

where y is $(n \times 1)$, X is $(n \times K)$, ε is $(n \times 1)$ and β is $(K \times 1)$. Assume
that $E(\varepsilon) = 0; V(\varepsilon) = V$, positive definite, and Cov $(X, \varepsilon) = 0$.

Case of a known V
In this case, since V is positive definite, we can find a non-singular P such
that $P'P = V^{-1}$ and premultiply (1.1) by P to obtain :

$$\tilde{y} = \tilde{X}\beta + \tilde{\varepsilon} \tag{2}$$

where $\tilde{y} = Py$; $\tilde{X} = PX$; $\tilde{\varepsilon} = P\varepsilon$ and $E(\tilde{\varepsilon}\tilde{\varepsilon}') = I$. Thus the BLUE of β is
given by the solution of the following system of normal equations :

$$\sum_{i=1}^{n} \tilde{x}_i \left(\tilde{y}_i - \tilde{x}_i'\hat{\beta} \right) = 0 \quad \text{or} \quad \sum_{i=1}^{n} s \left(\tilde{x}_i, \tilde{y}_i, \hat{\beta} \right) = 0 \tag{3}$$

Noting that $\hat{\beta}$ is an M-estimator and applying Theorem 1 of Hampel *et al.* (1986), we conclude that the optimal B-robust estimator of β is the solution of :

$$\sum_i h_c \left(A \left(s(\tilde{x}_i, \tilde{y}_i, \hat{\beta}_r) - a \right) \right) = 0 \quad \text{or} \quad \sum_i \psi_c \left(\tilde{x}_i, \tilde{y}_i, \hat{\beta}_r \right) = 0 \quad (4)$$

where, for a given bound c on the influence function, and writing $s(\tilde{x}_i, \tilde{y}_i, \hat{\beta}_r)$ compactly as s_i,

$$h_c \left(A(s_i - a) \right) = A(s_i - a) \min \left\{ 1, \frac{c}{\|A(s_i - a)\|} \right\} \quad (5)$$

and A and a are implicitly defined by

$$\int \psi_c s' dF = I; \quad \int \psi_c dF = 0 \quad (6)$$

respectively. The resulting estimator $\hat{\beta}_r$, which is a weighted generalised least-squares one in which "outlier" observations are automatically down-weighted, is the best compromise between efficiency at the model and bounded influence.

Case of an unknown V :
It is extremely rare that one knows all the elements of V. However, it is very common that we know its *structure*. In almost all econometric models, the known structure contains a finite number of unknown parameters considerably less than $\frac{n(n-1)}{2}$ (which is the maximum number of unknowns in any symmetric V). Let us denote the vector of these parameters by θ of dimension $p \le \frac{n(n-1)}{2}$ so that we can write $V = V(\theta)$. Further, appropriate classical procedures already exist for the estimation of θ and can be formally written, in general, as follows :

$$\phi_1 \left(\hat{\theta}, \hat{\varepsilon}, X, \hat{\beta} \right) = 0 \quad (7)$$

or

$$\hat{\theta} = \phi_2 \left(\hat{\varepsilon}, X, \hat{\beta} \right) \quad (8)$$

as the case may be, with $\hat{\varepsilon} = y - X\hat{\beta}$ being the residual vector. Note that $\hat{\theta}$ depends on $\hat{\beta}$ (the estimator of the coefficient vector of the model) and vice versa. Hence the above equation has to be solved simultaneously (or iteratively in practice) along with the normal equation for $\hat{\beta}$ and therefore the necessity to "robustify" both $\hat{\beta}$ and $\hat{\theta}$. Note also that it may not be always possible to write ϕ_1 or ϕ_2 as a sum of independent "scores" as in the case of $\hat{\beta}$ and that, in most cases especially in panel data models, they are generally quadratic forms in the residual vector (or increasing functions of squared residuals) and hence unbounded with respect to the same.

We propose to robustify $\hat{\theta}$ by replacing $\hat{\varepsilon}$ by $\psi_b(\hat{\varepsilon})$ in the ϕ functions, along the same lines as in Rocke (1983), Fellner (1986) and Richardson and Welsh (1993). By doing so, we limit the (empirical) influence of any single observation. The function $\psi_b(\cdot)$, which applies separately to individual elements of a matrix or a vector, is defined in the same way as $h_c(\cdot)$. Thus our "robust" estimating equations would be given by :

$$\phi_1\left(\hat{\theta}_r, \psi_b(\hat{\varepsilon}), X, \hat{\beta}_r\right) = 0 \qquad (9)$$

or

$$\hat{\theta}_r = \phi_2\left(\psi_b(\hat{\varepsilon}), X, \hat{\beta}_r\right) \qquad (10)$$

with $\hat{\varepsilon}$ calculated using $\hat{\beta}_r$, the solution of (1.4) with $\hat{\theta}_r$ instead of θ.

In case we can write ϕ_1 as a sum of independent scores $\phi_{1i}, i = 1, \cdots, n$, then $\hat{\theta}$ can also be robustified in an *optimal* manner by solving the following normal equations :

$$\sum_i \psi_b\left(A^*\left(\phi_{1i} - a^*\right)\right) = 0 \qquad (11)$$

(where A^* and a^* are defined similarly to A and a, and $\hat{\beta}_r$ is used for $\hat{\beta}$ in ϕ_{1i}) thus yielding optimal B-robust estimators $\hat{\beta}_r$ and $\hat{\theta}_r$. The same reasoning holds when ϕ_2 can be written as a sum of independent $\phi_{2i}, i = 1, \ldots, n$.

Asymptotic Properties

Consistency of $\hat{\beta}_r$ is implied in general by the definitions of A and a and similarly that of $\hat{\theta}_r$ by those of A^*, a^* in the case of *optimal* B-robust estimation of θ. In case $\hat{\theta}_r$ is not optimal but of bounded influence as defined by either (1.7) or (1.8), consistency can be achieved by introducing a correction factor $K(\theta)$, following Huber (1981), such that for $\hat{\theta}_r$ defined by ϕ_2, a typical element say $\hat{\theta}_j^r$ will be corrected as :

$$\hat{\theta}_{j,c}^r = \frac{\hat{\theta}_j}{K_j(\theta)}\phi_{2,j}\left(\psi_b(\hat{\varepsilon}), X, \hat{\beta}_r\right) \quad , j = 1, \ldots p \qquad (12)$$

where

$$K_j(\theta) = E_{\theta,\beta}\left[\phi_{2,j}(\cdot)\right]. \qquad (13)$$

In case $\hat{\theta}_r$ is defined implicitly by ϕ_1, the following correction may be made :

$$\hat{\theta}_c^r = \hat{\theta}_r + K(\theta) \tag{14}$$

where $K(\theta)$ is given by :

$$K(\theta) = \operatorname*{plim} \left[\frac{1}{T} \frac{\partial \phi_1}{\partial \theta'} \right]^{-1} \frac{1}{T} \phi_1(\theta) \tag{15}$$

In practice, the unknown values of parameters θ and β will be replaced by consistent estimates, in formulas (1.13) and (1.15). Thus we would have a bounded influence estimator which is also consistent. As for the implementation of the method, an iterative algorithm enabling the numerical computation of the estimators will be outlined shortly.

The limiting distribution of $\hat{\beta}_r$ is given by the following result :

Assuming that

$$\operatorname*{plim}_{n \to \infty} \left(\frac{1}{n} \sum_i \frac{\partial \psi_c}{\partial \hat{\beta}_r'} \right)_{\beta,\theta} = A(\beta, \theta); \tag{16}$$

$$\operatorname*{plim}_{n \to \infty} \left(\frac{1}{n} \sum_i \frac{\partial \psi_c}{\partial \hat{\theta}_r'} \frac{\partial \hat{\theta}_r}{\partial \hat{\beta}_r'} \right)_{\beta,\theta} = D(\beta, \theta) \tag{17}$$

and

$$\frac{1}{\sqrt{n}} \sum_i \psi_c(\beta, \theta(\beta)) \sim N(0, C(\beta, \theta)) \tag{18}$$

we obtain

<u>Lemma 1</u> : $\sqrt{n} \left(\hat{\beta}_r - \beta \right) \sim N \left(0, (A+D)^{-1} C (A+D)'^{-1} \right)$

The above result can be established by application of appropriate Central Limit Theorems after suitably rearranging the Taylor expansion of the estimating equations (1.4).

Numerical Algorithm

Step 1. Choose initial values of $\hat{\theta}$ say the classical estimators, and bounds b and c.

Step 2. Solve (1.4) for $\hat{\beta}_r$ using $\hat{\theta}$ to calculate $\tilde{y}_i, \tilde{x}_i, i = 1, \ldots, n$.

Step 3. Substitute $\hat{\beta}_r$ for $\hat{\beta}$ in (1.7) or (1.8) as the case may be, and solve for $\hat{\theta}_r$.

Step 4. Compute $K(\hat{\theta}_r)$ and in turn $\hat{\theta}_c^r$ using (1.12) or (1.14).

Step 5. Go back to Step 2 replacing $\hat{\theta}$ by $\hat{\theta}_c^r$ and repeat until convergence.

3   Two-Way Error Component Models

The two-way error component model (cf. Wallace and Hussain (1969) e.g.) can be written as :

$$y_{it} = x'_{it}\beta + \mu_i + \nu_t + \varepsilon_{it} \tag{19}$$

with

$$
\begin{aligned}
&E(\mu_i) = E(\nu_t) = E(\varepsilon_{it}) = 0, \quad \forall i, t\\
&V(\mu_i) = \sigma_\mu^2; V(\nu_t) = \sigma_\nu^2; V(\varepsilon_{it}) = \sigma_\varepsilon^2, \quad \forall i, t\\
&\text{Cov}\,(\mu_i, \nu_t) = \text{Cov}\,(\mu_i, \varepsilon_{jt}) = \text{Cov}\,(\nu_t, \varepsilon_{js}) = 0 \quad \forall i, j, t, s\\
&\text{Cov}\,(x_{it}, \mu_j) = \text{Cov}\,(x_{it}, \nu_s) = \text{Cov}\,(x_{it}, \varepsilon_{js}) = 0 \quad \forall i, j, t, s
\end{aligned} \tag{20}
$$

From these assumptions, we have (see Nerlove (1971))

$$V = V(y) = \lambda_1 M_1 + \lambda_2 M_2 + \lambda_3 M_3 + \lambda_4 M_4 \tag{21}$$

where

$$
\begin{aligned}
\lambda_1 &= \sigma_\varepsilon^2 + T\sigma_\mu^2; \quad \lambda_2 = \sigma_\varepsilon^2 + N\sigma_\nu^2;\\
\lambda_4 &= \sigma_\varepsilon^2; \quad \lambda_3 = \sigma_\varepsilon^2 + T\sigma_\mu^2 + N\sigma_\nu^2 = \lambda_1 + \lambda_2 - \lambda_4\\
M_1 &= \tfrac{1}{T}(I_N \otimes l_T l'_T) - \tfrac{1}{NT} l_{NT} l'_{NT}\\
M_2 &= \tfrac{1}{N}(l_N l'_N \otimes I_T) - \tfrac{1}{NT} l_{NT} l'_{NT}\\
M_3 &= \tfrac{1}{NT} l_{NT} l'_{NT}\\
M_4 &= I - M_1 - M_2 - M_3
\end{aligned}
$$

with l_ℓ denoting a vector of ones of size ℓ, I_M denoting an identity matrix of order M and $\otimes$ denoting the Kronecker product. The GLS normal equations for β are :

$$\sum_i \sum_t \tilde{x}_{it}\left(\tilde{y}_{it} - \tilde{x}'_{it}\hat{\beta}\right) = 0 \tag{22}$$

where $\tilde{y} = \sum_{j=1}^{4} \lambda_j^{-1/2} M_j y$; $\tilde{X} = \sum_{j=1}^{4} \lambda_j^{-1/2} M_j X$. In more general terms, we can write

$$\sum_i \sum_t s(\tilde{x}_{it}, \tilde{y}_{it}, \hat{\beta}) = 0 \quad \text{or} \quad \sum_i \sum_t s_{it} = 0 \tag{23}$$

Thus, for known $\lambda_j(s)$, the optimal B-robust estimator of β satisfies :

$$\sum_i \sum_t \psi_c\left(A\left(s_{it} - a\right)\right) = 0 \tag{24}$$

Now V depends only on three (linearly) independent parameters λ_1, λ_2 and λ_4 which form our θ. These $\lambda(s)$, along with λ_3, are the distinct eigenvalues of the variance-covariance matrix V of the model and are estimated by ANOVA method as follows (see Amemiya (1971)) [1]:

$$
\begin{aligned}
\hat{\lambda}_1 &= \tfrac{1}{N-1}\hat{u}'M_1\hat{u} &&= \tfrac{1}{N-1}\sum_i \left(\hat{u}_{i\cdot} - \hat{u}_{\cdot\cdot}\right)^2 \\
\hat{\lambda}_2 &= \tfrac{1}{T-1}\hat{u}'M_2\hat{u} &&= \tfrac{1}{T-1}\sum_t \left(\hat{u}_{\cdot t} - \hat{u}_{\cdot\cdot}\right)^2 \\
\hat{\lambda}_4 &= \tfrac{1}{(N-1)(T-1)}\hat{u}'M_4\hat{u} &&= \tfrac{1}{(N-1)(T-1)}\sum_i\sum_t \left(\hat{u}_{it} - \hat{u}_{i\cdot} - \hat{u}_{\cdot t} + \hat{u}_{\cdot\cdot}\right)^2
\end{aligned}
$$

$$(25)$$

with $\hat{\lambda}_3 = \hat{\lambda}_1 + \hat{\lambda}_2 - \hat{\lambda}_4$ and where $\hat{u} = y - X\hat{\beta}$. Thus we can write in general terms :

$$
\hat{\theta} = \left[\hat{\lambda}_1, \hat{\lambda}_2, \hat{\lambda}_4\right]' = \phi(\hat{u}) \tag{26}
$$

In fact, since ϕ is a function of the original observations only through the residuals $\hat{u}_{it}$, it suffices to bound the residuals directly and correct the asymptotic bias in the following manner :

$$
\hat{\lambda}_j^r = \frac{\hat{\lambda}_j}{K_j \cdot m_j}\psi_b\left(\hat{u}\right)' M_j \, \psi_b\left(\hat{u}\right), \quad j = 1, 2, 4 \tag{27}
$$

with $m_1 = N - 1$, $m_2 = T - 1$, $m_4 = (N - 1)(T - 1)$ (and $m_3 = 1$) and

where $K_j = \frac{1}{m_j}tr M_j E_\beta\left[\psi_b(u)\psi_b(u)'\right]$ for $u \sim N(0, \hat{V})$. Note that even though the variance component estimators can be written as a sum of squares of transformed residuals, it is not an *independent* sum as the elements of the sum are not independent across observations.

Combining equations (1.24) and (1.27) and solving them simultaneously yields our robust estimators $\hat{\beta}_r, \hat{\lambda}_j^r$, $j = 1, 2, 3, 4$.

In the case of error component models with only one specific effect, the full set of observations can be grouped to N independent sub-vectors, thus giving rise to M-estimators and enabling optimal robust versions, as in Stahel and Welsh (1992).

4   Random Coefficient Models

This model was introduced with one specific effect by Swamy (1970) and with two effects by Hsiao (1975) . Looking at the two-way structure straight away, the model is characterised by coefficients varying over time and cross-sections :

---

[1] Note that a dot in place of a subscript indicates the average taken over it.

$$y_{it} = x'_{it}\beta_{it} + \varepsilon_{it} \qquad (28)$$

with

$$
\begin{aligned}
&\beta_{it} = \overline{\beta} + \delta_i + \lambda_t \\
&E(\delta_i) = 0; \quad E(\delta_i \delta'_j) = \delta_{ij}\triangle \\
&E(\lambda_t) = 0; \quad E(\lambda_t \lambda'_s) = \delta_{ts}\Lambda \\
&E(\varepsilon_{it}) = 0; \quad E(\varepsilon_{it}\varepsilon_{js}) = \delta_{ij}\delta_{ts}\sigma^2_\varepsilon \\
&\text{Cov}(x_{it}, \delta_j) = \text{Cov}(x_{it}, \lambda_s) = \text{Cov}(x_{it}, \varepsilon_{js}) = 0 \\
&\text{Cov}(\delta_i, \lambda_t) = \text{Cov}(\delta_i, \varepsilon_{jt}) = \text{Cov}(\lambda_t, \varepsilon_{is}) = 0 \qquad \forall i, t, j, s
\end{aligned}
\qquad (29)
$$

and $\triangle$ and Λ being diagonal. Introducing the coefficient components in β_{it}, we get :

$$y_{it} = x'_{it}\overline{\beta} + (x'_{it}\delta_i + x'_{it}\lambda_t + \varepsilon_{it}) = x'_{it}\overline{\beta} + u_{it} \qquad (30)$$

or in matrix notation

$$y = X\overline{\beta} + u \qquad (31)$$

with $V(u) = X^*\triangle X^{*\prime} + X^{**}\Lambda X^{**\prime} + \sigma^2_\varepsilon I$ and X^* and X^{**} defined appropriately. The optimal robust estimator of $\overline{\beta}$ is once again given by :

$$\sum_i \sum_t \psi_c \left(s_{it}(\tilde{y}_{it}, \tilde{x}_{it}, \hat{\overline{\beta}}_r) \right) = 0 \qquad (32)$$

where $\tilde{y} = V^{-1/2}y$, $\tilde{X} = V^{-1/2}X$ and $s_{it} = \tilde{x}_{it}(\tilde{y}_{it} - \tilde{x}'_{it}\hat{\overline{\beta}})$.

The different unknowns in V are $\triangle, \Lambda$ and σ^2_ε i.e. $\triangle_{kk}, \Lambda_{kk}, k = 1, \ldots, K$ and σ^2_ε. Regrouping $\triangle_{kk}(s)$ and σ^2_ε in a vector $\delta = [\triangle_{11} + \sigma^2_\varepsilon, \triangle_{22}, \ldots, \triangle_{KK}]'$, the Hildreth-Houck type (1968) or MINQE type (cf. Rao (1970) , Hsiao (1975)) estimator is given by [2]:

$$\hat{\delta} = (F'F)^{-1} F' \dot{\hat{u}} = \left(\sum_i F_i F'_i \right)^{-1} \sum_i F'_i \dot{\hat{u}}_i \qquad (33)$$

where $F_i = M_i \dot{X}_i$, $M_i = I - X_i(X'_iX_i)^{-1}X'_i$, $\hat{u}_i = y_i - X_i(X'_iX_i)^{-1}X'_iy_i$, $F' = (F'_1, \ldots, F'_N)$ and $\hat{u}' = (\hat{u}'_1, \ldots, \hat{u}'_N)$.

The above estimator of the variances is unbounded in the residuals and can be made robust by modifying the equation as follows :

$$\hat{\delta}^r_k = f^{*\prime}_k \widehat{\psi_b(\hat{u})} \dot{\delta}_k K^{-1}_{\delta,k}, \quad k = 1, \ldots, K \qquad (34)$$

---

[2] A dot over a matrix (or a vector) indicates that all elements of the matrix (or the vector) are squared individually.

where $f_k^{*\prime}$ denotes the k-th row of $(F'F)^{-1} F'$ and $K_{\delta,k} = f_k^{*\prime} E\left(\widehat{\dot{\psi}_b(u)}\right)$ for $u \sim N\left(0, M\hat{V}M'\right)$ with $M = \mathrm{diag}\,(M_1, \ldots, M_N)$.

A similar procedure can be followed for robustifying the estimator of $\lambda = \left(\Lambda_{11} + \sigma_\varepsilon^2, \Lambda_{22}, \ldots, \Lambda_{KK}\right)'$, obtained in an analogous way using the residuals from regressions grouping all individuals for each time period. Finally, Hildreth-Houck (1968) also give a consistent estimator for σ_ε^2 :

$$\hat{\sigma}_\varepsilon^2 = \frac{1}{NT} \sum_i \sum_t \hat{u}_{it}\tilde{\hat{u}}_{it} \tag{35}$$

whose robust extension would be :

$$\hat{\sigma}_{\varepsilon,r}^2 = \frac{\hat{\sigma}_\varepsilon^2}{K_\varepsilon} \frac{1}{NT} \sum_i \sum_t \psi_b(\hat{u}_{it})\psi_b\left(\tilde{\hat{u}}_{it}\right) \tag{36}$$

with $K_\varepsilon = E\left(\psi_b(u)'\psi_b(\tilde{u})\right)$.

5 Concluding Note

This paper presents a general robust estimation approach for generalised regression models with special emphasis on panel data models. As a concluding note, we would like to say a word on the practical implementation of our procedure. As seen in the earlier sections, in most cases, the solution consists of explicit analytical expressions (for one sub-set of parameters in terms of another, enabling "zig-zag" iterative techniques) in which the scores or the residual values are replaced by their Huber functions. Hence the computational complexity of our algorithm should be in principle comparable to that of classical ML methods though the time taken to achieve numerical convergence may be longer. An important element in this respect is the calculation of the correction factor introduced for consistency purposes. As the expectation involved cannot be easily derived analytically, it has to be determined with the help of simulations using random variable generations.

Acknowledgments: The author is grateful to Elvezio Ronchetti for helpful comments.

References

Amemiya, T. (1971). The Estimation of Variances in a Variance - Components Model, *International Economic Review*, 12, 1-13.

Fellner, W.H. (1986). Robust Estimation of Variance Components, *Technometrics*, 28, 1, 51-60.

Hampel, F.R., E.M. Ronchetti, P.J. Rousseeuw and W.A. Stahel (1986). *Robust Statistics : The Approach Based on Influence Functions*, John Wiley and Sons, New York.

Hildreth, C. and J.P. Houck (1968). Some Estimators for a Linear Model with Random Coefficients, *Journal of the American Statistical Association*, 63, 584-95.

Hsiao, C. (1975). Some Estimation Methods for a Random Coefficients Model, *Econometrica*, 43, 305-25.

Huber, P.J. (1981). *Robust Statistics*, John Wiley & Sons, New York.

Huggins, R.M. (1993a). On the Robust Analysis of Variance Components Models for Pedigree Data, *Australian Journal of Statistics*, 35, 43-57.

Huggins, R.M. (1993b). A Robust Approach to the Analysis of Repeated Measures, *Biometrics*, 49, 715-720.

Matyas, L. and P. Sevestre (eds., 1992). *The Econometrics of Panel Data : Handbook of Theory and Applications*, Kluwer Academic Pulbishers, Netherlands.

Nerlove, M. (1971). A Note on Error Components Models, *Econometrica*, 39, 359-382.

Rao, C.R. (1970). Estimation of Heteroscedastic Variances in Linear Models, *Journal of the American Statistical Association*, 65, 161-172.

Richardson, A.M. and A.H. Welsh (1993). Robust Restricted Maximum Likelihood in Mixed Linear Models, Manuscript.

Rocke, D.M. (1983). Robust Statistical Analysis of Interlaboratory Studies, *Biometrika*, 70, 2, 421-431.

Stahel, W.A. and A.H. Welsh (1992). Robust Estimation of Variance Components, Manuscript.

Swamy, P. (1970). Efficient Inference in a Random Coefficient Regression Model, *Econometrica*, 38, 311-323.

Wallace, T.D. and A. Hussain (1969). The Use of Error Components Models in Combining Cross-Section with Time-Series Data, *Econometrica*, 37, 55-72.

Comparing Local Fitting to Other Automatic Smoothers

Michael Maderbacher
Werner G. Müller

ABSTRACT: Two automatic local fitting procedures are compared by simulation. They appear to exhibit similar small sample behaviour as other more popular smoothing methods.

KEYWORDS: Curve estimation; Simulation; Smoothing; Optimized Local Regression

1 Introduction

In a public service enterprise by Breiman and Peters (1992) various automatic smoothers, such as the supersmoother (SSMU), cross-validated smoothing splines (BART), delete-knot regression splines (DKS) and the cross-validated kernel smooth (KERNEL) were compared by simulation on a variety of sample sizes, noise levels and functions. The intention was to give practitioners guidelines when to use which type of smoother. The given work completes those simulations by including th e increasingly popular local fitting approach, that was introduced to the statistical literature by Cleveland (1979). Fedorov *et al.* (1993) have modified the technique in order to take account possible misspecification bias, termed 'optimized moving local regression', and here we use an automatic version (by cross-validation) of it as given in Fedorov *et al.* (1994).

At each point ξ in the regressor space the underlying local model

$$y = \theta + \delta\varphi(|\xi - x|) + \varepsilon \tag{1}$$

is assumed, where $\varphi(.)$ is a polynomial term. Weighted least squares are employed to estimate θ and hence provide a smoother for y. It turns out (see Fedorov *et al.* (1993)) that optimal weights have the general form

$$\lambda^* = a + b\varphi(.) \tag{2}$$

2   Simulation

In order to make the results of Breiman and Peters (1992) and our simulation study comparable, it was necessary to duplicate their data generation
mechanism

$$y_i = f(x_i) + \epsilon_i, \quad i = 1, \ldots, n$$

for three different sample sizes n, namely 25, 75 and 225. First the independent variable x is generated using three different designs: equispaced, normally distributed and gap distributed (at each x the distance to the neighbouring point is exponentially distributed with parameter 1 with probability p and with parameter μ with probability $(1-p)$), all on the interval $[0, 1]$.
For gap distribution the values for p and μ were: $N = 25, p = 0.75, \mu = 8$;
$N = 75, p = 0.60, \mu = 6$; $N = 225, p = 0.75, \mu = 20$ respectively.
In generating the dependent variable y seven different increasingly complex
underlying functions $f(x)$ are employed, see Figure 1.

- $f(x) = 2x$

- 'broken line': $f(x) = \begin{cases} -1 + 3x & 0 \leq x \leq 2/3 \\ 3 - 3x & 2/3 < x \leq 1 \end{cases}$

- $f(x) = \sin \pi x$

- $f(x) = \sin 2\pi x$

- 'spike': $f(x) = \begin{cases} -8 + 20x & 0.4 \leq x \leq 0.5 \\ 12 - 20x & 0.5 < x \leq 0.6 \\ 0 & \text{else} \end{cases}$

- Friedman's function: $f(x) = 3 \sin \left(2\pi (1 - x)^2 \right)$

- 'sawtooth': $f(x) = \begin{cases} 2x & 0 \leq x \leq 0.5 \\ -2 + 2x & 0.5 < x \leq 1 \end{cases}$

At each simulation run identically, independently and normally distributed
noise ϵ is added to the underlying function, using one of the three different
levels of variation ($\sigma = 0.5, 1, 2$). To make the smoothing procedure automatic the optimal smoothing parameter δ is determined by cross-validation.
For each setting $J = 500$ runs are simulated and six different performance
criteria are calculated:

- root mean squared error

$$rmse = \sqrt{\frac{1}{J} \sum_{j=1}^{J} mse_j}$$

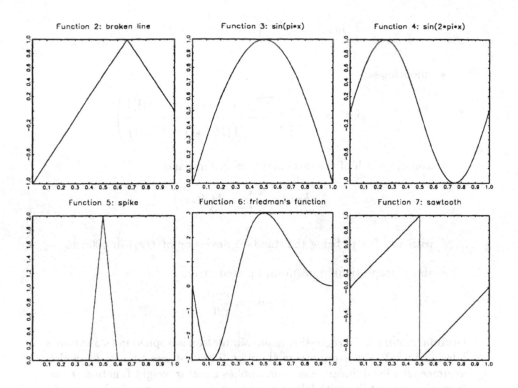

FIGURE 1. Functions $f(x)$ used in the simulation.

with

$$mse = \frac{1}{n} \sum_i \left(f(x_i) - \hat{f}(x_i) \right)$$

- root mean squared bias

$$rmsb = \sqrt{\frac{1}{n} \sum_i \left(f(x_i) - \bar{f}(x_i) \right)^2}$$

with

$$\bar{f}(x_i) = \frac{1}{J} \sum_{j=1}^{J} \hat{f}(x_i)$$

- maximum deviation

$$mxdv = \frac{1}{J} \sum_{j=1}^{J} \left(\max_i |(f(x_i) - \hat{f}(x_i)| \right)$$

- smoothness

$$smth = \frac{1}{J} \sum_{j=1}^{J} \left(\frac{\sum_{i=1}^{n-1} \left(|\hat{f}(x_{i+1}) - \hat{f}(x_i)| \right)}{\sum_{i=1}^{n-1} \left(|f(x_{i+1}) - f(x_i)| \right)} \right)$$

- average width of the 2-standard-deviation-band

$$band = \frac{2}{n} \sum_i sd\left(\hat{f}(x_i) \right)$$

with $sd\left(\hat{f}(x_i) \right)$ being the standard deviation of $\hat{f}(x_i)$ in 500 runs

- and a proportional root mean squared error

$$prmse = \frac{rmse}{sd(f)}$$

Optimal moving local regression in our simulation is applied in two variants refering to different assumptions about φ. The first one (model L), which assumes true local linear behaviour, applies a linear weight function in the form of a straight line with falling weights as distance increases. The second (model Q assumes a locally quadratic scheme) results in a quadratic form of the weight function. The parameter δ controls the width of the window and hence defines the amount of localit y. Since the two models guarantee a fixed shaped function the computational efficiency in searching for the optimal weight function can be enhanced. One only has to determine the window size with minimal mean squared error using cross-validation, i.e. finding

$$a^*, b^* = \arg\min_{a,b} \sum_i (y_i - \sum_{k \neq i} \lambda_k(a,b) y_k)^2 \qquad (3)$$

given (2), such that $\lambda_i \geq 0, \forall i$ and $\sum_i \lambda = 1$. Windows with 10, 20, ..., 100 percent of available datapoints are examined.

In addition to the results given by Breiman and Peters (1992) we computed the average of the degrees of freedom (defined as in Cleveland (1979)) and of the selected windowsize over the 500 runs to show that model Q tends to use smaller windows than model L.

All simulations were undertaken with the help of the program system GAUSS 3.15 (see Aptech (1993)). Further details and the corresponding software code can be found in Maderbacher (1994).

3 Results

As in Breiman and Peters (1992) results are shown in detail only for the medium noise level ($\sigma = 1$). For comparative reasons all the tables were produced in the same or a similar form like those in Breiman and Peters (1992).

3.1 Analysis of the *prmse*

Table (1), giving the *prmse* for the medium noise level, shows the advantage of Q over L in the case of sharp curvature. For both settings neither significant advantages nor drawbacks in comparison to alternative methods can be distinguished.

The average of *prmse* over all considered functions is given in table (2). It seems like there is no significant difference in performance between L and Q. On average over all functions and x-distributions they produce standard results with the best values for large sample sizes and in each case beating at least one of their competitors. Local fitting seems to be the safe option even with rather carelessly selected weight functions but with the opportunity to do extremely well if the opti mal weight function is used.

3.2 Analysis of Smoothness

Table (3) shows the average of *smth* over functions for the medium noise level. Unlike the others L and Q seem to get worse with increasing sample size but at very moderate speed. The only one to cope with local fitting in this respect is DKS.

3.3 Computational Speed

The average durations per run for the whole procedure of smoothing, i.e. finding optimal smoothing parameter and carrying out the estimation on a significantly slower computer (486DX-33, 4MB) than Breiman and Peters (1992) (SUN 3/160 with FPA), were 3, 20 and 187 seconds for the sample sizes 25, 75, 225 respectively.

4 Conclusions

The values of all the performance criteria are given for the two models L and Q and for the smoothers used by Breiman and Peters (1992) in the full table of Maderbacher and Müller (1995). These figures show that L and Q do considerably better on 'simple' functions, on larger sample sizes (where they do extremely well compared to the alternative smoothers) and on more unregular designs like 'normally' and 'gapped'. Comparing L and

sample size: 25					
	f(x)=2x			f(x)=broken line		
	equi	normally	gapped	equi	normally	gapped
L	0,44	0,40	0,38	0,52	0,46	0,46
Q	0,44	0,38	0,36	0,52	0,48	0,47
BART	0,36	0,38	0,84	0,46	0,47	0,53
DKS	0,32	0,33	0,30	0,48	0,49	0,45
SSMU	0,44	0,45	0,46	0,46	0,47	0,47
	f(x)=sin(π x)			f(x)=sin(2π x)		
	equi	normally	gapped	equi	normally	gapped
L	0,55	0,45	0,44	0,56	0,52	0,52
Q	0,52	0,47	0,49	0,55	0,50	0,54
BART	0,44	0,43	0,51	0,47	0,48	0,52
DKS	0,44	0,46	0,46	0,54	0,56	0,49
SSMU	0,44	0,46	0,49	0,46	0,50	0,49

sample size 75					
	f(x)=2x			f(x)=broken line		
	equi	normally	gapped	equi	normally	gapped
L	0,22	0,21	0,22	0,29	0,32	0,31
Q	0,22	0,24	0,24	0,30	0,32	0,29
BART	0,22	0,27	0,28	0,27	0,29	0,29
DKS	0,19	0,18	0,19	0,29	0,34	0,27
SSMU	0,33	0,34	0,34	0,33	0,38	0,33
KERNEL	0,46	0,49	0,59	0,34	0,38	0,40
	f(x)=spike			f(x)=sin(2π x)		
	equi	normally	gapped	equi	normally	gapped
L	0,44	0,50	0,48	0,30	0,31	0,32
Q	0,44	0,38	0,41	0,30	0,38	0,32
BART	0,42	0,41	0,43	0,27	0,28	0,29
DKS	0,44	0,38	0,55	0,32	0,35	0,32
SSMU	0,42	0,39	0,55	0,34	0,35	0,35
KERNEL	0,42	0,47	0,48	0,28	0,31	0,34
	f(x)=Friedman					
	equi	normally	gapped			
L	0,35	0,40	0,39			
Q	0,35	0,43	0,38			
BART	0,32	0,32	0,33			
DKS	0,33	0,36	0,35			
SSMU	0,34	0,38	0,37			
KERNEL	0,31	0,35	0,37			

sample size 225					
	f(x)=2x			f(x)=broken line		
	equi	normally	gapped	equi	normally	gapped
L	0,12	0,13	0,12	0,18	0,19	0,19
Q	0,12	0,13	0,13	0,18	0,18	0,18
BART	0,14	0,17	0,21	0,16	0,17	0,22
DKS	0,12	0,11	0,12	0,18	0,19	0,18
SSMU	0,16	0,14	0,15	0,18	0,17	0,16
	f(x)=spike			f(x)= sin(2π x)		
	equi	normally	gapped	equi	normally	gapped
L	0,27	0,28	0,28	0,18	0,23	0,19
Q	0,27	0,23	0,27	0,18	0,21	0,19
BART	0,27	0,25	0,27	0,17	0,17	0,21
DKS	0,27	0,23	0,35	0,21	0,23	0,20
SSMU	0,25	0,23	0,32	0,20	0,20	0,21
	f(x)=Friedman			f(x)=sawtooth		
	equi	normally	gapped	equi	normally	gapped
L	0,22	0,25	0,21	0,36	0,33	0,33
Q	0,22	0,23	0,21	0,36	0,33	0,40
BART	0,19	0,18	0,21	0,36	0,37	0,27
DKS	0,21	0,22	0,21	0,35	0,32	0,29
SSMU	0,21	0,21	0,24	0,33	0,29	0,27

TABLE 1. *prmse* for the medium noise level.

	sample size 25				sample size 75			
	equi	normally	gapped	average	equi	normally	gapped	average
L	0.52	0.46	0.45	0.48	0.32	0.35	0.34	0.34
Q	0.51	0.46	0.47	0.48	0.32	0.35	0.33	0.33
BART	0.43	0.44	0.60	0.49	0.30	0.31	0.32	0.31
DKS	0.45	0.46	0.43	0.45	0.31	0.32	0.34	0.32
SSMU	0.45	0.47	0.48	0.47	0.35	0.37	0.39	0.37
KERNEL					0.32	0.40	0.44	0.40
	sample size 225							
	equi	normally	gapped	average				
L	0.22	0.23	0.22	0.22				
Q	0.22	0.22	0.23	0.22				
BART	0.22	0.22	0.23	0.22				
DKS	0.22	0.22	0.23	0.22				
SSMU	0.22	0.21	0.23	0.22				
KERNEL								

TABLE 2. average of *prmse* over all functions

	sample size		
	25	75	225
L	1.11	1.11	1.12
Q	1.09	1.10	1.11
BART	1.32	1.22	1.26
DKS	1.09	1.08	1.07
SSMU	1.63	2.02	1.06
KERNEL		1.50	

TABLE 3. *smth* for the medium noise level

Q, the first one dominates the other in the case of medium sample sizes and on functions that are linear or show sharp but local changes (broken line, spike, sawtooth) or with low and steady curvature $(\sin(x\pi))$, whereas model Q dominates in the case of functions with relatively high and steady curvature $(\sin(2x\pi)$, Friedman's function).

Our simulations produced values of the criterion *band* of about two times the magnitude of the ones of Breiman and Peters (1992). However, from graphical output there is no evidence that the variability of our smoothers is that much larger. Therefore we suspect that Breiman and Peters (1992) omitted the factor '2' in their computations of *band*. Our results are corrected to reflect this discrepancy.

In addition to the simulation results the graphical output of Breiman and Peters (1992) was duplicated as well: The average of $\hat{f}(x)$ over the runs and the confidence interval curves are given in Maderbacher and Müller (1995). The graphic for model Q and Friedman's function shows that even with a rather bad design (due to large gaps) local fitting can give good smoothing results.

Summarizing, local fitting procedures such as 'optimized moving local regression' can, besides their theoretical advantages, also be regarded as serious competitors to more traditional smoothers from the point of view of practical performance.

References

Aptech Systems, Inc., Maple Valley (1993). *GAUSS Volume 2, Command Reference.*

Breiman, L. and Peters, S. (1991). Comparing automatic smoothers (a public service enterprise). *International Statistical Review*, 60(3):271–290.

Cleveland, W.S. (1979). Robust locally weighted regression and smoothing scatterplots. *Journal of the American Statistical Association*, 74(368): 829–836.

Fedorov, V.V., P.Hackl, Müller, W.G. (1993). Moving local regression: The weight function. *Journal of Nonparametric Statistics*, 2(4):355–368.

Fedorov, V.V., P.Hackl, Müller, W.G. (1994). Optimized moving local regression: Computational methods for the moving average case. In Dutter, R., Grossmann, W., editors, *COMPSTAT: Proceedings in Computational Statistics*, pages 476–481. Physica.

Maderbacher, M., Müller, W.G. (1995). Comparing local fitting to other automatic smoothers (with full tables). Technical Report #45, Institut für Statistik, Wirtschaftsuniversität Wien.

Maderbacher, M. (1994). Vergleich automatischer Glättungsverfahren und Implementierung der gleitenden lokalen Regression in GAUSS. Master's thesis, Wirtschaftsuniversität Wien.

Iterative Reweighted Partial Least Squares Estimation for GLMs

Brian D. Marx

ABSTRACT: We extend the concept of partial least squares (PLS) into the framework of generalized linear models. These models form a sequence of rank one approximations useful for predicting the response variable when the explanatory information is severely ill-conditioned or ill-posed. An Iterative reweighted PLS algorithm is presented with various theoretical properties. Connections to principal component and maximum likelihood estimation are made, as well as suggestions for rules to choose the proper rank of the final model.

KEYWORDS: Bias, Ill-conditioned, Latent, Principal Component

1 Introduction

We revisit and focus on modelling an exponential family response through generalized linear regression (GLR) with details and notation to follow in Section 2. Despite the popularity of Maximum Likelihood parameter estimation, the effects of ill-conditioning in training data can be nontrivial. In disciplines such as chemometrics, perhaps more importantly is that often the number of explanatory variables far exceeds the number of observations; thus we start with an ill-posed estimation problem. One useful tool for such data is partial least squares (PLS), due to H. Wold et al. (1984). PLS was initially developed for social science problems having scarce information, but more recently has received a great amount of attention in the chemometrics literature; Helland (1988) provided a nice overview and summary.

However, response data may be discrete, e.g. presence/absence or Poisson counts requiring generalized linear modelling. For example, a researcher may need to predict a discrete or non-Normal physical or chemical composition of a substance where the explanatory variables consist of several (hundred) signals of (collinear) wavelengths from spectroscopy. We propose an iterative reweighted partial least squares (IRPLS) estimation technique for generalized linear regression. IRPLS forms a sequence of rank one approximations useful to predict response variables in the exponential family.

Many biased estimation techniques for GLRs have surfaced in the last decade, but nearly all these efforts addressed alternatives to ML estimation when the information matrix is *near* singular. Often the motivation was to combat (weighted) collinearity in this more complex setting. Apart from producing alternative models through variable subset selection (VSS), research efforts in biased estimation include: the lasso (Tibshirani, 1994), ridge (Le Cessie & van Houwelingen, 1992);, principal component (Marx & Smith, 1990), among other penalized likelihood approaches. Both the PLS and the IRPLS algorithms, developed here, are variants of the conjugate gradient method of finding generalized inverses (Hestenes & Stiefel, 1952) that construct noniterfering directions to solve a maximum (minimum) of a multidimensional function.

2   Background and Notation for GLR

Let $X^* = (x_1^*, x_2^*, ..., x_p^*)$ be a N x p matrix of explanatory variables. We center and scale X^* into X using a weighted mean and weighted sum of squares, respectively. The weights are defined below, and standardization details are explicitly given in the algorithm defined in Section 3. Further, $X'X$ is in correlation form. If desired, the resulting solutions can transformed back into the variables' natural metric. Denote x_i' a $1 \times p$ row vector of X. The methods discussed include an intercept term β_0, thus X will be augmented by a N-vector of ones. The random response vector, $Y_{N \times 1}$, has independent entries Y_i following a distribution in the exponential family expressed, $f(y; \theta, \phi) = \exp[\{y\,\theta + b(\theta)\}/a(\phi) + c(y, \phi)]$, where a, b, c are known functions. The parameter θ is referred to as the canonical parameter. Let the nuisance parameter ϕ be constant for all Y_i. Given the set of p explanatory variables, generalized linear regression utilizes the relationship,

$$g(\mu_i) = \beta_0 + x_i'\beta = \eta_i,$$

satisfying: (i) $\mu_i = E(Y_i)$; (ii) g is a monotone, twice differentiable link function with $g^{-1} = h$; (iii) x_i' is a p x 1 row vector of standardized regressor variables; (iv) β is the $p \times 1$ unknown parameter vector and β_0 is the unknown intercept; (v) the estimation of β or β_0 does not depend on having knowledge of ϕ.

The loglikelihood equation can be expressed, using the canonical link, as

$$l(\beta_0, \ \beta; X) = \sum_{i=1}^{N}\{[y_i\eta_i + b(\theta)]/a(\phi) + c(y_i, \phi_i)\}.$$

The maximum likelihood estimation of the parameters is typically based on maximizing l through the method of scoring iterative equations

$$\tilde{\eta}_t = \tilde{\beta}_{0,t-1}\mathbf{1} + X(X'\tilde{V}_{t-1}X)^{-1}X'\tilde{V}_{t-1}\tilde{y}_{t-1}^*,$$

where, if convergence is attained, the estimated information matrix $\tilde{\Phi} = X'\tilde{V}X$, $\tilde{V} = \text{diag}(\tilde{v}_{ii}) = \text{diag}[\{h'(\tilde{\eta}_i)\}^2/\text{var}(Y_i)]$, $\tilde{y}_i^* = \tilde{\eta}_i + \tilde{e}_i(\partial\eta_i/\partial\mu_i)$, and $\tilde{e}_i = y_i - \tilde{\mu}_i$. The scalar intercept $\hat{\beta}_0$ is the weighted mean of the adjusted dependent vector $\tilde{y}^*$ using the weights in $\tilde{V}$ and is uncorrelated to estimation of β.

3 IRPLS Estimation

Much like principle component (PC) estimation, iterative reweighted partial least squares estimation in the GLR setting produces a sequence of models $\{\hat{\eta}_K^{PLS}\}_1^R$ that are constructed from the set of explanatory variables, where $R = \text{rank}(\tilde{\Phi})$. In Section 5, suggestions are provided for choosing the optimal value of $K \leq R$. The key difference between PLS and PC estimation is that once a latent explanatory variable is constructed it is immediately related to the response variable, then in some sense removed from the remaining predictor space. The construction of the next latent variable, in sequence, also takes this relationship into account. In a simultaneous fashion, a set response vectors are being constructed that too takes this relationship into account. Critics of the PLS estimation approach point out its highly nonlinear features and claim it is more of an algorithm than a linear model. However, Helland (1988) elegantly removed much of this algorithmic armor and showed equivalence between various PLS algorithms. These equivalences also hold for IRPLS, given below.

The beauty of (IR)PLS is that only two (iterated) matrix multiplications are needed for each desired rank estimate. Further, the (iterated) moment matrix calculations are not needed. Thus (IR)PLS can be an alternative estimation method when other techniques are prohibited by large matrix inversion or diagonalization. IRPLS utilizes details of both the PLS algorithm and the GLR method of scoring algorithm. In addition to iterating the observation weights and the adjusted dependent vector, the IRPLS algorithm also simultaneously iterates the latent variables, their loadings, along with their relationship to the response variable, until specified convergence. As in PLS, the explanatory variable subspace is carved out into R orthogonal components, in a weighted metric, i.e. the latent variables. The analog to the dependent variable in the PLS algorithm is the iterated adjusted dependent vector in the IRPLS algorithm. An important feature of IRPLS is that the following two decompositions, of the data matrix and adjusted dependent vector, are carried out together,

$$E_0 \equiv X = \sum_{j=1}^{K} t_j p_j' + E_K$$

$$f_0 \equiv y^* = \sum_{j=1}^{K} q_j t_j + f_K,$$

where the t_j are N-vector latent variables, p_j are the loadings, E_K is a residual matrix. When $K = R$, we have $E_R = 0$. The q_j are scalar coefficients, and f_K is a N-vector of residuals. The uniqueness of the t_j's and p_j's will come from imposing conditions of orthogonality.

We now provide one of several forms of the IRPLS algorithm for generalized linear regression. As we see from (line 2.a.i) the algorithm below, the adjusted dependent vector residuals, $\hat{f}_{k-1}$ (in step $k-1$), are partially regressed on the explanatory variable residuals, $\hat{E}_{k-1}$. Line 1 provides the initializations. This partial regression consists of computing the weighted covariance and using this vector to construct latent variables (line 2.a.ii). Next, the adjusted dependent vector residuals (in step $k-1$) are regressed on the current latent variable (in step k) (line 2.a.iv). The result of this fitted value is then subtracted from the residuals (in step $k-1$) to form the next sequence of adjusted dependent vector residuals (step k) (line 2.a.v). The explanatory variables residuals (in step k) are formed by subtracting from the residuals $\hat{E}_{k-1}$ its (weighted) projection on the estimated kth latent variable line 2.a.vii. The GLM method of scoring portion of the algorithm are given in lines 2.b-2.f. Once the estimated latent variables are constructed and converged, an appropriate GLR is performed in lines 3-4. The details follow.

Algorithm IRPLS.

1. Initialize $\hat{E}_0 \leftarrow X$; $\epsilon \leftarrow$ suitable disturbance ; $\hat{f}_0 \leftarrow g(y+\epsilon)$; $\hat{V} \leftarrow \{h'(y+\epsilon)\}^2/\mathrm{var}(Y)$

2. Iterate until $\Delta\hat{\eta}$ small

   (a) For k=1 to R

      i. $\hat{w}_k \leftarrow \hat{E}'_{k-1}\hat{V}\hat{f}_{k-1}$ #orthogonal vector loadings

      ii. $\hat{t}_k \leftarrow \hat{E}_{k-1}\hat{w}_k$ #latent variables st $\hat{V}^{1/2}\hat{t}_k$ orthogonal

      iii. $\hat{t}_k \leftarrow$ scale{ $\hat{t}_k$, ctr = wt.mean($\hat{t}_k$, wt = $\hat{V}$), scale = SS($\hat{t}_k$)}

      iv. $\hat{q}_k \leftarrow$ coefficient lsfit ($\hat{f}_{k-1}$ on $\hat{t}_k$, wt = $\hat{V}$, no intercept)

      v. $\hat{f}_k \leftarrow \hat{f}_{k-1} - \hat{t}_k\hat{q}_k$

      vi. $\hat{p}_k \leftarrow$ coeffs lsfit ($\hat{E}_{k-1}$ on $\hat{t}_k$, wt = $\hat{V}$, no intercept)

      vii. $\hat{E}_k \leftarrow$ resids lsfit ($\hat{E}_{k-1}$ on $\hat{t}_k$, wt = $\hat{V}$, no intercept)

   (b) end For

   (c) $\hat{\eta} \leftarrow$ wt.mean ($\hat{f}_0$, wt = $\hat{V}$) + $\sum_{k=1}^{R} \hat{q}_k\hat{t}_k$

   (d) $\hat{V} \leftarrow \{h'(\hat{\eta})\}^2/\mathrm{var}(Y)$

   (e) $\hat{f}_0 \leftarrow \hat{\eta} + \mathrm{diag}(\partial\eta_i/\partial\mu_i)\{y - h(\hat{\eta})\}$

   (f) $\hat{E}_0 \leftarrow$ scale{ X, center = wt.mean(X, wt = $\hat{V}$), scale = SS(X)}

3. Choose $s \ni \|\hat{f}_{s+1}\|$ small, $s \leq R$

4. $\texttt{glm}(y \sim \hat{t}_1 \dots \hat{t}_s)$

In addition to the loading vectors w_k being orthogonal, the vectors $\hat{V}^{1/2}\hat{t}_k$, $(k = 1, 2, \dots)$ build up an orthogonal basis of the matrix of explanatory variables $X = E_0$. The matrix E_k is the projection of $X = E_0$ orthogonal to $\hat{V}^{1/2}\hat{t}_k$. Perhaps this is best seen through the following formulation,

$$\hat{E}_k = \hat{E}_{k-1} - \frac{\hat{t}_k \hat{t}_k' \hat{V} \hat{E}_{k-1}}{\hat{t}_k' \hat{V} \hat{t}_k} = \prod_{i=1}^{k} \left(I - \frac{\hat{t}_i \hat{t}_i' \hat{V}}{\hat{t}_i' \hat{V} \hat{t}_i} \right) X,$$

since $\hat{t}_i' \hat{V} \hat{t}_{i'} = 0$ for $i \neq i'$. Note that $\hat{t}_k$ in the algorithm's *line 2.a.ii* above is a gradient step, $\partial l / \partial \beta$. This is particularly enlightening since it draws a connection between IRPLS and the conjugate gradient algorithm applied to the GLR normal equations, $X'\{y - h(\eta)\} = 0$. A similar recursive formula exists for the adjusted dependent vector $\hat{y}^* = \hat{f}_0$, i.e.

$$\hat{f}_k = \prod_{i=1}^{k} \left(I - \frac{\hat{t}_i \hat{t}_i' \hat{V}}{\hat{t}_i' \hat{V} \hat{t}_i} \right) \hat{y}^*.$$

The above two equations will be particularly useful in drawing a connection to between IRPLS and principal component estimation for GLR. An alternative, but equivalent IRPLS algorithm can be extended from work of Martens (1985), such that the orthogonal loading vectors are now found by multiple regression.

For all $k \leq R$, it can be shown by induction that

$$S(k) \equiv \text{span}(\hat{t}_1, \dots, \hat{t}_k) = \text{span}(X\hat{w}_1, \dots, X\hat{w}_k).$$

Perhaps it is useful to also view

$$\hat{E}_k = \{I - P_{S(k)}\}X$$
$$\hat{f}_k = \{I - P_{S(k)}\}\hat{f}_0,$$

where $P_{S(k)}$ is the projection operator onto the space spanned by $S(k)$. Given converged estimates of $\tilde{w}$, $\tilde{V}$ and the converged adjusted dependent vector $\tilde{y}_s^{*PLS}$ based on $s \leq R$ components, we have

$$\tilde{w}_{k+1} = E_k' \tilde{W} \tilde{f}_k$$
$$= X'\{I - P_{S(k)}\}\tilde{V}\{I - P_{S(k)}\}\tilde{y}_s^{*PLS}$$
$$= \{I - \tilde{\Phi}\tilde{W}_k(\tilde{W}_k'\tilde{\Phi}\tilde{W}_k)^{-1}\tilde{W}_k'\}X'\tilde{V}\tilde{y}_s^{*PLS},.$$

where $\tilde{W}_k = (\tilde{w}_1 \| \dots \| \tilde{w}_k)$. In conjunction with the span equivalence above, the above weight result is particularly important since it allows us to re-express IRPLS in a modified form of the converged iterate from the method of scoring algorithm,

$$\tilde{\beta}_s^{PLS} = \tilde{W}_s(\tilde{W}_s'\tilde{\Phi}\tilde{W}_s)^{-1}\tilde{W}_s'X'\tilde{V}\tilde{y}_s^{*PLS}$$
$$= \tilde{W}_s(\tilde{P}_s'\tilde{W}_s')^{-1}\tilde{q}_s,$$

where $\tilde{P} = (\tilde{p}_1||...||\tilde{p}_s)$ and $\tilde{q}_s = (\tilde{q}_1, ..., \tilde{q}_s)'$. It can be shown that $\tilde{T}_s = (\tilde{t}_1||...||\tilde{t}_s) = XW_s(\tilde{P}_s\tilde{W}_s)^{-1}$. Given a future or new observation $x_{new}^{*\prime}$, prediction could be obtained by standardizing $x_{new}^{*\prime}$ into x_{new}', then constructing $\tilde{\eta}_{new} = \tilde{\beta}_0 + x_{new}'\tilde{\beta}_s^{PLS}$ and using $\tilde{\mu}_{new} = h(\tilde{\eta}_{new})$.

4   Connecting IRPLS to PC Estimation

An interesting connection exists between Maximum Likelihood, Principal Component and Iteratively Reweighted Partial Least Squares for GLR parameter estimation. It will be useful to define GLR principal components for each observation, $Z = XM$, where the (i, j)th element of Z is the score of the jth principal component for the ith observation. Define M as the $p \times p$ matrix whose jth column is the jth eigenvector of the information matrix (without the intercept), $\Phi = X'VX$. Hence M is an orthogonal matrix and $M'\Phi M = \text{diag}(\lambda_j) = \Lambda$, where λ_j are the corresponding eigenvalues of Φ. In many chemometrics applications, Φ can be semi-positive definite. Denote T as the number of iterations until convergence. The method of scoring maximum likelihood algorithm can be expressed

$$\tilde{\eta}^{ML} = \sum_{t=1}^{T}\{\tilde{\beta}_0 1 + X\sum_{j=1}^{p}\tilde{\lambda}_j^{-1}\tilde{m}_j\tilde{m}_j'X'\tilde{V}\tilde{y}^*\}_t,$$

which is undefined in the presence of zero eigenvalues. Again $\tilde{\beta}_0$ is the iterated weighted mean of the GLR adjusted dependent vector, and $\tilde{V}$ is the usual updated GLR weight matrix. Based on $R = \text{rank}(\Phi)$ components, it will be useful to define the matrices $\tilde{\Lambda}_R$ and $\tilde{M}_R$ corresponding to the diagonal matrix of nonnull eigenvalues and the associated matrix of eigenvectors, respectively, of the converged truncated maximum likelihood estimate of the information matrix, if they exist.

One of several strategies to reduce the effects of ill-conditioned information is to further delete, in sequence, terms in the sum corresponding to the $r = R - s$ smallest nonnull $\tilde{\lambda}_j$. Utilizing results of the converged truncated ML solution, e.g. $\tilde{\beta}_0$ and $\tilde{\Phi}$, the principal component estimation technique, based on $s \leq R$ components, can be expressed as

$$\hat{\eta}_s^{PC} = \{\tilde{\beta}_0 1 + X\sum_{j=1}^{s}\tilde{\lambda}_j^{-1}\tilde{m}_j\tilde{m}_j'X'\tilde{V}\sum_{t=1}^{T}\hat{y}_{s,t}^{*PC}\},$$

where $\hat{y}_s^{*PC}$ is again the adjusted dependent vector but this time updated using only s terms in $\hat{\eta}^{PC}_{s,(t-1)}$. If all R terms are used in PC estimation, then truncated ML estimation is achieved.

Lastly, IRPLS estimation can be expressed in a similar fashion. One iterative scheme based on s latent variables can be expressed as

$$\hat{\eta}^{PLS}_s = \{\tilde{\beta}_0 \mathbf{1} + X\tilde{W}_s \sum_{j=1}^s \tilde{\phi}_j^{-1} \tilde{\gamma}_j \tilde{\gamma}_j' \tilde{W}_s' X' \tilde{V} \sum_{t=1}^T \hat{y}^{*PLS}_{s,t}\},$$

where similarly to PC estimation, $\hat{y}^{*PLS}_s$ is the adjusted dependent vector utilizing $s \leq R$ latent variables in $\hat{\eta}^{PLS}_s$ and again $\tilde{W}_s = (\tilde{w}_1 || \dots || \tilde{w}_s)$. The $\tilde{\gamma}_j$ and the $\tilde{\phi}_j$ correspond to the eigenvectors and eigenvalues of $\tilde{W}_s' \tilde{\Phi} \tilde{W}_s$, respectively.

5   Choosing Model Rank

In applications requiring the identity link and approximately Normal errors, the choice of the meta parameter K is commonly made through ordinary cross-validation (CV) (Stone, 1974). In the GLR framework, the square residual is replace with the ith observation deviance, and an analog of CV can be constructed using first order approximations. It is often convenient to work with Akaike's Information Criterion (AIC), where now choose K based on,

$$\hat{K}^{AIC} = \arg\min_{0 \leq K \leq R} \{\text{deviance}(\hat{\eta}_K) + 2\dim(\beta_0, \beta; K)\}.$$

However, the problem with using AIC in this setting is that the dimension of the optimally constructed variables is not clear, perhaps intractable. We can also choose the *meta parameter s* based on the estimation criterion

$$\hat{K}^* = \arg\min_{1 \leq s \leq R} \text{trace}\{\text{MSE}(\hat{\beta}^{PLS}_s)\}$$

or the prediction criterion

$$\hat{K}^{**} = \arg\min_{1 \leq s \leq R} \text{MSE}(c'\hat{\beta}^{PLS}_s),$$

for all nonnull c of proper dimension. The variance component of $\hat{K}^{**}$ can be expressed $\sum_{j=1}^p \text{var}(\hat{\beta}^{PLS}_s) = \text{trace}\{(\tilde{W}_s' \tilde{\Phi} \tilde{W}_s)^{-1} \tilde{W}_s' \tilde{W}_s\}$, asymptotically.

6   Discussion

We have extended the concept of PLS estimation into the generalized linear model framework. PLS estimation has been criticized for being highly nonlinear and algorithmic in nature; IRPLS does not circumvent this feature. However as seen in previous sections, IRPLS does have elegant properties and theoretical underpinnings to the conjugate gradient algorithm.

Further, it is not uncommon that ML solutions do not exist, and IRPLS can be computationally more efficient than PC or ridge approaches. Penalized likelihood extensions of adaptive schemes in the GLR setting appear promising, but too are highly nonlinear and computationally intensive.

References

Helland, I.S. (1988). On the structure of partial least squares regression. *Communications in Statistics - Simulation and Computation* **17**, 581-607.

Hestenes, M.R. and E. Stiefel (1952). Methods of conjugate gradients for solving linear systems. *Journal Res. NBS* **49**, 409-436.

Le Cessie, S. and J.C. van Houwelingen (1992). Ridge estimators in logistic regression. *Applied Statistics* **41**(1), 191-201.

Martins, H. (1985). *Multivariate Calibration.* Dr. Techn. Thesis. Technical University of Norway, Trondheim.

Marx, B.D. and E.P. Smith (1990). Principal component estimators for generalized linear regression. *Biometrika* **77**(1), 23-31.

Stone, M. (1974). Cross-validatory choice and assess of statistical descriptions (with discussion). *Journal of the Royal Statistical Society* **B 36**, 111-147.

Tibshirani, R. (1994). Regression shrinkage and selection via the lasso. Technical report no. 465, Department of Statistics, Stanford University.

Wold, S., A. Ruhe, H. Wold and W.J. Dunn, III. (1984). The collinearity problem in linear regression. The partial least squares (PLS) approach to generalized inverses. *Siam Journal on Scientific and Statistical Computing* **5**, 735-743.

Protective Estimation of Longitudinal Categorical Data With Nonrandom Dropout

Bart Michiels
Geert Molenberghs

ABSTRACT: Longitudinal categorical data, subject to dropout, are analyzed using a *protective estimator*, assuming dropout depends on the unobserved outcomes only. Necessary and sufficient conditions are derived in
order to have an estimator in the interior of the parameter space. A variance
estimator is proposed. The method is illustrated using data taken from a
psychiatric study.

KEYWORDS: Contingency Table; Dropout; Longitudinal Data; Missing
Values; Pattern-Mixture Models; Protective Estimator; Selection Models.

1 Introduction

Recent methodological research in incomplete data focuses on both longitudinal studies with dropout and on nonignorable missingness. Diggle and
Kenward (1994) propose a selection model for multivariate normally distributed data with informative dropout. A similar model for ordinal data
is presented in Molenberghs, Kenward, and Lesaffre (1994). An overview
of methods for missing data in longitudinal data is given in Laird (1988).
Baker and Laird (1988) consider models for non-ignorable non-response
with categorical outcomes. A general framework is provided by Fay (1986).
Most methods are formulated within the selection modelling frame (Little
and Rubin, 1987) as opposed to pattern-mixture modelling (Little, 1993). It
is generally accepted that as soon as missingness is informative, the mechanism has to be modelled explicitly. Recurring problems are inestimable
parameters (Little, 1993) or fundamentally untestable assumptions (see
discussion to Diggle and Kenward (1994)).
Brown (1990) has shown that progress can be made when the missingness is
allowed to depend on the unobserved values, but not on the observed measurements. For multivariate Gaussian data, Brown constructs an estimator
for the mean and covariance parameters of the joint normal distribution:

the *protective estimator*. He allows for the presence of all possible response patterns, conditional on the fact that each subject is measured at the first occasion.

Here, we consider repeated categorical measurements, where each subject is observed at the first occasion, and missingness is due to attrition. A protective estimator will be constructed, which can be used both in a selection model, as well as in a pattern-mixture framework. Necessary and sufficient conditions for a unique solution in the interior of the parameter space are derived. An intuitive and appealing interpretation of these conditions is given. A simple algorithm is presented. A connection between likelihood based estimation and the proposed protective estimator is established. Our findings are illustrated with data from a longitudinal psychiatric study.

2   Longitudinal Categorical Data With Dropouts

For each subject s $(s = 1, \ldots, N; t = 1, \ldots, T)$ in a study a complete set of measurements consists of a sequence of outcomes Y_{st}. We assume that there are T measurement occasions (indicated by t). Outcomes Y_{st} can take on r distinct levels. In the presence of dropout, the full data consist of the responses Y_{st}, and a random variable D_s, which takes T levels indicating the last observed measurement, exhibited by subject s.

All information about the response on the sth unit is contained in a cross-classification of the outcomes D_s and Y_{st} into a $(T + 1)$ dimensional table of indicator variables

$$Z^c_{s d k_1 \ldots k_T} = I(D_s = d, Y_{s1} = k_1, \ldots, Y_{sT} = k_T). \tag{1}$$

where $I(.)$ stands for the usual indicator function and the superscript c refers to "complete". Accordingly, the multinomial cell probabilities are

$$\nu^c_{d k_1 \ldots k_T} = \text{pr}(D_s = d, Y_{s1} = k_1, \ldots, Y_{sT} = k_T), \tag{2}$$

Let $Z^c_{d k_1 \ldots k_T} = \sum_{s=1}^{N} Z^c_{s d k_1 \ldots k_T}$ and group these quantities in a vector $\mathbf{Z}^c$. A vector $\boldsymbol{\nu}^c$ is defined similarly. Thus $\mathbf{Z}^c$ represents a multi-way contingency table.

The probabilities $\boldsymbol{\nu}^c$ can be factorized in essentially two different ways. A *selection model* (Little and Rubin, 1987) uses the factorization

$$\nu^c_{d k_1 \ldots k_T} = \mu^c_{k_1 \ldots k_T} \, \phi^c_{d | k_1 \ldots k_T}. \tag{3}$$

Here, $\boldsymbol{\mu}^c$ are complete data measurement probabilities, marginalized over all possible dropout patterns and $\phi \equiv \phi^c$ are the conditional dropout probabilities. Alternatively, in a *pattern-mixture model* (Little, 1993) the probabilities are factorized as

$$\nu^c_{d k_1 \ldots k_T} = \phi^c_d \, \mu^c_{k_1 \ldots k_T | d}. \tag{4}$$

Now, μ^c are complete data measurement probabilities, conditional on the observed pattern of dropout, while the dropout pattern probabilities are expressed in marginal form.

3   The Protective Estimator For Categorical Data

In this section a protective estimators for repeated categorical data is presented.
There are T possible dropout patterns, indexed by $d = 1, \ldots, T$. Complete data for pattern d consist of the cells $Z^c_{dk_1\ldots k_T}$. However, we only observed the first d measures, hence the observed data are $Z_{dk_1\ldots k_d}$. Recall that the first variable is always observed. We can estimate the probabilities

$$\mu^c_{k_1\ldots k_d \mid d}, \tag{5}$$

directly from the data. These probabilities are of a pattern-mixture nature and correspond to a "partial classification" of model (4). The aim is to construct selection probabilities $\mu^c_{k_1\ldots k_T}$ or pattern-mixture probabilities $\mu^c_{k_1\ldots k_T \mid d}$.
Protective estimation is based on the assumption that dropout solely depends on the unobserved outcome. This implies

$$\nu^c_{k_1 \mid k_2\ldots k_t; (D \geq d)} = \nu^c_{k_1 \mid k_2\ldots k_t}. \tag{6}$$

Choosing $t = d$, the left hand side of (6) are the probabilities directly observed through the patterns for which at least d measurements are available; they can be written as $\mu_{k_1 \mid k_2\ldots k_d}$. The right hand side are the complete data measurement probabilities, marginalized over all dropout patterns and hence (6) can be rewritten as

$$\mu^c_{k_1 \mid k_2\ldots k_d} = \mu_{k_1 \mid k_2\ldots k_d}. \tag{7}$$

This result means that the conditional probability of the first outcome, given the outcomes on the $d - 1$ measures that follow can be estimated directly through the observed data.
To summarize, from the available data we immediately estimate (5), i.e. table d yields $\mu^c_{k_1\ldots k_d \mid d}$; tables d through T provide the conditional probabilities

$$\mu^c_{k_1 \mid k_2\ldots k_d}. \tag{8}$$

We will now outline the computation of $\mu^c_{k_1\ldots k_d}$, for all d. Denote by $\mu_{k_1\ldots k_d}$ the cell probabilities, estimated from tables d through T. For $d = 1$, $\mu^c_{k_1}$ follows from the data. Consider the construction of $\mu^c_{k_1 k_2}$. Then, $\mu^c_{k_1 k_2}$ is determined by solving the system of equations

$$\sum_{k_2=1}^{r} \mu^c_{k_1 \mid k_2} \mu^c_{k_2} = \mu^c_{k_1}, \qquad k_1 = 1, \ldots r. \tag{9}$$

Here, $\mu_{k_2}^c$ has to be determined. Solving this system yields $\mu_{k_2}^c$ and hence $\mu_{k_1 k_2}^c$. Writing this system as $M_{1|2} M_2 = M_1$, it clearly follows that a unique solution can be found if and only if the determinant of the matrix $M_{1|2}$ is nonzero. This is equivalent with $\det(M_{12}) = \det(\mu_{k_1 k_2}) \neq 0$. Further, in order to obtain a valid solution, one has to guarantee that all components of M_2 are nonnegative. Necessary and sufficient conditions are given in Theorem 1.

We proceed by induction. Suppose that we have constructed all marginal probabilities, up to order $d - 1$. We will construct $\mu_{k_1 \ldots k_d}^c$. For a fixed multi-index $(k_2, \ldots, k_{d-1})$, consider the system of equations

$$\sum_{k_d=1}^{r} \mu_{k_1|k_2 \ldots k_d}^c \mu_{k_d|k_2 \ldots k_{d-1}}^c = \mu_{k_1|k_2 \ldots k_{d-1}}^c, \qquad k_1 = 1, \ldots r. \qquad (10)$$

Solving this system yields $\mu_{k_d|k_2 \ldots k_{d-1}}^c$ and hence $\mu_{k_1 k_d|k_2 \ldots k_{d-1}}^c$, resulting in

$$\mu_{k_1 k_2 \ldots k_{d-1} k_d}^c = \mu_{k_1 k_2 \ldots k_{d-1}}^c \frac{\mu_{k_1 k_d|k_2 \ldots k_{d-1}}^c}{\mu_{k_1|k_2 \ldots k_{d-1}}^c}; \qquad (11)$$

all quantities on the right hand side being determined. Writing (10) as

$$M_{1|d}^{(k_2 \ldots k_{d-1})} M_d^{(k_2 \ldots k_{d-1})} = M_1^{(k_2 \ldots k_{d-1})}, \qquad (12)$$

we obtain a family of equal systems of equations, one for each combination $(k_2, \ldots, k_{d-1})$. We now state the conditions for a valid solution.

Theorem 2 *The system of equations (10) has a unique, valid (i.e. non-negative) solution if and only if*

1. *$\det(\mu_{k_1 \ldots k_d}) \neq 0$, for $k_2, \ldots, k_{d-1}$ fixed;*

2. *the column vector $M_1^{(k_2 \ldots k_{d-1})} = (\mu_{k_1|k_2 \ldots k_{d-1}})_{k_1}$ is an element of the convex hull of the r column vectors, indexed by k_d, $((\mu_{k_1|k_d;k_2 \ldots k_{d-1}})_{k_1})$*

Note that the vectors $((\mu_{k_1|k_d;k_2 \ldots k_{d-1}})_{k_1})$ are the columns of $M_{1|d}^{(k_2 \ldots k_{d-1})}$. By requiring that this theorem holds for all $d = 2, \ldots T$ and for all $1 \leq k_2, \ldots, k_{d-1} \leq r$, one assures the existence of an overall solution.

These conditions can be interpreted as follows. The model implies that the column distributions in both tables are the same, as indicated by (6). Therefore, the single column we observe on the right hand side of (10) must be a convex linear combination of the set of column distributions we observe on the left hand side. An interesting consequence is that a negative solution points to a violation of the assumptions: the data can contradict the model. A nonzero determinant is equivalent to a set of column distributions which is of full rank.

In order to complete the estimation procedure, we have to compute an estimate of the variance. Write (12) in short-hand notation as $M_{1|d} M_d =$

M_1. Let $V(.)$ indicate the covariance function. An estimator follows based on the following facts. First, $V(M_{1|d})$ and $V(M_1)$ follow immediately from the data. Secondly, $V(M_d)$ is found by applying the delta method to $M_d = M_{1|d}^{-1} M_1$. Thirdly, $V(M_{1d})$ follows from the fact that the entries of M_{1d} are given by a pointwise multiplication of entries of $M_{1|d}$ and entries of M_d. Probabilities of the form $\mu_{k_1 \ldots k_d}^c$ can be written as a product of probabilities that have been determined:

$$\mu_{k_1 \ldots k_d}^c = \mu_{k_1 k_d | k_2 \ldots k_{d-1}}^c \prod_{t=2}^{d-1} \mu_{k_t | k_2 \ldots k_{t-1}}^c. \tag{13}$$

Note that in each matrix (set of probabilities) a sum constraint applies. This fact needs to be discounted in computing the derivatives, needed for the delta method.

4 Example

Our example is a multicentre study involving 315 patients that were treated by fluvoxamine for psychiatric symptoms described as possibly resulting from a dysregulation of serotonine in the brain. The data are described in Molenberghs and Lesaffre (1994). After recruitment of the patient in the study, he or she was investigated at three visits. The therapeutic effect and the side-effects were scored at each visit in an ordinal manner. Side effect is coded as (1) = no; (2) = not interfering with functionality of patient; (3) = interfering significantly with functionality of patient; (4) = the side-effect surpasses the therapeutic effect. Similarly, the effect of therapy is recorded on a four point ordinal scale: (1) no improvement or worsening; (2) minimal improvement; (3) moderate improvement and (4) important improvement. Thus, a side effect occurs if new symptoms occur while there is therapeutic effect if old symptoms disappear. These data were analyzed by Molenberghs and Lesaffre (1994).

In this paper we analyze a dichotomized version of the data, were category 1 is contrasted with the others for both outcomes. 299 patients have at least one measurement, including 242 completers. Table 4 shows estimated cell probabilities under an MAR and a protective assumption, together with the probabilities estimated from the completers only. Standard errors lie between 0.01 and 0.03. To assess the fit of the models, a deviance statistic was computed. The deviance is 11.23 for the protective estimator and 9.40 for the MAR model, both on 4 degrees of freedom. P values are 0.024 and 0.052 respectively, pointing to a similar (lack of) fit for both models.

An alternative strategy consists of estimating both the measurement parameters *and* the dropout model, using the model advocated by Molenberghs, et al. (1994). We describe the dropout probability, given the out-

TABLE 1. Estimated Cell Probabilities For The Psychiatric Study.

	Side Effects			Therapeutic Effect		
Cell	MAR	Prot.	Comp.	MAR	Prot.	Comp.
111	355	342	388	50	44	45
112	23	31	25	5	12	4
121	17	18	17	0	0	0
122	34	37	33	8	7	8
211	129	113	128	177	184	190
212	21	26	21	12	37	12
221	117	115	107	217	265	215
222	305	318	281	530	449	525

(Note: all quantities were multiplied by 1000.)

TABLE 2. Dropout Models For The Psychiatric Study, Side Effects.

Par.	MCAR	MAR	Inf.(1)	Inf.(2)
β_0	-2.19	-3.56	-4.33	-5.59(0.24)
β_d			1.35	2.71(0.10)
β_{d-1}		0.86		-0.70(0.27)
deviance(overall)	1236.32	1227.71	1227.37	1227.09
deviance(dropout)	369.96	361.34	348.66	320.15

comes, by a logistic regression

$$\text{logit}(\phi_{d|k_{d-1}k_d}) = \beta_0 + \beta_1 k_d + \beta_2 k_{d-1}.$$

Estimates for the dropout parameters are given in Table 2 (standard errors between parentheses). The measurement probabilities, for which a saturated model is assumed, are not shown. The overall deviances (corresponding to the likelihood of measurement and dropout processes simultaneously) convey a different message than the deviances of the dropout process. The overall deviances do not show a clear distinction between MAR and informative models. Indeed, although both terms in the informative model Inf.(2) are significant, the MAR model and the informative model Inf.(1) with dependence on the current outcome only, describe the data equally well. It clearly shows that the likelihood is very flat (as it heavily relies on assumptions) and similar likelihoods are obtained for conceptually very different models. When either the previous measurement only or the current observation only are included to describe the dropout process, the latter is the clear winner from the view-point of describing the dropout process. It should be remembered, however, that it uses the current (possibly unobserved) value as a covariate and hence has to be considered jointly with the measurement model. We conclude that at least one outcome should be included in the dropout model. This is in agreement with the results by Molenberghs, et al (1994) who postulated that dropout mainly depends on the *size* of side effects, whereas a *decrease* in the therapeutic outcome seems to be responsible for dropout. From the models in Table 4 we would infer that the previous measurement is a slightly better candidate to describe dropout with respect to the side effects outcome.

For therapeutic effect, Molenberghs, et al. (1994) found that MAR and MCAR fitted equally well, but the fit was much improved by allowing for informative dropout. In our analysis, the deviance for the MAR model is 5.57 on 4 degrees of freedom, which has to be contrasted with a deviance of 20.28 on 4 degrees of freedom for the protective estimator. Interpretation of these statistics should be done with caution, as the frequencies in some cells are very small and the protective estimator lies on the boundary (another indication for a bad fit). Note that a boundary solution can occur by chance alone for sparsely filled tables. This problem can be overcome by reparametrizing the cell probabilities properly.

5   Discussion

A protective estimator for a longitudinal categorical data table, subject to dropout, has been proposed, assuming that dropouts depends on the unobserved outcomes only. As under a MAR assumption, the dropout mechanism need not be explicited. It can be used in both selection and pattern-mixture models.

Our procedure extends to a modelling approach where covariates are measured, along with outcomes. Given parameter values, the observed data probabilities can be computed and hence they can be extended to complete data tables. This yields a complete dataset to which complete data methods can be applied. This issue is the subject of further research.

The protective estimator can also be derived through likelihood estimation, by considering a saturated measurement model and a dropout model which only depends on the unobserved outcome, i.e. the likelihood based on the factorization $\nu^c_{dk_1...k_T} = \mu^c_{k_1...k_T} \phi^c_{d|k_{d+1}...k_T}$. In the special case of two binary measurements, maximizing the likelihood

$$L = \prod_{k_1,k_2} (\mu^c_{k_1 k_2} \phi^c_{1|k_2})^{z_{1k_1k_2}} \prod_{k_1} (\mu^c_{k_1 1} \phi^c_{2|1} + \mu^c_{k_1 2} \phi_{2|2})^{z_{2k_2}}$$

yields the protective estimator which is also equal to the estimator for the model (XY, YR), discussed in Baker and Laird (1988). The likelihood estimator is seemingly flexible in allowing for extensions to covariates, but explicit and often untestable assumptions about the dropout process have to be made. This was shown clearly in the analysis of the example. Using a protective estimation technique, one can assume a parsimonious model to describe the influence of predictor variables on the measurement probabilities without having to model the dropout process explicitly. Of course, one has to have evidence that either MAR or protective assumptions are plausible, e.g. when measurements lie sufficiently apart in time. Preferably, contextual information should be included.

References

Baker, S.G. and Laird, N.M. (1988). Regression analysis for categorical variables with outcome subject to nonignorable nonresponse. *J. Amer. Statist. Assoc.*, **83**, 62–69.

Brown, C.H. (1990). Protecting against nonrandomly missing data in longitudinal studies. *Biometrics*, **46**, 143–155.

Diggle, P.J. and Kenward, M.G. (1994). Informative dropout in longitudinal data analysis (with discussion). *Appl. Statist.*, **43**, 49–93.

Fay, R.E. (1986). Causal models for patterns of nonresponse. *J. Amer. Statist. Assoc.*, **81**, 354–365.

Laird, N.M. (1988). Missing data in longitudinal studies. *Statistics in Medicine*, **7**, 305–315.

Little, R.J.A. (1993). Pattern-mixture models for multivariate incomplete data. *J. Amer. Statist. Assoc.*, **88**, 125–134.

Little, R.J.A. and Rubin, D.B. (1987). *Statistical Analysis with Missing Data*. New York: Wiley.

Molenberghs, G., Kenward, M.G., & Lesaffre, E. (1994). The analysis of longitudinal ordinal data with informative dropout. *Submitted*.

Molenberghs, G. and Lesaffre, E. (1994). Marginal modelling of correlated ordinal data using a multivariate Plackett distribution. *J. Amer. Statist. Assoc.*, **89**, 633–644.

Analysis of Counts Generated by an Age-dependent Branching Process

Salomon Minkin

ABSTRACT: Analytical expressions for the implementation of likelihood inference for data generated by an age-dependent branching process are available for only very few special cases. On the other hand, the generation of simulated data from such a process is extremely simple. This suggests the use of Monte-Carlo methods for the implementation of likelihood inference. This work compares the results obtained by Monte-Carlo methods with those obtained with the aid of symbolic algebra software, in the context of an application from colon cancer prevention. The study suggests that Monte-Carlo analysis can provide accurate approximations in an environment of flexible interactive analysis using standard software.

KEYWORDS: Likelihood; Monte-Carlo; Colon Cancer; Gamma lifetime

1 Introduction

This research is concerned with the analysis of data representing the size of clusters at a given point in time, when cluster growth is governed by an age-dependent branching process. The motivation stems from the use of the aberrant crypt assay for assessing the effect of compounds on promoting the growth of preneoplastic lesions in the colon. In other biological applications (e.g. clonogenic assays) similar data is produced, so developments in this area are likely to have an impact beyond the motivating application. The project builds on some recently reported work (Minkin 1994).

The model used here is the Bellman-Harris age-dependent branching process. The main theoretical results for this branching process have been available for many years (see the monographs by Harris (1963) and Jagers (1975)). Estimation methods have been proposed when either the full record of the population size in the interval $[0, t]$ is available (Athreya and Keiding 1977), or at least observations at different time points have been made (Hoel and Crump 1974). Nevertheless, likelihood inference for the situation when independent observations are available only at a fixed point t has seldom been implemented unless the lifetime distribution is exponential. This is due to the complicated formulas needed to evaluate the resulting probability distributions. For the richer family of Erlangian (i.e. Gamma with integer shape parameter) lifetimes, Minkin (1994) indicates how likelihood inference can be implemented with the help of symbolic algebra software. However, extensions to other families will probably require a different approach.

The objective of this work is to evaluate the use of Monte-Carlo techniques to compute the likelihood function (Diggle and Gratton 1984). This evaluation is done using data presented by Minkin (1994), which was generated by an assay to evaluate the colon cancer promotion effect of a compound often found in human diets. Specifically, the research tried to answer two questions. 1) To what extent can accurate Monte-Carlo results be obtained within reasonable constraints on implementation and computation time? 2) Can the added flexibility of Monte-Carlo techniques result in the identification of better fitting models for this application?

2   The Data

An assay used to evaluate potential colon cancer promoters consists of determining the size of all the aberrant crypt clusters present in the colon of rodents after the animals have received one dose of a potent colon carcinogen followed by 100 days of exposure to the putative promoter. Minkin (1994) lists the data obtained from 27 rats; nine were exposed to 1% HMF, a compound formed when sugar is caramelized; and the remaining 18 animals served as controls. The data consists of 27 frequency distributions listing the number of clusters with $1, \ldots, 9$, and 10 or more aberrant crypts. Also, the paper provides the results of fitting a model that assumes that the number of aberrant crypts in each cluster identified in the i-th animal is the result of an age-dependent branching process with a Gamma$(3, \theta_i)$ lifetime, and that clusters behave independently. Maximum likelihood estimates of the scale parameters θ_i and their standard errors, obtained using the symbolic manipulation program Maple, are listed there.

3   Implementation of Monte Carlo Likelihood

The log-likelihood for the i-th animal is $L_i(\theta) = \sum x_{ij} \log P_j(t \mid \theta)$, where x_{ij} denotes the number of clusters in the i-th animal with j aberrant crypts, and $P_j(t \mid \theta)$ is the probability that the number of crypts in a cluster at time t will be j when the crypt lifetimes follow a distribution with density $f_\theta(u)$. For the special case when the lifetime distribution $f_\theta(u)$ belongs to the Gamma family with known integer shape parameter α, the expressions for $P_j(t \mid \theta)$ are of the form $\sum_{k=1}^{j} \exp(-k\theta t) \text{poly}[\theta t, k(\alpha - 1)]$, where $\text{poly}[u, m]$ denotes an m-th degree polynomial in u. Clearly, even for small values of

j and α, the expressions become very difficult to manage. On the other hand, $P_j(t \mid \theta)$ can easily be estimated as the observed relative frequency of clusters of size j at time t when an age-dependent branching process with Gamma(α, θ) lifetime is simulated a large number of times. Such a simulation is easy to implement using computer software such as the S language which allows recursive functions. The function *size*, given below, provides a suitable implementation to generate one observation from the relevant branching process:

```
function size(u, t, j, α, θ)
          v ← random number from Gamma(α, θ)
          j* ← j
          if (u + v < t) then {u* ← u + v
                    j* ← j + 1
                    j* ← size(u*, t, j*, α, θ)
                    j* ← size(u*, t, j*, α, θ)}
          return(j*)
```

For a fixed α, the functions $P_j(t \mid \theta)$ and thus $L_i(\theta)$ can be evaluated at time $t = 100$ for a grid of values of θ, e.g. $\theta = .03(.001).07$. Such a naive approach would require one simulation study for each value of θ in the grid, i.e. 41 in the example given. Since each simulation study can involve several thousand replications, it is important to look for alternatives. Fortunately, in this case θ is a scale parameter in such a way that the likelihood functions $L_i(\theta)$ are functions of the product $\theta \times t$. Therefore, the same information can be obtained by running only one simulation study with θ fixed at some convenient value, say $\theta = 1$, with $t = 7$, but keeping track of the realizations at intermediate times $t = 3(0.1)7$. Clearly, the savings in computation time would be substantial as long as the overhead from keeping track of realizations at intermediate times is not too demanding. In our implementation in Splus version 3.2, taking into account this overhead, we were able to obtain the profile log-likelihood 20 to 30 times faster in terms of CPU time, by running one single simulation study with intermediate times, instead of several simulation studies with a single time.

4 Maximum Likelihood Estimates

To estimate $L_i(\theta)$ at time $t = 100$, 10,000 simulations were generated from the age-dependent branching process with a Gamma$(3, 1)$ lifetime, and the population sizes were recorded for times $t = 3(0.1)7$. These simulations, which provided estimates for $P_j(100 \mid \theta), \theta = .03(.001).07$, took 41 seconds of CPU time on a Sparcstation-10. Using the data from the i-th animal, the log-likelihood was evaluated over the grid of values of θ. To smooth these values and to carry out the interpolation needed for a more accurate evaluation of the maximum likelihood estimate of θ_i, a second order polynomial was fitted by least-squares to the estimated values of $L_i(\theta)$, using all the values of θ whose log-likelihood differed from the observed maximum in the grid by less than half of the upper 5% percentile of a χ_1^2 distribution. Figure 1 shows that the estimates obtained in this way are extremely close to the 27 values reported by Minkin (1994) . Specifically, the maximum relative error was $MRE = 1.58\%$ while the root mean square error was $RMSE = 0.00041$, where $RMSE = [\sum_{i=1}^{27}(\hat{\theta}_i^{MC} - \hat{\theta}_i^{EX})^2/27]^{1/2}$, the superscripts MC and EX used to denote Monte-Carlo and exact values, respectively. The estimated standard errors described in the next section were a by-product of these calculations.

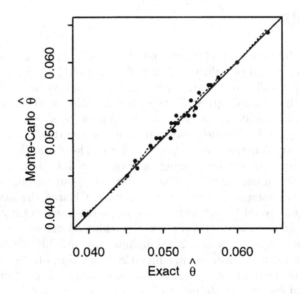

FIGURE 1. Monte Carlo estimates of $\hat{\theta}_i$. The plotted points correspond to the maximum observed in the grid. The broken line joins the estimates obtained by interpolation with a second degree polynomial in a neighborhood of the observed maximum. The continuous line is the reference 45 degree line representing the exact estimates.

We also considered some modifications to the procedure just described, including changing the χ^2 upper percentile which defines the smoothing neighborhood from 5% to 10%, and introducing the bias correction to the log of the estimated P_j suggested by Diggle and Gratton (1984) . These modifications had practically no impact on the RMSE and MRE of the estimates. We also conducted another Monte-Carlo using a much finer grid, recording realizations at times $t = 3.5(.01)7$. This finer partition resulted in a 14.6% reduction in $RMSE$ at the expense of a 6.7-fold increase in CPU time. Notice that the exact values reported by Minkin are also subject to rounding and calculation error. The rounding error is easy to determine, since the reported results have four decimals and three significant digits, which induce a rounding error with $RMSE$ and MRE one order of magnitude smaller than those obtained with Monte-Carlo. To put the accuracy of these Monte-Carlo estimates into perspective, note that the standard errors for $\hat{\theta}_i$ reported by Minkin range from .0014 to .0026, so $[s.e.^2(\hat{\theta}_i) + .00041^2]^{1/2}$ would range from .00146 to .00263. Thus, there is a very small increase in variability due to replacing an exact estimation by one based on Monte-Carlo.

5   Standard Errors

Having a closed-form expression for the log-likelihood function permitted Minkin (1994) to compute standard errors for the $\hat{\theta}_i$ as the reciprocal of the square root of the observed Fisher information. An approximation to the observed Fisher information is provided by the negative of the second derivative of the second order polynomial fitted in the neighborhood of the MLE as described in the previous subsection. In this way, an estimate of $s.e.(\hat{\theta}_i)$ is given by $[-2\beta_{2i}]^{-1/2}$, where β_{2i} is the coefficient of the quadratic term in the fitted second-order polynomial to the log-likelihood for the i-th animal. A comparison of the values obtained in this way with those reported by Minkin is given in Figure 2. Note that in absolute terms, these Monte-Carlo estimates for $s.e.(\hat{\theta}_i)$ are as accurate as the Monte-Carlo estimates of $\hat{\theta}_i$ (since the scale in the plot has been reduced by two orders of magnitude) with $RMSE = .000155$, which corresponds to 8% of the average of the 27 values listed by Minkin.

6   Shape Parameter

The calculations carried out by Minkin (1994) are only possible when the shape parameter of the Gamma lifetime is an integer. The Monte-Carlo analysis does not have this limitation. To maximize the profile log-likelihood $\sum_{i=1}^{27} L_i(\hat{\theta}_i(\alpha), \alpha)$ for the shape parameter α, a simple search was carried out first using four sets of 2,500 simulations, followed by four additional sets of 5,000 simulations as the region for the maximum was narrowed down, and finally 4 sets of 10,000 simulations. This search identified the MLE for α as 2.89. The profile log-likelihood at 2.89 exceeded by 6.7 the value

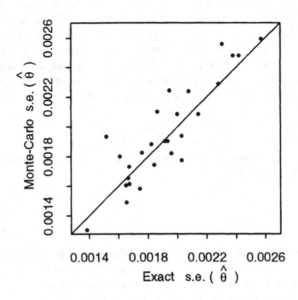

FIGURE 2. Monte Carlo estimates of $s.e.(\hat{\theta}_i)$.

obtained at $\alpha = 3$. This difference would easily be considered significant compared to half of the χ_1^2 percentile. However, one should also take into account the error due to Monte-Carlo. Although it is not difficult to compute the estimated variance of $L_i(\hat{\theta}_i(\alpha), \alpha)$ using properties of the multinomial distribution (Bishop, Fienberg and Holland, 1975, p 495), these terms are clearly positively correlated. A lower bound to the standard error is obtained if the covariances are assumed to be zero. This gives standard errors of 1.72 and 1.67 for the maximum of the profile log-likelihood for α equal to 3 and 2.89, respectively. An upper bound can be obtained by noting that the maximum possible covariance for $L_i(\hat{\theta}_i)$ and $L_j(\hat{\theta}_j)$ occurs when $\hat{\theta}_i = \hat{\theta}_j$ (ignoring here the effects of the quadratic smoothing, so the value of $\hat{\theta}_i$ is assumed to be one of those defining the original grid). This gives an upper bound for the standard error of the profile log-likelihood at $\alpha = 3$ and $\alpha = 2.89$ of 3.30 and 2.99, respectively. Although the lower bound for the standard errors would indicate that the increase in the profile log-likelihood is significantly higher than $\chi_{1;.05}^2/2$, the use of the upper bound would lead to different conclusions. Thus, it is not clear that a considerably better fit results from the added flexibility of estimating α.

7 Conclusions

These findings indicate that, as suggested by Thompson, Atkinson and Brown (1987), the use of Monte-Carlo techniques represents a feasible option even in the context of interactive analysis, without the need to develop specialized software. The approach is flexible enough to permit the consideration of more complicated and biologically sensible models, including those which incorporate crypt death or mixtures in the lifetime distribution.

Acknowledgments: The author would like to thank Mark Ossip for computational assistance and to Kathy Sykora for valuable comments. This work was partially supported by a grant from the Natural Sciences and Engineering Research Council of Canada.

References

Athreya, K.B. and Keiding, N. (1977). Estimation theory for continuous-time branching processes. *Sankhya* **39A**, 101-123

Bishop, Y.M.M., Fienberg, S.E., and Holland, P.W. (1975). *Discrete Multivariate Analysis.* MIT, Cambridge

Diggle, P. J. and Gratton, R. J. (1984). Monte Carlo methods of inference for implicit statistical models. *Journal of the Royal Statistical Society* **B46**, 193-227

Harris T.E. (1963). *Theory of Branching Processes.* Springer-Verlag, Berlin

Hoel, D.G. and Crump, K.S. (1974). Estimating the generation-time distribution of an age-dependent branching process. *Biometrics* **30**, 125-135

Jagers, P. (1975). *Branching Processes with Biological Applications.* Wiley, London

Minkin S. (1994). Statistical analysis of aberrant crypt assays for colon cancer promotion studies. *Biometrics*, **50**, 279-288

Thompson, J.R., Atkinson, E.N., and Brown, B.W. (1987). Simest: an algorithm for simulation-based estimation of parameters characterizing a stochastic process. In: Thompson, J.R. and Brown, B.W. (eds). *Cancer Modelling.* Marcel Dekker, 387-415

Quantitative Risk Assessment for Clustered Binary Data

Geert Molenberghs
Lieven Declerck
Marc Aerts

ABSTRACT: The effect of misspecifying the parametric response model for a univariate clustered binary outcome from a toxicological study on the assessment of dose effect and on estimating a virtually safe dose is investigated. Marginal and conditional models are contrasted.

KEYWORDS: Clustered Data; Dose-Response Models; Likelihood Estimation; Risk Assessment.

1 Introduction

Quantitative risk assessment refers to the task of predicting safe exposure levels for humans. Laboratory studies in rodents play an important role in testing and regulating substances with potential danger. Applications are found in developmental toxicity studies. A standard study includes a control group and several exposed groups, each involving 20 to 30 pregnant dams. Exposure usually occurs early in gestation, during the period of major organogenesis. Outcomes include fetal deaths and resorbtions, and malformation indicators. An important problem involves the calculation of a virtually safe dose (VSD), it is defined as the dose at which the excess risk over the background rate is small (Crump and Howe, 1985).

For normally distributed data, multivariate methods can yield much more powerful results. In our setting, some evidence has been gathered that the efficiency of both estimation and testing is not always higher when it is based on a multivariate outcome, rather than on a collapsed version (Lefkopoulou and Ryan 1994).

Several models, such as the Bahadur, the beta-binomial, and the conditional model proposed by Molenberghs and Ryan (1994) can be specified for univariate clustered outcomes. Emphasis is placed on estimating a dose effect parameter, on testing the null hypothesis of no dose effect, and on determining a VSD. We address the effect of misspecifying the model. Methods are compared on real developmental toxicity data, through asymptotic calculations and by means of small sample simulations.

2   Models for Clustered Data

Consider an experiment involving N clusters, the ith of which contains m_i implants, with r_i deaths/resorbtions and n_i viable fetuses. Suppose Y_{ij} indicates whether the jth individual in cluster i is abnormal. A covariate of interest is dosing d_i. Assuming exchangeability, it is useful to define $Z_i^{(1)} = Z_i = \sum_{j=1}^{n_i} Y_{ij}$, the number of malformations in cluster i, and $Z_i^{(2)} = \sum_{j<k} Y_{ij} Y_{ik}$, the number of pairs where both are malformed.

We will first give Bahadur's description of the joint distribution (Cox, 1972). We will describe it for the case of a univariate clustered outcome, adopting exchangeability and restricting the association to pairwise effects. With π_i the marginal malformation probability and ρ_i the pairwise correlation coefficient,

$$f(z_i|\pi_i,\rho) = \binom{n_i}{z_i} \pi_i^{z_i}(1-\pi_i)^{n_i-z_i}$$

$$\times \left[1 + \rho \left\{ \binom{z_i}{2} \frac{1-\pi_i}{\pi_i} + \binom{n_i-z_i}{2} \frac{\pi_i}{1-\pi_i} - z_i(n_i-z_i) \right\} \right].$$

A convenient model is

$$\begin{pmatrix} \ln(\frac{\pi_i}{1-\pi_i}) \\ \ln(\frac{1+\rho}{1-\rho}) \end{pmatrix} = X_i\beta = \begin{pmatrix} 1 & d_i & 0 \\ 0 & 0 & 1 \end{pmatrix} \begin{pmatrix} \beta_0 \\ \beta_d \\ \beta_a \end{pmatrix}$$

Maximum likelihood estimation is straightforward.

Rather than modelling marginal functions directly, a popular approach is to assume a random effects model in which each litter has a random parameter. An example is the beta-binomial model (Kleinman, 1973):

$$f(z_i \mid \pi_i, \rho) = \binom{n_i}{z_i} \frac{B(\pi_i(\rho^{-1}-1)+z_i, (1-\pi_i)(\rho^{-1}-1)+(n_i-z_i))}{B(\pi_i(\rho^{-1}-1), (1-\pi_i)(\rho^{-1}-1))}$$

where B denotes the beta function. The beta-binomial model and the Bahadur model have the same first and second order moments.

Molenberghs and Ryan (1994) proposed a conditional model which for a univariate clustered outcome becomes:

$$f(z_i|\theta_i,\beta_a) = \binom{n_i}{z_i} \exp\left\{\theta_i z_i + \beta_a z_i^{(2)} - A(\Theta_i)\right\},$$

with $A(\Theta_i)$, $\Theta_i = (\theta_i, \beta_a)$, a normalizing constant. Assume $\theta_i = \beta_0 + \beta_n n_i + \beta_d d_i$, θ_i can be interpreted as a conditional logit and β_a is the log odds ratio for a pair of fetuses, given all others observations are healthy. The "covariate" n_i ensures independence of coding reversal.

3 Dose Effect Estimation

We want to investigate the effect of misspecifying the response model on the assessment of dose effect. We contrast two models: the Bahadur and the conditional model. When the model chosen to fit the data is inappropriate, the true is inefficiently estimated and/or bias occurs.

A concern investigators have is whether a dose effect can be found. Likelihood ratio (LR) and Wald tests are considered. Small sample simulations and asymptotic calculations (Rotnitzky and Wypij, 1994) are carried out. For the latter, an artificial sample can be constructed, where each realization is weighted according to its true probability.

We assume there are 4 dose groups, with one control group ($d_i = 0$) and three active groups ($d_i = 0.25, 0.5, 1.0$); 30 clusters are assigned to each. The number n_i of viable fetuses per cluster is chosen at random from a local linear smoothed version of the relative frequency distribution given in Kupper et al. (1986).

For the Bahadur (conditional) model we choose intercepts -5.5 and -4.5 (-3.5 and -2.5), dose effects in the range $(0.0; 8.0)$, $((0.0; 5.5))$ and associations 0.0 and 0.1 (0.0 and 0.25). A correlation of 0.1 in the Bahadur model is considerable, given the range restrictions. β_n is chosen to be zero.

The tests statistics follow a noncentral χ^2 distribution with 1 degree of freedom and noncentrality parameter equal to the population value of the statistic $\lambda_d = \beta_d^2 W_{dd}^{-1}$, where W_{dd} is the population variance of β_d The LR test is characterized by $2(\ell_1 - \ell_0)$. The behaviour of the Wald test is often aberrant (Hauck and Donner 1977): it decreases to zero as the distance between the parameter estimate and the null value increases, and the power decreases to the significance level for alternatives far from the null. Fitting the incorrect model, we compute population values of likelihood ratio and Wald statistics. Care has to be taken with interpreting these values, but they are useful to assess the effect of misspecification. The intercepts closest to zero were used.

We studied 5 population values for test statistics (LR and Wald for Bahadur, LR, Wald and robust Wald for the conditional model), in four different settings, arising from Bahadur versus conditional model and zero versus nonzero association. Under the Bahadur model, the LR test under the correct model performs best. The Wald test deteriorates when the dose effect increases. The LR test under the conditional model increases monotonically, its value is comparable to Bahadur's Wald. The worst performance is seen with the Wald tests under the conditional model. It hardly increases beyond a dose effect parameter of about 3.5. Similar behaviour is seen for the conditional model. These findings are confirmed by the real data example (Table 3). All of them show highly significant dose effects.

We performed a limited small sample simulation study. Table 1 show estimated rejection probabilities. According to the asymptotic study, one could expect that these probabilities reflect the same ordering as in the test statistics, but differences are much less pronounced. There is no systematic order and most tests show comparable results. If Bahadur is true, all have a 100 % rejection probability for $\beta_d = 4.0$. The main reason for this apparant discrepancy is the fact that very modest sample sizes are sufficient to obtain a certain prescribed power. A sample size of 4×30 will lead to very high powers for all test statistics. This is comforting because it shows that even a small dose effect is detected with customary sample sizes.

Focusing on the type I error, we observe that, if data are generated from a Bahadur model, estimated error probabilities vary in the range $(0.8; 13.1)$. H_0 is rejected too often with an LR test when there is no association. For baseline $\beta_0 = -5.5$, the Wald rejection probability is very low. For the higher level, the picture is less clear. If data are generated from the conditional model, the estimated type I error probabilities stay in the acceptable range $(4.00; 8.27)$ except when $\beta_0 = -2.5$ and $\beta_a = 0.35$ and using the incorrect Bahadur model. The enormous error probabilities 51.84 % and 47.07 % are confirmed by intermediate simulations. For $\beta_0 = -3.0$, they are about 28 %.

4   Estimating the Virtually Safe Dose

Suppose we wish to estimate a safe level of exposure. The standard approach requires the specification of $p(d)$, the probability that an adverse event occurs at dose d. A measure for the excess risk over background is $r(d) = (p(d) - p(0))/(1 - p(0))$. The VSD at which the excess risk is acceptably low, can then be defined and estimated in several ways (Crump and Howe, 1985).

We use $p(d) = \pi(d; \beta)$, the marginal probability of malformation at level d and define the VSD by $r(d; \beta) = 10^{-4}$. For setting confidence limits in low dose extrapolation, Crump and Howe (1985) prefer the asymptotic distribution of the likelihood ratio. An approximate $100(1-\alpha)\%$ lower limit is

$$\min\{d(\beta) : r(d; \beta) = 10^{-4} \text{ over all } \beta \text{ s.t. } 2(\ell(\hat{\beta}) - \ell(\beta)) \le \chi_p^2(1 - 2\alpha)\}.$$

The probability of malformation for the conditional model has no simple expression and the distribution of the cluster sizes is involved; we used the local linear smoothed frequencies. An alternative procedure (Ryan 1992) first determines a benchmark dose, corresponding to an excess risk of 1 %, which is then extrapolated to a VSD. The main advantage is that is is less model dependent.

Results from an asymptotic simulation study are now summarized. When the true model is Bahadur and there is no association between littermates, there are only two curves: the model based and the extrapolated version. The latter ones yields the lowest VSD. When association between littermates is introduced, all curves separate. The highest VSD comes from

TABLE 1. Simulation Study. Each Dataset Consisted of 4 Dose Groups With 30 Litters in Each. Each Simulation Result is Based on 500 Replications. Estimated Rejection Probabilities of $H_0 : \beta_d = 0$ Are Shown. The Significance Level Is 0.05.

	Bahadur		Conditional		
β_d	Wald	LR	Wald(n)	Wald(r)	LR
True Model is Bahadur					
$\beta_0 = -4.5; \beta_a = 0.0$					
0.0	8.25	8.93	6.45	5.65	7.26
0.5	13.80	13.40	11.38	11.59	11.85
1.0	54.21	53.78	45.61	48.18	47.85
$\beta_0 = -5.5; \beta_a = 0.0$					
0.0	2.81	7.32	2.00	9.09	6.31
0.5	6.58	9.98	7.04	9.38	10.78
1.0	27.56	29.49	28.00	29.33	32.00
$\beta_0 = -4.5; \beta_a = 0.1$					
0.0	3.04	4.66	5.30	6.72	7.98
0.5	11.44	12.95	15.77	13.39	15.58
1.0	35.07	36.27	37.60	28.20	40.76
$\beta_0 = -5.5; \beta_a = 0.1$					
0.0	0.82	4.41	3.65	13.09	9.05
0.5	3.75	8.43	7.25	9.73	10.90
1.0	10.69	14.47	17.22	15.13	21.56
True Model is Conditional					
$\beta_0 = -2.5; \beta_a = 0.0$					
0.00	4.20	4.80	4.60	4.80	5.60
0.25	16.20	14.80	14.60	14.80	15.20
0.50	56.20	55.40	53.60	53.00	55.40
$\beta_0 = -3.5; \beta_a = 0.0$					
0.00	8.27	7.88	5.06	5.67	6.30
0.25	12.22	12.03	11.00	10.06	11.20
0.50	32.46	31.86	28.06	28.06	30.04
$\beta_0 = -2.5; \beta_a = 0.35$					
0.00	51.84	47.07	4.20	5.60	4.80
0.25	82.51	83.18	78.80	78.40	80.00
0.50	99.17	98.97	100.00	100.00	100.00
$\beta_0 = -3.5; \beta_a = 0.35$					
0.00	5.81	6.01	4.00	5.40	5.80
0.25	17.40	17.00	14.40	14.40	15.00
0.50	55.00	54.40	44.20	43.60	46.40

the conditional model (model based), the lowest one by extrapolating the Bahadur estimator. The difference is seen most prominently for very small dose effects. When dose effect increases, two pairs of curves are seen: model based and extrapolated. When the true model is conditional and no association is assumed, all procedures yield similar results. A different conclusion is obtained for associated outcomes: two pairs of curves are found: Bahadur versus conditional. Overall, the conditional model yields a higher VSD.

5   Examples

We applied the models to three developmental toxicity studies conducted by the Research Triangle Institute under contract to the National Toxicology Program (NTP). The studies investigated the effects in mice of three different chemicals: d-(2-ethyhexyl)-phtalate (DEHP) ethylene glycol (EG) and diethylene glycol dimethyl ether (DYME). References are found in Molenberghs and Ryan (1994).

Each study involved a control group and 3 or 4 dosed groups, each including 20 to 30 dams with between 2 and 17 offspring per litter. For these experiments, malformations were classified as being external, visceral and skeletal.

We will present analyses for the DEHP, EG, and DYME data. A collapsed outcome (ANY) was considered as well, which is 1 if a fetus has at least one malformation, and zero otherwise. Table 2 contains maximum likelihood estimates (results for external outcome omitted). Although Bahadur and beta-binomial yield similar intercepts and dose effect parameters, it is observed that the association parameter β_a is often larger in the beta-binomial model. This can be explained through range restrictions on the correlation in the Bahadur model. The beta-binomial model features all positive correlations and there are no restrictions on the parameters in the conditional model.

Wald and LR statistics are presented in Table 3. The Bahadur model almost always gives larger statistics than the beta-binomial model, while the conditional model yields smaller values. LR tests are larger than Wald type tests (cf. asymptotic findings). This holds for both Bahadur's and the conditional model. In both cases, the (higher order) association can be misspecified. We conjecture that omitting the higher order association might be one of the sources for the observed discrepancies.

VSD's were computed as in the previous section. In agreement with the simulation results, the conditional model in general yields higher values for both VSD and the lower limit. This effect is seen clearer in the model based version than with the extrapolation method. Exceptions are the EG study, where the conditional model yields the smallest values for external and visceral outcomes. The extrapolation method yields much smaller values in all cases.

TABLE 2. Maximum Likelihood Estimates (Standard Errors).

Outcome	Parameter	DEHP	EG	DYME
		Bahadur Model		
Visceral	β_0	-4.42(0.33)	-7.38(1.30)	-6.89(0.81)
	β_d	4.38(0.49)	4.25(1.39)	5.49(0.87)
	β_a	0.11(0.02)	0.05(0.08)	0.08(0.04)
Skeletal	β_0	-4.67(0.39)	-2.49(0.11)	-4.27(0.61)
	β_d	4.68(0.56)	2.96(0.18)	5.79(0.80)
	β_a	0.13(0.03)	0.27(0.02)	0.22(0.05)
Collapsed	β_0	-3.83(0.27)	-2.51(0.09)	-5.31(0.40)
	β_d	5.38(0.47)	3.05(0.17)	8.18(0.69)
	β_a	0.12(0.03)	0.28(0.02)	0.12(0.03)
		Beta-Binomial Model		
Visceral	β_0	-4.38(0.36)	-7.45(1.17)	-6.21(0.83)
	β_d	4.42(0.54)	4.33(1.26)	4.94(0.90)
	β_a	0.22(0.09)	0.04(0.09)	0.45(0.21)
Skeletal	β_0	-4.88(0.44)	-2.89(0.27)	-5.15(0.47)
	β_d	4.92(0.63)	3.42(0.40)	6.99(0.71)
	β_a	0.27(0.11)	0.54(0.09)	0.61(0.14)
Collapsed	β_0	-3.83(0.31)	-2.51(0.09)	-5.42(0.45)
	β_d	5.59(0.56)	3.05(0.17)	8.29(0.79)
	β_a	0.32(0.10)	0.28(0.02)	0.33(0.10)
		Conditional Model		
Visceral	β_0	-2.04(0.60)	-4.61(1.70)	-2.82(1.10)
	β_d	2.55(0.58)	3.96(1.41)	2.88(0.93)
	β_n	-0.19(0.04)	-0.26(0.11)	-0.31(0.06)
	β_a	0.30(0.12)	0.09(0.73)	0.54(0.15)
Skeletal	β_0	-3.19(0.71)	-1.22(0.36)	-1.11(0.60)
	β_d	2.85(0.61)	0.99(0.20)	2.36(0.53)
	β_n	-0.12(0.05)	-0.15(0.03)	-0.27(0.05)
	β_a	0.42(0.08)	0.41(0.03)	0.49(0.06)
Collapsed	β_0	-1.47(0.45)	-0.98(0.33)	-2.53(0.67)
	β_d	3.11(0.53)	0.97(0.20)	5.05(0.75)
	β_n	-0.20(0.03)	-0.17(0.03)	-0.20(0.04)
	β_a	0.29(0.06)	0.41(0.03)	0.37(0.06)

TABLE 3. Wald and Likelihood Ratio Test Statistics.

		DEHP		EG		DYME	
		LR	Wald	LR	Wald	LR	Wald
Visceral	Bah.	71	79	16	9	65	40
	BB	60	67	17	12	43	30
	Cond.	34	19	14	8	14	9
Skeletal	Bah.	76	71	182	261	311	52
	BB	59	61	64	75	117	96
	Cond.	41	22	53	24	62	20
Collapsed	Bah.	165	130	190	314	421	140
	BB	92	99	65	72	152	109
	Cond.	74	34	52	24	118	46

References

Cox, D. R. (1972) The analysis of multivariate binary data. *Applied Statistics* **21**, 113–120.

Crump, K. S. and Howe, R. B. (1985) A review of methods for calculating statistical confidence limits in low dose extrapolation. In: Toxicologiacal Risk Assessment, Volume I: Biologicial and Statistical Criteria, Clayson, D. B., Krewski, D., and Munro, I. (eds.) CRC Press.

Hauck, W.W. and Donner, A. (1977) Wald's test as applied to hypotheses in logit analysis. *J. Amer. Stat. Assoc.*, **72**, 851–853.

Kleinman, J.C. (1973) Proportions with extraneous variance: single and independent samples. *J. Amer. Stat. Assoc.*, **68**, 46–54.

Kupper, L. L., Portier, C., Hogan, M. D., and Yamamoto, E. (1986) The impact of litter effects on dose-response modeling in teratology. *Biometrics*, **42**, 85–98.

Lefkopoulou, M. and Ryan, L. (1994) Global tests for multiple binary outcomes. *Biometrics*, to appear.

Molenberghs, G. and Ryan, L.M. (1994) Likelihood inference for clustered multivariate binary data. Submitted.

Rotnitzky, A. and Wypij, D. (1994) A note on the bias of estimators with missing data. *Biometrics*, **50**, in press.

Ryan, L. (1992) Quantitative risk assessment for developmental toxicity. *Biometrics*, **48**, 163–174.

Optimal Design and Lack of Fit in Nonlinear Regression Models

Timothy E. O'Brien

ABSTRACT: This paper points out that so-called optimal designs for non-linear regression models are often limited when the assumed model function is not known with complete certainty and argues that robust designs - near optimal designs but with extra support points - can be used to also test for lack of fit of the model function. A simple robust design strategy - which has been implemented with a popular software package - is also presented and illustrated.

KEYWORDS: D-optimality; Model mis-specification ; Nonlinear models; Robust designs

1 Introduction

Researchers often find that nonlinear regression models are more applicable for modelling various biological or chemical processes than are linear ones since they tend to fit the data well and the models and model parameters are more scientifically meaningful. These researchers are often in a position of obtaining optimal or near-optimal designs for a given nonlinear model. A common shortcoming of most optimal designs for nonlinear models used in practical settings, though, is that these designs often have only p support points where p is the number of model parameters. Such designs may present no problem when the model function is assumed to be known with complete certainty, but researchers typically desire designs which are near-optimal for the assumed model but which contain "extra" design points that can be used to test for model adequacy. This paper introduces and illustrates such a "robust" design procedure which also has been implemented using a popular software package.

2 Optimal Design Theory

The design problem for the nonlinear model

$$y_i = \eta(\mathbf{x}_i, \theta) + \varepsilon_i \qquad i = 1, ..., n \tag{1}$$

typically involves choosing an n-point design, ξ, to estimate some function of the p-dimensional parameter vector, θ, with high efficiency. Here ξ can be written as

$$\xi = \left\{ \begin{array}{c} \mathbf{x}_1, \mathbf{x}_2, ..., \mathbf{x}_n \\ \omega_1, \omega_2, ..., \omega_n \end{array} \right\}$$

where the design vectors (or points), $\mathbf{x}_1$, $\mathbf{x}_2$, ..., $\mathbf{x}_n$, are elements of the designs space Δ and are not necessarily distinct, and the associated weights, $\omega_1, \omega_2, ..., \omega_n$, are non-negative real numbers that sum to one. Alternatively, ξ can be expressed in terms of its m (m $\leq$ n) distinct support points s_1, s_2, ..., s_m, and their associated weights, $\lambda_1, \lambda_2, ..., \lambda_m$.

When the residuals in (1) are uncorrelated Gaussian random variables with zero mean and constant variance (without loss of generality taken to equal one), the Fisher information matrix is given by $\mathbf{M}(\xi, \theta) = \mathbf{V}^T \Omega \mathbf{V}$, where $\mathbf{V}$ is the n×p Jacobian of η and $\Omega = \mathrm{diag}\{\omega_1, \omega_2, ..., \omega_n\}$, and the corresponding variance function of η for the design ξ is given by

$$d(\mathbf{x}, \xi, \theta) = \frac{\partial \eta(\mathbf{x})}{\partial \theta^T} \mathbf{M}^{-1}(\xi, \theta) \frac{\partial \eta(\mathbf{x})}{\partial \theta} \tag{2}$$

where $\partial \eta(\mathbf{x}) / \partial \theta$ is of dimension p×1 and a generalized inverse is used whenever $\mathbf{M}$ is singular.

Optimal designs typically minimize some convex function of $\mathbf{M}^{-1}$. For example, designs which minimize the determinant $|\mathbf{M}^{-1}(\xi, \theta^o)|$ are called locally D-optimal, and those that minimize the maximum (over all $\mathbf{x} \in \Delta$) of the variance function in (2) are called locally G-optimal; the term "locally" is used here to emphasize that the design is based on an initial estimate of the parameter vector, θ^o. Further, the General Equivalence Theorem of Kiefer and Wolfowitz (1960) establishes the equivalence between locally D- and G-optimal designs; a corollary to this theorem states that the variance function in (2) evaluated using a D-optimal design achieves its maximum value at the support points of this design.

To illustrate, consider the two-parameter intermediate product model function (IP2),

$$\eta(x, \theta_1, \theta_2) = \frac{\theta_1}{\theta_1 - \theta_2} \left(e^{-\theta_2 x} - e^{-\theta_1 x} \right),$$

and the initial parameter estimates $\theta_1^o = 0.70$ and $\theta_2^o = 0.20$. In this case the locally D-optimal design, ξ_D, associates the weight $\lambda = \frac{1}{2}$ with each of the support points $s_1 = 1.229$ and $s_2 = 6.858$. D-optimality of this design can be established by noting that the corresponding variance function, $d(x, \xi_D, \theta^o)$, reaches its maximum value at $x = 1.229$ and $x = 6.858$. An algorithm is presented in O'Brien (1995) which uses the SAS$^{(R)}$ software package and the results of the General Equivalence Theorem to obtain and verify locally D-optimal designs.

For the previous example, note that the number of support points of the D-optimal design is equal to the number of parameters (2). Thus, regardless of the final sample size chosen, half of the observations are to be taken at x = 1.229 and the other half at x = 6.858. Although this design is "optimal" for estimating the two parameters of the IP2 model function, it provides no opportunity to check the validity of the assumed model function. The design strategy introduced below is suggested for situations where robust designs, or designs with "extra" support points, are desired.

3   Lack of Fit in Regression Models

When replicates are taken at at least one of a design's support points, lack of fit of the assumed model function (to the means model) can be tested using the F-statistic

$$F_{m-p,n-m} = \frac{SSLF \, / \, (m-p)}{SSPE \, / \, (n-m)}$$

where SSLF and SSPE are the lack-of-fit and pure-error sums of squares, respectively, m is the number of support points, n is the final sample size and p is the number of model parameters. This statistic may be used to test lack of fit for nonlinear models when intrinsic curvature is negligible, and can be adjusted using the methods given in Hamilton and Wiens (1987) when it is not. Obviously, if the number of support points of a given design is equal to the number of model parameters, no test for inadequacy of the assumed model can be made using this test.

For a given final sample size (n) and model function with p parameters, we are interested in determining (in some sense) the optimal number of support points to maximize the power of this lack-of-fit test. One such measure is to choose m to minimize the quantile F statistic, $F_{\alpha,m-p,n-m}$, for, say, $\alpha = 0.05$. Interestingly, our preliminary research has shown that, at least when n > 2p, this "optimal" m is approximately $\frac{n+2p}{3}$. For example, if a sample size of n = 20 is used to fit the IP2 model function (where p = 2), then indeed $F_{.05,m-p,n-m}$ is minimized for m = 8. The main point here is that to test for lack of fit of the assumed model function, we usually wish to choose designs with more than p support points.

4   A Robust Design Strategy

Strategies to obtain robust designs, or designs with "extra" support points, include Bayesian D-optimality discussed in Chaloner and Larntz (1989), Q-optimality introduced in O'Brien (1992) and the nesting design approach given in O'Brien (1994). As none of methods are intended to obtain designs to test for general departures of the assumed model function, we recommend the following strategy be used in these situations.

4.1   The Algorithm

- Find the (locally) D-optimal design, ξ_D. Often this will have only p support points (see Gaffke, 1987), and the following assumes that this is indeed the case.

- Choose the number "de" between 0 and 1 (typically around .90), and find all (t) values of x such that

$$d(x, \xi_D, \theta^o) = p \left[\left(\frac{p+1}{p} \, de \right)^p - 1 \right] \tag{3}$$

- Choose as a final design r_1 replicates of the p support points of ξ_D and r_2 replicates of the t points obtained from the previous step.

4.2   Justification

Our motivation for suggesting the above algorithm is the following. Let ξ_x represent a one-point design putting all weight at the support point x. Then the design $\xi_N = \frac{p}{p+1}\xi_D + \frac{1}{p+1}\xi_x$ associates the weight $\frac{1}{p+1}$ with each of s_1, s_2, ...,s_p (the support points of ξ_D) and x. One measure of the "distance" between ξ_D and ξ_N is the D-efficiency (Atkinson and Donev, 1992)

$$de = \left[\frac{|\mathbf{M}(\xi_N)|}{|\mathbf{M}(\xi_D)|} \right]^{1/p}$$

which, in the current case, is equal to $\frac{p}{p+1}[1 + d(x, \xi_D, \theta^o)/p]^{1/p}$. Solving this expression for $d(x, \xi_D, \theta^o)$ in terms of "de" yields the expression in (3).

4.3   Implementation

Programs to obtain and verify D-optimal designs for nonlinear regression models and to graph the corresponding variance function using the SAS$^{(R)}$ software package are given in O'Brien (1994). These programs have also been adapted to obtain the t values of x which satisfy (3) for a given D-efficiency, "de". Also, these programs (available from the author) have also been extended to incorporate Bayesian D-optimality, subset D-optimality, and heteroskedastic error structures.

4.4   An Example

By way of illustration, consider again the IP2 model function with initial parameter estimates $\theta_1^o = 0.70$ and $\theta_2^o = 0.20$. Taking de $= 0.90$, we include those values of x such that the corresponding variance function equals 1.645; this yields the t $= 4$ values x $= 0.761, 1.909, 4.890$, and 9.366. The final design would then consist of r_1 replicates of the D-optimal support points 1.229 and 6.858 and r_2 replicates of the check points 0.761, 1.909, 4.890, and 9.366. This design has the advantage of having "extra" support points to test for model mis-specification, yet in being "near optimal" in the sense of having a reasonably high final efficiency.

4.5   Final Efficiencies

Denote

$$\xi_x = \left\{ \begin{array}{c} x_1, x_2, ..., x_t \\ \frac{1}{t}, \frac{1}{t}, ..., \frac{1}{t} \end{array} \right\}$$

where $x_1, x_2, ...,x_t$ are the t points that satisfy (3) for given values of "de" and p, and note that $\mathbf{M}(\xi_x) = \frac{1}{t}\mathbf{V}_x^T\mathbf{V}_x$, where $\mathbf{V}_x$ is the t $\times$ p Jacobian matrix evaluated at $x_1, x_2, ...,x_t$. Then the final D-efficiency of the design with r_1 replicates of the D-optimal support points and r_2 replicates of the support points of ξ_x is given by

$$de_F = \frac{(r_1 p)^{1-\frac{t}{p}}}{r_1 p + r_2 t} \, |r_1 p \mathbf{I}_t + r_2 \mathbf{D}(\mathbf{x}, \xi_D, \theta^o)|^{1/p}, \qquad (4)$$

where $\mathbf{D}(\mathbf{x}, \xi_D, \theta^o) = \mathbf{V}_x \mathbf{M}^{-1}(\xi_D)\mathbf{V}_x^T$ is the corresponding *variance-covariance matrix function* (c.f., equation (2)). Note that for the IP2 model function and the six-point design comprising one replicate of ξ_D and one replicate of ξ_x, the final D-efficiency is 88%, and this number can be increased (up to 100%) by increasing the number of observations chosen at the support points of ξ_D.

5   Discussion

The above robust design strategy has been used for sixteen data sets and model functions used in practical settings including the four-parameter log-logistic model used by Vølund (1978) to model a process with one explanatory variable and a five-parameter growth model used by Gerig, *et al.* (1989) to detect synergy of two chemicals. In all settings, near-optimal designs were easily obtained with the SAS software package, a package commonly used by practitioners. The key to this design procedure is its simplicity for obtaining near-optimal designs with extra support points for situations where the model function is not known with complete certainty.

References

Atkinson, A.C. and Donev, A.N. (1992). *Optimum Experimental Designs.* Oxford: Clarendon Press.

Chaloner, K. and Larntz, K. (1989). Optimal Bayesian design applied to logistic regression experiments. *J. Stat. Plann. Inf.*, **21**, 191-208.

Gaffke, N. (1987). On D-optimality of exact linear regression designs with minimum support. *J. Stat. Plann. Inf.*, **15**, 189-204.

Gerig, T.M., Blum, U., Meier, K. (1989). Statistical analysis of the joint inhibitory action of similar compounds. *J. Chem. Ecol.*, **15**, 2403-2412.

Hamilton, D., Wiens, D. (1987). Correction factors for F ratios in nonlinear regression. *Biometrika*, **74**, 423-5.

Kiefer, J., Wolfowitz, J. (1960). The equivalence of two extremum problems. *Can. J. Math.*, **12**, 363-6.

O'Brien, T.E. (1992). A note on quadratic designs for nonlinear regression models. *Biometrika*, **79**, 847-9.

O'Brien, T.E. (1994). A new robust design strategy for sigmoidal models based on model nesting. In Dutter, R. and Grossmann, W., eds., *Proceedings in Computational Statistics: Compstat, 1994*, Heidelberg: Physica-Verlag, 97-102.

O'Brien, T.E. (1995). Obtaining and verifying optimal designs for nonlinear regression models using SAS software. To appear in *Proceedings of SUGI 20*.

Vølund, A. (1978). Application of the four-parameter logistic model to bioassay: comparison with slope ratio and parallel line models. *Biometrics*, **34**, 357-65.

Nonparametric Regression, Kriging and Process Optimization

M. O'Connell, P. Haaland
S. Hardy and D. Nychka

ABSTRACT: Thin plate splines and kriging models are proposed as methods for approximating unknown response functions in the context of process optimization. Connections between the methods are discussed and implementation of the models using S-PLUS is described. Results are presented from a simulation study comparing the methods and further necessary work is identified.

KEYWORDS: Thin plate splines; Kriging; Process optimization

1 Introduction

The goal of process optimization is to explore and present relationships between the process response and design variables with the view to identifying settings of the design variables that provide an optimum response. These settings may be associated with a maximum, minimum or target/window and may also be robust to extraneous sources of variability.

Response surface methods (Box and Draper, 1987), even as augmented with notions of robustness (Myers *et al.* 1992), rely on polynomial models for the response surface. Such characterization, while convenient and generally useful, is unappealing primarily due to the global nature of the fit and the necessary extrapolation in corners of the design.

Nonparametric regression and kriging provide alternative approximations to the unknown response function that capture local structure. A feature of both these approaches is the more data-faithful prediction of the unknown response function over the design space.

In this note, implementation of thin plate splines and kriging methods for approximating response surfaces is described. The close correspondence of the two methods is discussed and the methods are compared by way of a simulation study involving two design variables. Based on the simulation study, some recommendations for further studies are made.

2   Thin plate splines

We consider thin plate spline functions f of a variable $\boldsymbol{x} \in \Re^d$ with data model $y_i = f(\boldsymbol{x}_i) + \epsilon_i$, $i = 1, ..., n$, where y_i is the observed response at the ith combination of design variables $\boldsymbol{x}_i$, f is a smooth function and ϵ_i are independent, zero-mean random variables.

The thin plate spline is a generalization of the usual cubic smoothing spline with a penalty function $J_m(f)$ of the form

$$J_m(f) = \int_{\Re^d} \sum \frac{m!}{\alpha_1! ... \alpha_d!} \left(\frac{\partial^m f}{\partial x_1^{\alpha_1} ... \partial x_d^{\alpha_d}} \right)^2 d\boldsymbol{x}$$

with the sum over all non-negative integers α such that $\sum \alpha_1 + ... + \alpha_d = m$, and with $2m > d$.

The estimator of f is the minimizer of the penalized sum of squares

$$S_\lambda(f) = \frac{1}{n} \sum_{i=1}^{n} (y_i - f(\boldsymbol{x}_i))^2 + \lambda J_m(f) \qquad (1)$$

As discussed in Bates *et al.* (1988), Green and Silverman (1994, section 7.9) and Wahba (1990, section 2.4) the function f satisfying this minimization is of the form

$$f(\boldsymbol{x}) = \sum_{i=1}^{n} \delta_i E_{md}(\boldsymbol{x} - \boldsymbol{x}_i) + \sum_{i=1}^{t} \beta_i \phi_i(\boldsymbol{x})$$

where δ and β are vectors of coefficients; $t = \begin{pmatrix} m + d - 1 \\ d \end{pmatrix}$; ϕ_i, $i = 1, ..., t$, are linearly independent polynomials spanning the t-dimensional space of polynomials in $\Re^d$ of total degree $\leq m - 1$; and E_{md} are the radial basis functions

$$
\begin{aligned}
E_{md}(\boldsymbol{r}) &= a_{md} \| \boldsymbol{r} \|^{(2m-d)} \log(\| \boldsymbol{r} \|) \quad d \text{ even} \\
&= a_{md} \| \boldsymbol{r} \|^{(2m-d)} \qquad\qquad d \text{ odd}
\end{aligned}
$$

The coefficients δ and β are functions of the smoothing parameter λ and basis matrices K and T where $K_{ij} = E_{md}(\boldsymbol{x}_i - \boldsymbol{x}_j)$ and $T_{ij} = \phi_j(\boldsymbol{x}_i)$. The thin plate spline is thus completely determined once the smoothing parameter λ is estimated. This is achieved by generalized cross validation (GCV) using the GCVPACK routines (Bates *et al.*, 1987) called from an S-PLUS function. The GCV estimate of λ and the coefficients δ and β are then used by an S-PLUS function predict() to obtain the optimum response value and to construct contour and surface plots.

3 Kriging

The kriging models considered are described in Zimmerman and Harville (1991) and Sacks *et al.* (1989). In general terms, consider

$$y(x) = X\beta + Z(x) + \epsilon$$

where $y(x)$ is an observed response with argument $x \in \Re^d$, X is a mean model matrix and typically contains just an intercept, β is a corresponding vector of coefficients, $Z(x)$ is a zero mean stochastic process and ϵ denotes additive measurement error with $E(\epsilon) = 0$, $cov(\epsilon) = \sigma^2 I$.

Zimmerman and Harville (1991) refer to this as a random field linear model and use the notation $\mathcal{F}_Z \equiv \{Z(x); \ x \in \Re^d\}$ to denote an unobservable random field. In keeping with Sacks *et al.* (1989) we consider $\mathcal{F}_Z$ to be second-order stationary, geometrically anisotropic and separable with covariance function of the form

$$C(s, t; \theta) \equiv C(r; \theta) = \theta_0 \prod_{k=1}^{d} \exp(-\theta_k r_k) = \theta_0 C^*(r; \theta^*)$$

where $r_k = \mid s_k - t_k \mid^{p_k}$ for $p_k \in (0, 2)$ is a measure of distance in the kth design variable dimension.

The additive measurement error, σ^2, is often referred to as a nugget effect. If there is replication at some design runs one might consider a log-linear variance model $\sigma^2 \mathrm{diag}[\exp(U\psi)]$, where ψ is a variance function parameter and U is a variance model matrix. In this setting one could identify a robust process.

The variance of y is conveniently written as $V = \sigma^2(I + \alpha C^*) \equiv \sigma^2 W$ for $\alpha = \theta_0/\sigma^2$. To compute the predicted surface note that the MLE of β is the GLS estimate $\hat{\beta} = (X^T V^{-1} X)^{-1} X^T V^{-1} Y$ and the MLE of σ^2 is $\hat{\sigma}^2 = (1/N) Y^T P_W Y$ for $P_W = W^{-1} - W^{-1} X (X^T W^{-1} X)^{-1} X^T W^{-1}$. Estimation of α and θ^* is then achieved by maximizing the concentrated restricted likelihood

$$l_R^*(\theta^*, \alpha, \hat{\beta}, \hat{\sigma}^2 \mid y) = -\frac{1}{2} \log |W| - \frac{1}{2} \log |X^T W^{-1} X| - \frac{n-p}{2} \log(Y^T P_W Y)$$

The predicted (kriging) surface is then calculated using the estimated values of θ via the familiar (EBLUP) expression for prediction at a new point g

$$\hat{y}(g) = X\hat{\beta} + c^T(g, x) V^{-1} (y - X\hat{\beta})$$

where $c(g, x)$ is the vector of correlations between the Z's at g and the design points x_i, $i = 1, ..., n$,

$$c(g, x)[i] = \theta_0 \prod_k \exp(-\theta_k r_{k[i]})$$

where $r_{k[i]} = \mid g - x_i \mid^{p_k}$.

4   Connections between the methods

Several authors have discussed the connection between thin plate splines and kriging models (e.g. Wahba 1990, section 2.4-2.5 and Cressie 1991, section 3.4.5). In essence, a kriging estimate is identical to a thin plate spline if an appropriate covariance function is used. For example, in one dimension the cubic smoothing spline may be written in value-2nd derivative form as the minimizer in f of $(y-f)^T(y-f) + \lambda f^T K f$, and the estimated function is $\hat{f} = SY$ for $S = (I + \lambda K)^{-1}$. It turns out that kriging with covariance matrix K^- produces the same estimate.

The choices of covariance function and spline basis thus distinguish the predicted functions. The polynomial basis of the thin plate spline corresponds to the mean model matrix in the kriging model and these specify the 'mean' or trend surface. The 'stochastic' part of the predicted response function is driven by the functional form $r^{2m-d} \log r$ for d even and r^{2m-d} for d odd as represented by the matrix E in the thin plate spline, and by the function $\prod_k \exp(-\theta_k r_k)$ for the kriging model. The coefficients, δ, of this 'stochastic' part of the thin plate spline are completely determined by λ which is estimated by GCV. This λ corresponds to the ratio of variances, α, in the kriging model. The coefficients, θ_k, of the kriging model form part of the basis for the prediction. The thin plate spline may thus be viewed as a deterministic analogue to the kriging model characterized by a base polynomial surface that is isotropically smoothed towards the observed data.

Laslett (1994) shows empirically that for data that may be considered as a locally stationary stochastic process with dominant short range correlations decreasing monotonically to zero with increasing distance, kriging may outperform splines if : (a) data are on a grid : small nugget effect, correlations between adjacent values are weak and positive (e.g. 0.2); or (b) data are irregularly spaced : close data with high correlation (e.g. 0.8), prediction required between data that are far apart. For (a) the difference is small and depends on extrapolation of the fitted covariance function; for (b) the difference is more apparent.

Laslett suggests the reason for the performance difference may be that the smoothing parameter in splines is estimated by GCV and is thus dependent on the geometry of the training set, and that splines may not be flexible enough to use information on covariation at distances needed for prediction even if available. However, as pointed out by Handcock et al. (1994), the better performance may be simply due to the kriging model using the mean function for prediction at points that have no near neighbor data.

The important issue for the study presented herein is whether it is best to predict the response surface, particularly in the region of the optimum, by smoothing based on neighboring data (splines), or by distance-based correlation between points and data as determined from the entire set of responses (kriging), over the chosen design.

5 Comparison of thin plate splines and kriging for process optimization

A small Monte Carlo study was used to compare fits obtained from thin plate splines ($m = 3$, i.e. quadratic base polynomial) and the kriging model for two generating functions of differing complexity and one 'space-filling' design. Results are scaled to those from a 2nd-order polynomial model fit to the same data.

The generating functions used were:

$$y(x_1, x_2) = 1 + [x_1 \sin(1.2\pi x_1)] * [x_2 \sin(2.6\pi x_2)]$$
$$0 \le x_1 \le 1, \quad 0 \le x_2 \le 10 \tag{2}$$

$$y(x_1, x_2) = [30 + x_1 * \sin(x_1)] * [4 + \exp(x_2)]$$
$$0 \le x_1, x_2 \le 5 \tag{3}$$

The more complex function 2 is shown in Figure 1. The smoother function 3 was considered by Welch *et al.* (1992).

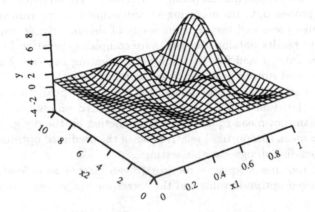

FIGURE 1. Product sine generating function

The designs used were 9, 13 and 17 run maximin designs on the unit circle as obtained from Gosset (Hardin and Sloane, 1994). These designs were augmented with 3 center points and rescaled over the specified domains. The 13 run design on the unit circle is shown in Figure 2.

For each combination studied, each of 100 simulated designs were rotated through a random angle θ, $0 < \theta < 2\pi$, and Gaussian realizations of the generating functions on the design were obtained using a standard deviation of 0.1 times the range of the generating function over the specified domain. For each realization, the 3 models were fit and used to obtain a prediction of the generating function on the design and on a 20 x 20 grid of evenly spaced points over the specified domain.

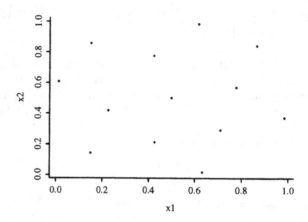

FIGURE 2. 13 run minimax design on the unit circle.

The goal of the study was to assess and compare performance of the thin plate spline and kriging models (using 2nd-order polynomial models for reference) for process optimization using a small number of experimental runs, relatively high noise and varying complexity of the underlying response. Comparative results obtained for the more complex generating function 2 are given in Table 1 and for the smoother generating function 3 in Table 2. The tabulated summary statistics are as follows:

ermse : median empirical root mean square error of prediction over the domain calculated on a 20 x 20 grid of evenly spaced points.

rmse : median root mean square error of prediction on the design.

bias.(x,y) = mean (proportion) relative bias of the predicted optimum value and corresponding design variable settings.

rmse.(x,y) : root mean square error (sqrt(mean bias^2 + sample variance)) of the predicted optimum value and the corresponding design variable settings.

The ermse, rmse, rmse.y and rmse.x are all expressed as proportions of the same quantity calculated for the 2nd-order polynomial model so that a value of less than 1 indicates improvement over that model.

6   Discussion

For both generating functions both the kriging model and the thin plate spline approximated the true function better than the polynomial model on the design and on the finer grid. The thin plate spline appeared to do a little better than the kriging model on the design but a little worse on the grid. Also, the thin plate spline showed less bias in predicting the

TABLE 1. Comparative simulation results for thin plate spline and kriging models with generating function 2 and a minimax design on the unit circle.

	runs=9		runs=13		runs=17		
	Krig	TPS	Krig	TPS	Krig	TPS	TPS*
ermse	0.754	0.789	0.797	0.858	0.812	0.902	0.832
rmse	0.616	0.528	0.938	0.592	0.998	0.522	0.046
bias.y	-0.812	-0.703	-0.785	-0.422	-0.835	-0.315	-0.123
rmse.y	1.278	1.110	1.532	1.035	1.713	1.042	0.645
bias.x1	-0.466	-0.445	-0.115	0.045	-0.092	0.031	0.183
rmse.x1	0.981	1.076	0.689	0.889	0.654	0.994	1.082
bias.x2	-0.895	-0.892	-0.276	-0.274	-0.163	-0.162	-0.024
rmse.x2	1.002	1.000	0.881	0.935	0.745	0.864	0.328

optimum than the kriging model although neither model did at all well for the more complex surface of generating function 1.2. Bias did improve for the thin plate spline with number of runs and also appeared to be due to the relatively high noise used since an additional simulation run at 1/10th the noise of the others showed quite low bias (see TPS*, the final column of Table 1). Neither mechanism was evident for the kriging model. The rmse of the predicted optimum was smaller than that of the polynomial surface only for the thin plate spline run with the small noise and was particularly poor for the kriging model in all cases. The rmse of the corresponding design variables showed little improvement over the polynomial model.

TABLE 2. Comparative simulation results for thin plate spline and kriging models with generating function 3 and a minimax design on the unit circle.

	runs=9		runs=13		runs=17	
	Krig	TPS	Krig	TPS	Krig	TPS
ermse	0.801	0.933	0.769	1.005	0.847	0.954
rmse	0.813	0.677	0.827	0.697	0.817	0.640
bias.y	-0.056	-0.028	-0.074	-0.037	-0.048	-0.023
rmse.y	1.808	0.955	1.703	1.007	1.310	0.913
bias.x1	0.046	-0.058	0.032	-0.154	0.034	-0.062
rmse.x1	0.918	1.091	0.884	1.145	0.751	0.966
bias.x2	0.103	0.072	0.157	0.113	0.132	0.101
rmse.x2	1.714	1.000	1.682	1.000	1.434	1.166

The kriging model and the thin plate spline are conceptually attractive approaches for function approximation in the context of process optimization. The simulation studies indicate, however, potential problems, particularly for complex surfaces observed with relatively high noise on small experimental designs. In these cases thin plate splines appear to do a little better in distinguishing signal from noise in the region of the optimum. Clearly, more work is required to compare and assess suitability of the methods for process optimization. This should include studies involving design sequences in which data are irregularly spaced and response functions with a range of noise.

References

Bates, D.M., Lindstrom, M.J., Wahba, G. and Yandell, B.S. (1987). GCV-PACK - Routines for generalized cross validation. *Comm. Stat. Sim. Comp.* 16, 263-297.

Box, G.E.P., Draper, N.R. (1987). *Empirical Model Building and Response Surfaces.* Wiley, New York, NY.

Cressie, N.A.C. (1991). *Statistics for Spatial Data.* Wiley, New York, NY.

Green, P.J., Silverman, B.W. (1994). *Nonparametric Regression and Generalized Linear Models.* Chapman and Hall, London, UK.

Handcock, M.S., Meier, K. and Nychka, D. (1994). Comment on Laslett. *J.A.S.A.* 89, 401-403.

Hardin, R.H., Sloane, N.J.A. (1994). *Operating manual for Gosset: A general purpose program for constructing experimental designs.* AT&T Bell Laboratories, Murray Hill, NJ.

Laslett, G.M. (1994). Kriging and splines: an empirical comparison of their predictive performance in some applications. *J.A.S.A.* 89, 391-400.

Myers, R.H., Khuri, A.I. and Vining, G. (1992). Response surface alternatives to the Taguchi robust design approach. *The American Statistician* 46, 131-139.

Sacks, J., Welch, W.J., Mitchell, T.J. and Wynn, H.P. (1989). Design and analysis of computer experiments. *Statistical Science* 4, 409-435.

Wahba, G. (1990). *Spline Models for Observational Data.* Society for Industrial Applied Mathematics. Philadelphia PA.

Zimmerman, D.L., Harville, D.A. (1991). A random field approach to the analysis of field-plot experiments and other spatial experiments. *Biometrics* 47, 223-239.

Frailty In Multiplicative Intensity Models

GH.R. Oskrochi

ABSTRACT: In multiplicative intensity models, using the counting process approach to survival analysis, the intensity function may depend on covariates (risk variables). However not all risk variables may be known or measurable. An unknown or unmeasurable covariate is usually called individual heterogeneity or frailty. When more than one failure time is obtained for each individual, frailty is a common factor among such repeated failure times. Here, multiplicative intensity models including random frailty effects are constructed. The method is programmed via a GLIM macro, and a real data set is analyzed.

KEYWORDS: Counting process; Frailty model; Multiplicative intensity model; Poisson process; GLIM

1 Introduction

Suppose in an experiment, for each of I individuals, n_i of one particular type of event occur. Specifically the problem that motivates this study is the repeated times to infection at the point of insertion of the catheter for each kidney patient using a portable dialysis machine. The infection occurs at the point of insertion of the catheter and, when it occurs, the catheter must be removed, the infection cleared up, and then the catheter re-inserted. Repeated times are times from insertion until the next infection. Sometimes the catheter must be removed for other reasons, so that there may be right censoring of the data. The risk variables (covariates) are age, sex, and three different type of disease. These variables form the regression vector **X** (McGilchrist and Aisbett 1991). The assumption of independent event times within each individual is questionable in the analysis of such a data set. Moreover, ignoring this particular type of dependency may lead to erroneous conclusions. We suppose that this dependency (*frailty*) is due to a common unobservable covariate, R_i, which acts multiplicatively on the hazard rate of the individual i. Frailty is different from individual to individual, or in other words, it is a between subject (individual) effect and can be assigned a distribution with possibly unknown parameters.

Estimation of the parameters of such frailty models has been carried out by several authors, for example based on a gamma distribution for frailty by Clayton 1978, Clayton and Cuzick (1985), Oakes (1982). Based on a proportional hazards model and partial likelihood by Lee and Klein (1989), Klein (1992). Hougaard (1986a, 1986b, 1987) criticized the gamma model for frailty, and proposed estimators based on a positive stable distribution for frailty.

In the above approaches the traditional conditional hazard model as described in Cox (1972) has been used. Thus, not surprisingly ties in observed failure times and time varying covariates create problems. Here, an alternative approach, using a multiplicative intensity model and counting process techniques, is given. The intensity for some life history event depends partly on observed covariates and partly on unobserved frailty effects, supposed to act multiplicatively on the intensity. Time varying covariates and tied observed failure times add no complexity to this method. The frailty distribution does not need to be gamma (Nielsen et al 1992) nor any other conjugate distribution.

2   Frailty in Multiplicative Intensity Models

Let us indicate by T_{ij} the failure time for the jth observation from the ith individual. We suppose possible dependence may occur among failure times from the same individual. Therefore there might be a problem of inference with unobserved subject−specific heterogeneity or correlated failure times for each individual.

In the multiplicative intensity model terminology $N = \{N_{ij} \quad : \quad i = 1, \ldots, I$ and $j = 1, \ldots, n_i\}$ denotes the multivariate counting process, such that $N_{ij}(t) = I(T_{ij} \leq t, w_{ij} = 1)$, where w_{ij} is as usual the censoring indicator. Take $Y = (Y_{ij} : i = 1, \ldots, I; j = 1, \ldots, n_i)$, defined by $Y_{ij}(t) = I(T_{ij} \geq t)$ to be an observable non−negative predictable process indicating whether or not the ith individual is at risk at time t for the jth observation. We now suppose that the unobservable specific individual quantity (frailty) R_i, is independently and identically distributed for $i = 1, \ldots, I$. Therefore, the n_i observations from individual i are correlated and $\mathrm{cov}(T_{ij}, T_{il})$ is a function of the parameter(s) of R_i's distribution (the frailty distribution). The conditional independence model suggests that, conditional on the frailty R_i, the set of $\{N_{ij}, Y_{ij}\}$ are independent, i.e. $\lambda_{ij}(t|r_i) = r_i \lambda_{ij}(t)$, where

$$\lambda_{ij}(t) = \lim_{h \downarrow 0} h^{-1} \Pr[N_{ij}(t+h) - N_{ij}(t) = 1|\mathcal{F}_{t-}] = \lambda_{0j}(t) Y_{ij}(t) \psi(X_{ij}(t))$$

$$(1)$$

is the intensity process of N_{ij} with time varying covariates $(X_{ij}(t))$. The intensity model given by (1) has the form of a stratified multiplicative intensity model, e.g. baseline intensity (λ_{0j}) is assumed to be the same for the jth observation of all individuals, given the random effects. This means that, the baseline risk of failure for the jth times depends on the number of times the same individual failed in the past (i.e., depends on j). If one believes that the calendar time is also important, this my be modelled via a covariate (e.g. age/cohort model).

2.1 Likelihood

For statistical analysis of the model in which $\lambda_{0j}(t)$ is supposed to be arbitrary, we assume that, conditional on the frailty, i.e., $R = r$, censoring is independent and noninformative about T. The likelihood contribution of the ith profile with intensity given by (1), and the random effect R, is given by,

$$L_i = \int \left[\prod_j \left(\prod_t (r_i \lambda_{ij}(t))^{\Delta N_{ij}(t)} \right) \exp\left(-r_i \int_0^{t_{ij}} \lambda_{ij}(s)ds \right) \right] f(r_i)dr_i$$

(2)

where $f(r)$ is the distribution of the random effects. Performing a change of variable by taking $\Lambda_{0j}(t) = \int_0^t \lambda_{0j}(s)ds$ with Jacobian $\lambda_{0j}(t)$, leads to a monotone transformation of the time axis, so that the same transformation applies to each risk set characterized by the baseline intensity λ_{0j}. Therefore the value of the triple $\{N_{ij}, Y_{ij}, X_{ij}\}$ remains unchanged under this transformed time for each separate risk set. Now, we divide the time axis into small units (the unit in which we measured the responses, e.g. one day, one week, ... etc.) intervals $(t^-, t]$, such that the information stays unchanged in (t^-, t) and the time varying covariates $\psi(X_{ij}(s))$ are constant when $s \in (t^-, t]$. Therefore, assuming that the jth observation of the ith individual is at risk in interval $(t^-, t]$, if it was at risk in the beginning of that interval (at t^-). One can easily show that,

$$\exp\left(-r_i \int_0^{t_{ij}} \lambda_{ij}(s)ds \right) = \prod_{t>0} \exp\left(-r_i Y_{ij}(t^-)\psi(X_{ij}(t))[\Delta \Lambda_{0j}(t)] \right).$$ (3)

Therefore, using this, we can rewrite (2) as $\int \ell_i f(r_i)dr_i$ where ℓ_i is,

$$\prod_j [\lambda_{0j}(t_{ij})]^{-1} \prod_{t>0} \left[(r_i \lambda_{ij}(t))^{\Delta N_{ij}(t)} \exp\left(-r_i Y_{ij}(t^-)\psi(X_{ij}(t))\Delta \Lambda_{0j}(t) \right) \right]$$

(4)

and $\Delta \Lambda_{0j}(t^-) = \Lambda_{0j}(t) - \Lambda_{0j}(t^-)$. The easiest choice for t^- when the information stays unchanged between ordered failure times, is the preceding ordered failure time less than t, this reduces the computational cost in model fitting.

In the uncensored case, the value of $\Delta N_{ij}(t) = 1$ whenever $t = t_{ij}$ and at this time point $\frac{\lambda_{ij}(t)}{\Lambda_{0j}(t)} = Y_{ij}(t^-)\psi(X_{ij}(t))$. Therefore, (4) can be reduced to

$$\prod_j \prod_{t>0} \left[r_i Y_{ij}(t^-)\psi(X_{ij}(t))\right]^{\Delta N_{ij}(t)} \exp\left[-r_i Y_{ij}(t^-)\psi(X_{ij}(t))\Delta\Lambda_{0j}(t)\right] \quad (5)$$

Here, the time axis has been divided into unit sub−intervals $(t^-, t]$. The definition of the censored and uncensored observations is different here. The jth event time of the ith individual in each small interval $(t^-, t]$ contributes a right censored time epoch if $t_{ij} > t$ and a failure time epoch if $t^- < t_{ij} \leq t$ and $w_{ij} = 1$. Maximizing the logarithm of the inner part of (5) with respect to $\eta_{ij} = r_i Y_{ij}(t^-)\psi(X_{ij})$ gives,

$$\sum_j \sum_{t>0} \left[\frac{\Delta N_{ij}(t)}{r_i Y_{ij}(t^-)\psi(X_{ij})} - \Delta\Lambda_{0j}(t)\right], \quad (6)$$

which is Poisson processes. Its solution in the random effect case gives the ML estimate of the parameters. Hence, the problem of inference in multiplicative intensity models reduces to the assumption of Poisson processes for $\Delta N_{ij}(t)$ with parameter $r_i Y_{ij}(t^-)\ \psi(X_{ij}(t))\ \Delta\ \Lambda_{0j}(t)$ and $\Delta\Lambda_{0j}(t) = \{\Lambda_{0j}(t) - \Lambda_{0j}(t^-)\}$ as prior weights.

2.2   Ties in Observed Failure Times

To extend the multiplicative intensity models to a setting which allows for ties in the observed failure times, we adapt the following modifications,

1. If $\Delta N_{ij}(t) = 0$. Then,
   i) $\Delta^n N_{ij}(t) = \Delta N_{ij}(t) = 0$.   ii) Put counter $D_{ij}(t) = 1$.
   iii) Put $\eta_{ij}^n(t) = \eta_{ij}(t) = r_i Y_{ij}(t^-)\psi(X_{ij}(t))$

2. If $\Delta N_{ij}(t) = 1$. Then,
   i) Replace $\Delta N_{ij}(t)$ by $\Delta^n N_{ij}(t) = \sum_{l=1}^I \Delta N_{lj}(t)$.
   ii) Put the counter $D_{ij}(t) = \Delta^n N_{ij}(t)$.
   iii) Replace $\eta_{ij}(t) = r_i Y_{ij}(t^-)\psi(X_{ij}(t))$ by $\eta_{ij}^n(t) = \sum_{l=1}^I \Delta N_{lj}(t)\eta_{lj}(t)$.
   iv) Replace the prior weights $\Delta\Lambda_{0j}(t)$ by $\Delta^n\Lambda_{0j}(t) = \frac{\Delta\Lambda_{0j}(t)}{D_{ij}(t)}$.

$\Delta^n N_{ij}(t), \eta_{ij}^n(t)$, and $\Delta^n\Lambda_{0j}(t)$, stay as the new values for, $\Delta N_{ij}(t), \eta_{ij}(t)$ and $\Delta\Lambda_{0j}(t)$ respectively. Now, each $\Delta^n N_{ij}(t)$ can take the values $0, 1, 2, \ldots$. Again we have a Poisson process for $\Delta^n N_{ij}(t)$ with new parameter $\eta_{ij}^n \Delta\Lambda_{0j}(t)$ and prior weights $\Delta^n\Lambda_{0j}(t) = \frac{\Delta\Lambda_{0j}(t)}{D_{ij}(t)}$. This is an attractive result, because not only do tied observations create no problems, but their contributions to the global likelihood are also very accurate. This nice property is not shared by alternative methods.

Method of Approach

Rewrite the intensity model (1) as

$$\lambda_{ij}(t|r_i) = \lambda_{0j}(t|r_i)Y_{ij}(t^-)\exp(\beta^t X_{ij}(t) + \gamma u_i),$$

where $\log R_i$ is replaced by γU_i and $U_i \sim N(0,1)$. For simplicity we assume that $n_i = n$ for all i.

Using the rank regression, we specify a marginal likelihood based upon the rank vector of the survival time, consistent with the sample. After some algebra, the partial likelihood in case of censoring becomes:

$$L_m = \underbrace{\int \cdots \int}_{\Lambda_{0j}(t_{ij}) \in \mathcal{K}_j^*} \left\{ \prod_{i=1}^{I} \left[\int \prod_{j=1}^{n_i} \prod_{t>0} \exp(\Delta N_{ij}(t)Y_{ij}(t^-)(\beta^t X_{ij} + \gamma u_i) \right. \right.$$

$$\left. \left. - \Delta\Lambda_{0j}(t)Y_{ij}(t^-)\exp(\beta^t X_{ij} + \gamma u_i))f(r_i)dr_i] \right\} d\mathcal{A}^0 \qquad (7)$$

where $\mathcal{K}_j^*$ for $j = 1, \ldots, n$ are the generalized rank vectors (Prentice 1978) containing the labels of the uncensored failure time in the jth risk set (for illustration see Oskrochi (1994) page 130-140).

It is easy to show that both in the univariate and multivariate case this integral is identical to the presentation of the random effect Cox's partial likelihood in the multiplicative intensity model, and its integrand has the same structure as the likelihood of the multiplicative intensity model with association parameter γ_i discussed by Nielsen et al. (1992).

The inner part of the above likelihood is a parametric random effect likelihood for the Poisson processes model where $\Lambda_{0j}(t)$ are known transformations of the survival times.

2.3   Estimation

Maximization of likelihood (7) to estimate β and γ is quite difficult. But the use of an approximate marginal likelihood for estimation suggested by Clayton and Cuzick (1985) can be useful here. The approximate marginal likelihood is given by the inner part of the above likelihood with $\Lambda_{0j}(t)$ replaced by (scores),

$$\Lambda_{0j}^-(t) = E(\Lambda_{0j}(t)|\mathcal{K}_1^*, \mathcal{K}_2^*, \ldots, \mathcal{K}_n^*, \beta, \gamma)$$

Therefore the marginal likelihood has the form:

$$\prod_{i=1}^{I} \int \prod_{j=1}^{n} \prod_{t>0} \exp[\ell]f(r_i)dr_i$$

$$\ell = \Delta N_{ij}(t)Y_{ij}(t^-)(\beta^t X_{ij}(t) + \gamma u_i) - \Delta\Lambda_{0j}^-(t)Y_{ij}(t^-)\exp(\beta^t X_{ij}(t) + \gamma u_i)$$

$$(8)$$

Since estimation of $\Lambda_{0j}^-(t)$ implicitly depends on γ and β, ML estimation of γ, β proceeds by iteration, using a two step algorithm. The first step calculates $\Delta\Lambda_{0j}^-(t)$ and the second step maximizes the above marginal likelihood (8). This marginal likelihood can be maximized in the GLM framework, using a Poisson process random effect model fitted to the process $(\Delta N_{ij}(t))$ with the transformed time $\Delta\Lambda_{0j}(t)$ as weight. The prepared macro and its related programmes are available from the author.

2.4   Calculation of Scores

The assumption of conditional independence implies that $E[\Lambda_{0j}(t)|\mathcal{K}_j^*,\beta,\gamma]$ $=E[\Lambda_{0j}(t)|\mathcal{K}_1^*,\mathcal{K}_2^*,\ldots,\mathcal{K}_n^*,\beta,\gamma]$. Thus, $\Lambda_{0j}^-(t)$ will be the estimate of the conditional cumulated baseline intensity (hazard) of the jth risk set corresponding to the rank set $\mathcal{K}_j^*$ up to time t, which is given by :

$$\Lambda_{0j}^-(t) = \int_0^t \frac{\sum_{i=1}^I dN_{ij}(s)}{\{\sum_{i=1}^I Y_{ij}(t^-)\exp(\beta^t X_{ij}(t)+\gamma u_i)\}} \tag{9}$$

for $j=1,2,\ldots,n$. This estimate is a modified extended Nelson−Aalen estimator or Breslow's estimate of the cumulated baseline intensity. Note that in this setting the estimate of $\Delta\Lambda_{0j}^-(t)$ is:

$$\Delta\Lambda_{0j}^-(t) = \frac{\sum_{i=1}^I \Delta N_{ij}(t)}{\{\sum_{i=1}^I Y_{ij}(t^-)\exp(\beta^t X_{ij}(t)+\gamma u_i)\}}. \tag{10}$$

2.5   Algorithm

1. Find an initial estimate for β by ignoring u and put $\Lambda_{0j}^-(t)=1$ for all j and t.

2. Maximize (8) to find (β,γ) as a random effects Poisson processes model (Oskrochi 1995).

3. Calculate $\Lambda_{0j}^-(t)$ for all j and t (using (10)).

4. Repeat steps 2 and 3 with the current value for $\Lambda_{0j}^-(t), \gamma$ and β until convergence.

2.6   Example

The described multiplicative intensity model with additional frailty term were fitted by a GLIM macro to the repeated times to infection at point of insertion of the catheter for each kidney patient. Estimation of the parameters together with standard errors are:

Parameters Estimation				
μ_1	$\mu_2 - \mu_1$	age	sex	γ
1.06	0.053	-.0007	0.99	$1.17e^{-16}$
Standard Errors				
0.59	0.43	0.013	0.44	0.21

μ_i is the mean of the increment of the cumulated baseline intensity for the ith repeated time. This analysis suggest that the distribution of T_{i1} is identical to that of T_{i2} for all i. Inclusion/exclusion of the type of disease in the model makes no sense, and also age has almost no effect. These data set do not show any sign of random effects, since the estimation of γ is very close to zero. This suggests that the first failure time is uncorrelated (or independent) from the second one. The only covariate that gives some effect is the sex; it shows the risk of infection for males is higher than females. The result of the analysis of data set in McGilchrist and Aisbett (1991) is somewhat different. They showed that the association parameter is not one (i.e., γ is not zero). This happened because they used a model for recurrent events analysis, which is not appropriate for this type of the data.

2.7   Advantages:

- The distribution of the random effects is not restricted to be normal. The normal distribution with finite variance for random effects which is used here is in the class of stable distributions discussed by Hougaard (1986, 1987). Moreover, in the absence of any parametric form for random effects, it seems to be reasonable to use a normal distribution (white noise) according to the central limit theorem.

- The model is in the form of a multiplicative intensity. The definition of the risk set and thus the calculation of the estimates are much easier than the traditional conditional hazard.

- Tied failure times in observed data and time dependent covariates add no complexity to this method, and do not need any further caution.

- **Graphical interpretation of the data**

  In figure 1, the first and the second failure times for each individual are plotted vs individuals. A common (positive) correlation between these two failure times would correspond to seeing a pattern in this graph. This means in the case of positive correlation that, long failure time for the first observation of an specific individual results in expecting a longer failure time in the second observation compared to the others. This phenomenon happened for the first five individuals. These individuals should have a rather high within correlation. But the violation of this pattern for the rest of the individuals can not result in an over all within correlation among individuals.

Failure Times

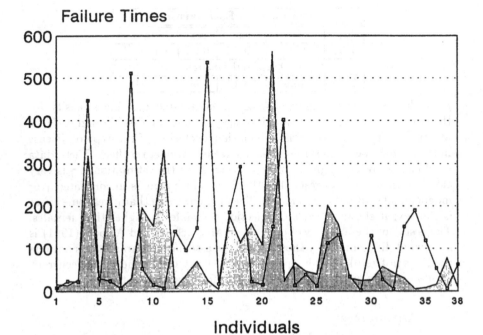

Individuals

FIGURE 1. Plot of the original data from the kidney patients

Acknowledgments: The most part of this work was done while the author was in the Biostatistical Centre for Clinical Trials of K.U. Leuven in Belgium. The author would like to acknowledge the comments and supervision of Prof. J.K. Lindsey and the support and help of Prof. E. Lesaffre and his group. The author is grateful to the ESRC for funding his current post under research grant number R000233850. I also acknowledge the Scholarship of the Ministry of health and medical education of I.R. Iran.

References

Clayton, D. (1978). A model for association in bivariate life tables. Biometrika 65, 141−151.

Clayton, D.G. and Cuzick, J. (1985). Multivariate generalization of the proportional hazards model (with discussion); J.Royal Statistical Society A−148 part 2.

Cox, D. R. (1972). Regression models and life tables (with discussion). J.Royal Statistical Society B 34, 187−220.

Hougaard, P. (1986a). A class of multivariate failure time distributions. Biometrika, 73 671−678.

Hougaard, P. (1986b). Survival model for heterogeneous populations derived from stable distributions, Biometrika, 387−396.

Hougaard, P. (1987). Modelling multivariate survival. Scan.J.Stat. 14, 291-304.

Klein, J, P. (1992). Semiparametric estimation of random effects using the Cox model based on EM algorithm. Biometrics, 48 795−806

Lee, S. and Klein,J. P. (1989). Statistical method for combining laboratory and field data ..., Recent Developments in Statistics and their application.

Nielsen, G. Gill, R.D., Sorensen, P.K. (1992), A counting process approach to maximum likelihood estimation in frailty model, Scan.J.Stat. 19

Oakes, D. (1982). A model for association in bivariate survival data. J.Royal Statistical Society B 44, 414−422.

Oskrochi, GH.R. (1995). Analysis of censored correlated observations. To appear in Journal of Statistical planning and inference, 1995.

Oskrochi, GH.R. (1994) Analysis of censored correlated observations. PhD. thesis K.U. Leuven Belgium

Prentice, R. L. (1978). Linear rank tests with right censored data. Biometrika, 65, 167−179.

Methods for Assessing the Adequacy of Probability Prediction Models

Joseph G. Pigeon
Joseph F. Heyse

ABSTRACT: There is an increasing interest in the development of statistical models which use a set of predictor variables in order to estimate the probability that a subject has a particular health outcome. Validation of these probability prediction models requires an assessment of how well the model is able to predict the outcomes of interest. Much of the work to date has considered the case of $K = 2$ distinct outcomes and a large candidate set of predictors. We consider the more general case with $K \geq 2$ outcomes and argue that the model needs to be validated on an individual basis as well as on a marginal basis. We propose some methods for these purposes and describe a new chi-square statistic for assessing goodness of fit. Two examples are given; one example applies these methods to a probability prediction model for predicting sets of symptoms associated with benign prostatic hyperplasia and the other example uses a logistic regression model.

KEYWORDS: Goodness of Fit; Probability Models; Logistic Regression

1 Introduction

Many areas of medical research, are becoming increasingly dependent upon statistical models which use a set of predictor variables in order to estimate the probability that a subject has a particular health outcome. The quality of inferences drawn from these models greatly depends on the validity of the underlying prediction model. Though it is impossible to specify the "true" prediction model, the ability to make predictions from "approximate" prediction models is still useful and highly desirable. Consequently, probability prediction models will continue to be developed and evaluated for their adequacy. Historically, standard goodness of fit tests such as X^2, G^2, and F^2 (see e.g. Read and Cressie, 1988; Agresti, 1990) have been used for this purpose. However because the standard goodness of fit tests rely on a multinomial assumption and only examine the overall or marginal fit of the model, they are not entirely satisfactory for the compete evaluation of a model's adequacy. We believe that probability prediction models require a different approach to the assessment of marginal fit as well as an assessment of the internal fit of the model. This internal assessment should consist of determining the extent to which individuals having certain outcomes are associated with high probabilities of those same outcomes.

This paper proposes a simple way to assess the internal fit and also describes a new test statistic recently suggested by Pigeon and Heyse (1995) to assess the marginal fit of probability prediction models. This new test statistic accounts for the failure of the multinomial assumption in a probability prediction model and has an approximate chi-square distribution with $K - 1$ degrees of freedom where K is the number of distinct outcome states. Two examples are provided to illustrate these methods; one example applies these methods to a probability prediction model for predicting sets of symptoms associated with Benign Prostatic Hyperplasia (BPH) and the other to a logistic regression model.

2   Problem Description and Proposed Methodology

The situation considered in this paper is one of determining whether a model which predicts the probability of certain outcome states provides an adequate fit to the data. Specifically, each of N subjects may be classified into one of K outcomes and we define π_{ij} as the true probability that subject i has outcome j. The model provides an estimate of this probability called $\hat{\pi}_{ij}$ which has been computed based on a set of covariates, $\mathbf{X}_i$. Consequently, the available data consists of a vector of covariates, $\mathbf{X}_i$, a vector of estimated probabilities, $\hat{\boldsymbol{\pi}}_i$ and a vector of outcomes, $\mathbf{Y}_i$ where $y_{ij} = 1$ if subject i has outcome j and $y_{ij} = 0$ otherwise.

Decisions about the adequacy of a probability prediction model should consider a comparison of the vector of estimated probabilities, $\hat{\boldsymbol{\pi}}_i$ to the vector of observed outcomes, $\mathbf{Y}_i$ on both an internal and a marginal basis. The internal evaluation compares the vectors $\hat{\boldsymbol{\pi}}_i$ and $\mathbf{Y}_i$ on an individual subject basis to assure that the model is generally estimating high probabilities for those outcome states that are actually observed for the subjects. The marginal evaluation compares the vectors $\sum \hat{\boldsymbol{\pi}}_i$ and $\sum \mathbf{Y}_i$ on an summary basis to assure that the model is estimating predicted numbers of subjects that are in agreement with the observed numbers of subjects for each outcome state.

We propose measuring internal closeness of fit through an analysis of the N estimated probabilities, $\hat{\pi}_{ij'}$, where j' is the observed outcome for subject i. Since these estimated probabilities are bounded by 0 and 1, an arcsine transformation should be used prior to analysis. An assessment of the model's internal fit may be made by describing the distribution of $\hat{\pi}_{ij'}$'s using measures such as, the mean, median, quartiles, etc. That is, if the distribution of $\hat{\pi}_{ij'}$'s is sufficiently "high", then we may infer that the model is generally estimating a high probability for the outcome actually observed for each subject.

While we think that this simple descriptive approach for assessing internal fit is sufficient for most purposes, we recognize the need for an inferential approach in some situations. For these situations, we suggest a simple t test comparing the mean transformed values of the $\hat{\pi}_{ij'}$'s to the transformed value of $1/K$, the expected value of the $\hat{\pi}_{ij}$'s under a null model.

Marginal assessment of fit should compare the K components of the $\sum \hat{\pi}_i$ vector, P_j, to the K components of the $\sum Y_i$ vector, O_j. This comparison might include plots of O_j versus P_j, a generalized R^2 statistic (Anderson-Sprecher, 1994) of the form

$$R^2 = 1 - \frac{\sum_{j=1}^{K}(O_j - P_j)^2}{\sum_{j=1}^{K}(O_j - P_j')^2}$$

or the familiar and popular statistic

$$X^2 = \sum_{j=1}^{K}\frac{(O_j - P_j)^2}{P_j}$$

where $O_j = \sum_{i=1}^{N} y_{ij}$ is the jth observed count, $P_j = \sum_{i=1}^{N} \hat{\pi}_{ij}$ is the jth predicted count and P_j' is the jth predicted count under some null model . When the $\hat{\pi}_{ij}$'s are the same for all subjects, then X^2 is the usual Pearson's chi-square statistic for multinomial outcomes which has an approximate chi-square distribution with $K-1$ degrees of freedom. However, when the $\hat{\pi}_{ij}$'s vary among the subjects, as is the case in a probability prediction model, then X^2 is no longer distributed as chi-square with $K-1$ degrees of freedom. When $\hat{\pi}_{ij}$ is an unbiased estimate of π_{ij}, then $E(O_j) = \sum_{i=1}^{N} \hat{\pi}_{ij} = P_j$ and $\mathrm{Var}(O_j) = \sum_{i=1}^{N} \hat{\pi}_{ij}(1 - \hat{\pi}_{ij})$ which is less than the usual multinomial variance, $N\bar{\pi}_j(1 - \bar{\pi}_j)$ where $\bar{\pi}_j = \sum_{i=1}^{N} \hat{\pi}_{ij}/N$. Consequently, the O_j's are under dispersed relative to a multinomial situation and the X^2 statistic will be biased downward. We propose to assess the adequacy of the model's fit by

$$G = \sum_{j=1}^{K}\frac{(O_j - P_j)^2}{\phi_j P_j}$$

where $\phi_j = \sum_{i=1}^{N} \hat{\pi}_{ij}(1 - \hat{\pi}_{ij})/N\bar{\pi}_j(1 - \bar{\pi}_j)$. Since ϕ_j is just the ratio of the true variance to the multinomial variance, the G statistic represents an adjustment to the usual X^2 statistic and it seems reasonable to expect its distribution to be chi-square with $K-1$ degrees of freedom. Finally, some algebra yields the alternate expression $\phi_j = 1 - (N-1)S_j^2/N\bar{\pi}_j(1 - \bar{\pi}_j)$ where $S_j^2 = \sum_{i=1}^{N}(\hat{\pi}_{ij} - \bar{\pi}_j)^2/(N-1)$ from which we easily observe that if

$\hat{\pi}_{ij}$ is the same for all subjects, then $S_j^2 = 0$ and $G = X^2$ but if $\hat{\pi}_{ij}$ differs for some subjects, then $S_j^2 > 0$ and $G > X^2$.

Pigeon and Heyse (1995) performed extensive simulations using various numbers of subjects (N), various numbers of outcomes (K) and various numbers of distinct covariate patterns (C). These simulations were used to compare the empirical distributions of the G and X^2 statistics as well as another test statistic based on Freeman-Tukey deviates, $Z^2 = \sum_{j=1}^{K}(\sqrt{O_j} + \sqrt{O_j + 1} - \sqrt{4N\bar{\pi}_{ij} + 1})^2$ and a corrected version of this test statistic, Z_C^2 obtained by dividing each squared deviate by ϕ_j. The simulations demonstrate that the G statistic has much less bias than the other three statistics and is very nearly distributed as chi-square with $K - 1$ degrees of freedom.

3   Examples

A probability prediction model was developed to examine the progression of six symptoms associated with benign prostatic hyperplasia (BPH). The model estimates the probability that a subject will be in each of the $2^6 = 64$ possible symptom states at time $t + 2$ years based upon the subject's age and symptom state at times t and $t - 2$ years. The 64 symptom states are defined by the presence or absence of the six BPH symptoms and the model was developed from longitudinal data obtained from the Baltimore Longitudinal Study of Aging (NIH, 1984). The model was constructed from 1585 records representing the cross-sectional pooling of the visit data from 700 men who reported no previous surgery for BPH, no history of prostatic cancer and who had at least 3 visits between 1.5 and 2.5 years apart. For illustrative purposes, the model has been reduced to consideration of only the four most prevalent symptoms which results in 16 possible symptom states. The probability of each of these 16 symptom states was estimated for each of the available 1585 subject-visits and the marginal data are presented in Table 1.

Calculation of the X^2 and G statistics yields $X^2 = 43.73$ and $G = 57.86$ both of which are significant at the $\alpha = 0.01$ level, suggesting that the model might not be adequate as a probability prediction model. However, this highly significant result may not be due to a poor fit of the model, but rather to the very large sample size. In fact, when a 50% random sample of the 1585 subject-visits was used to assess model fit, a nonsignificant result was obtained. Further justification of the model's fit is provided by the generalized R^2 value of 0.942 comparing the proposed model to a null model consisting of the multinomial product of the four symptom prevalences. This value of 0.942 may be interpreted as a measure of the proportion of "variability" between observed and predicted counts which is accounted for by the proposed model relative to a simple (null) model based on symptom prevalences.

Table 1
Marginal Assessment of Fit
Reduced BPH Symptom Model

Symptom State (j)	Predicted Number with State j	Observed Number with State j	ϕ_j
0000	458.06	500	0.526
0001	289.65	307	0.695
0010	30.14	16	0.946
0011	50.94	48	0.867
0100	147.71	132	0.758
0101	122.68	106	0.832
0110	24.11	26	0.875
0111	61.74	53	0.787
1000	32.54	16	0.948
1001	75.81	59	0.796
1010	5.78	7	0.947
1011	34.75	21	0.860
1100	29.59	33	0.840
1101	86.88	95	0.747
1110	12.78	18	0.895
1111	121.85	148	0.502
Total	1585	1585	

The internal goodness of fit was assessed by examination of the distribution of predicted probabilities of the observed cells. Employing an arcsine transformation, the observed cells were found to have an average predicted probability of 0.3869, which seems quite high considering that each subject had 16 possible symptom states. Furthermore, about 85% of the predicted probabilities of the observed cells exceeded $1/16 = 0.0625$. The t test comparing the average predicted probability (transformed) to 0.0625 (transformed) yields a test statistic of 50.96, which is highly significant. Thus, there seems to be ample evidence that the observed cells are indeed associated with "high" probabilities.

Considering both the marginal and internal goodness of fit assessments, we conclude that the model provides an adequate fit to the data and is in fact useful for understanding the development and progression of symptoms associated with BPH.

A common application arises for the case when there are only two outcome states ($K = 2$) and a logistic regression model is used as a fit to the data. Strictly marginal goodness of fit tests cannot be used in this instance since $\sum \hat{\pi}_{i1} = \sum y_{i1}$ and $\sum \hat{\pi}_{i2} = \sum y_{i2}$ which implies that $G = X^2 = 0$. To circumvent this problem, Hosmer and Lemeshow (1980, 1989) recommend forming g groups of subjects based on the values of $\hat{\pi}_{i1}$. That is, the N subjects are first sorted by $\hat{\pi}_{i1}$ and then combined into g groups of approximately equal size. For each group and outcome, estimates of expected frequencies are calculated by summing the $\hat{\pi}_{ij}$'s over all subjects in that group and a goodness of fit statistic, $\hat{C}$, is obtained by calculating the Pearson chi-square from the $2 \times g$ table of observed and estimated expected frequencies. Hosmer and Lemeshow note that the distribution of $\hat{C}$ is not chi-square with $g - 1$ degrees of freedom because of the varying $\hat{\pi}_{ij}$'s within each group and outcome but their extensive simulations indicate that the distribution of $\hat{C}$ is well approximated by the chi-square distribution with $g - 2$ degrees of freedom.

The G statistic can also be used in this setting, since it accounts for the varying $\hat{\pi}_{ij}$'s within the cells of the $2 \times g$ table and no adjustment to the degrees of freedom is necessary. An important advantage to the use of G is that groups can be formed using criteria based on values of the covariates rather than deciles of risk. This strategy allows inspection of the goodness of fit of the model within selected subgroups, a potentially useful diagnostic technique.

Table 2
Low Birth Weight Data Set (Hosmer-Lemeshow Example)
Grouping based on Estimated Probabilities of Events (Deciles of Risk)

Group	Group Size	Estimated Number of Events	Observed Number of Events	ϕ_j
1	19	14.12	15	0.9575
2	19	10.52	9	0.9924
3	19	8.57	10	0.9985
4	18	6.82	6	0.9976
5	18	5.57	6	0.9987
6	20	5.02	6	0.9978
7	20	3.65	2	0.9976
8	18	2.25	4	0.9968
9	20	1.64	1	0.9993
10	18	0.85	0	0.9951
Total	189	59	59	

$$\hat{C} = 5.23, \quad df = 8, \quad P = 0.7327$$
$$G = 5.26, \quad df = 9, \quad P = 0.8111$$

For a numerical example, we consider the low birth weight data set used by Hosmer and Lemeshow (1989) to illustrate the assessment of fit using $\hat{C}$. The raw data appear in Appendix 3 and details of the data set and the fitted model may be found in Chapters 2, 3 and 4. The results reported in Table 2 using a deciles of risk grouping strategy show that the ϕ_j's are quite close to 1.0 causing the values of $\hat{C}$ and G to be very similar. In addition, both statistics indicate that the model fits reasonably well.

A curious investigator might also wish to know if the model fits equally well for younger subjects as well as older subjects. The results reported in Table 3 using a grouping strategy based upon age show the ϕ_j's to be much less than one causing quite different values of $\hat{C}$ and G. However, because of the adjustment to the degrees of freedom in $\hat{C}$, the associated P-Values are very similar and both statistics indicate that the model fits reasonably well across all age groups.

Table 3
Low Birth Weight Data Set (Hosmer-Lemeshow Example)
Grouping based on Age Groups

Group	Group Size	Estimated Number of Events	Observed Number of Events	ϕ_j
1	12	4.31	5	0.9393
2	22	7.98	7	0.8621
3	34	12.82	11	0.8887
4	25	8.77	7	0.7798
5	26	6.91	10	0.8209
6	23	8.82	10	0.6795
7	27	5.73	7	0.7072
8	19	3.67	2	0.7253
Total	189	59	59	

$$\hat{C} = 4.76, \quad df = 6, \quad P = 0.5755$$
$$G = 5.99, \quad df = 7, \quad P = 0.5405$$

4 Concluding Remarks

We have suggested some simple techniques for assessing the adequacy of probability prediction models both internally and marginally. Specifically, we suggest that internal assessment of a model be made by examining the distribution of predicted probabilities associated with each subject's observed outcome. This assessment may be performed using descriptive statistics or through a test comparing the observed mean of the predicted probabilities to an appropriate hypothesized mean. We showed that the marginal assessment of the model should not be made using a Pearson chi-square statistic, since the observed counts are actually under dispersed relative to a true multinomial situation. However, the G statistic provides an unbiased test statistic for marginal assessment with an approximate chi-square distribution and the usual $K - 1$ degrees of freedom.

In addition, we showed how the G statistic can also be used to test the goodness of fit of a logistic regression model. In fact, the use of G seems to offer several advantages over the Hosmer Lemeshow goodness of fit test.

First, the groups may be formed using any strategy of interest to the investigator such as age, gender or risk factors and the G statistic will adjust to accommodate the predicted probabilities which may vary over a wide range. However, since the Hosmer Lemeshow test is based upon a grouping strategy which results in minimizing the variance of the predicted probabilities within each group, it is not known whether the same degrees of freedom adjustment will continue to work as well when other grouping strategies are used which might result in greater variances within groups. It is also not known whether the same adjustment of degrees of freedom works when there are more or less than 10 groups.

Finally, we suggest that no single statistic should be used to assess the adequacy of a probability prediction model. As demonstrated in our BPH symptom model example, a highly significant value of G does not necessarily indicate lack of marginal fit, since large sample sizes will usually yield a significant goodness of fit statistic. Plots of predicted counts versus observed counts and generalized R^2 statistics are also of great help in marginal goodness of fit assessments. All of these tools need to be used together in a cohesive manner in order to decide on the adequacy of probability prediction models.

References

Agresti, A. (1990). *Categorical Data Analysis*. Wiley, New York.

Anderson-Sprecher, R. (1994). Model comparisons and R^2. *Am. Statistician*, Vol. 48, No. 2, pp.113-117.

Hosmer, D.W. and Lemeshow, S. (1980). Goodness-of-fit tests for the multiple logistic regression model. *Communication is Statistics, Theory and Methods*, Vol. 9, pp. 1043-1069.

Hosmer, D.W. and Lemeshow, S. (1989). *Applied Logistic Regression*. Wiley, New York.

National Institutes of Health, Normal Human Aging. (1984). The Baltimore Longitudinal Study of Aging. November 1984. Washington, DC: US GPO (NIH publication No. 84-2450).

Pigeon, J.G. and Heyse, J.F. (1995). An improved goodness of fit statistic for probability prediction models. (submitted).

Read, T.R.C. and Cressie, N.A.C. (1988). *Goodness-of-Fit Statistics for Discrete Multivariate Data*. Springer-Verlag, New York.

Forecast Methods in Regression Models for Categorical Time Series

Helmut Pruscha

ABSTRACT: We are dealing with the prediction of forthcoming outcomes of a categorical time series. We will assume that the evolution of the time series is driven by a covariate process and by former outcomes and that the covariate process itself obeys an autoregressive law. Two forecasting methods are presented. The first is based on an integral formula for the probabilities of forthcoming events and by a Monte Carlo evaluation of this integral. The second method makes use of an approximation formula for conditional expectations. The procedures proposed are illustrated by an application to data on forest damages.

KEYWORDS: Forecasting; Categorical Time Series; Regression Model; Cumulative Model; Monte Carlo Method; Forest Damage Data

1   Introduction

We are concerned with the problem of predicting forthcoming outcomes Y_{T+l} of a categorical time series Y_t, $t = 1, 2, \ldots$, if the history of the process up to time T was observed.

The evolution of the categorical response variable Y_t, where $Y_t \in J = \{1, 2, \ldots, m\}$, is assumed to be driven by (1) an r–dimensional covariate process Z_t, $t = 1, 2, \ldots$, (2) the last response Y_{t-1}, (3) a vector summarizing the history before $t - 1$.

Thus, we are dealing with a transition type of regression model. The conditional probabilities

$$p_{t,j} = \mathbb{P}(Y_t = j \mid \mathcal{H}_t)$$

are modelled in the form $h_j(\eta_t)$, where h_j, $j \in J$, are response functions, η_t is a regression term depending on the regressors introduced in (1) to (3) above, and $\mathcal{H}_t$ comprises the variables $Z_1, Y_1, \ldots, Z_{t-1}, Y_{t-1}, Z_t$. Special attention is given to a cumulative regression model in the case where Y_t is measured on an ordinal scale.

To tackle the forecast problem we have to assume that the covariate process Z_t, $t = 1, 2, \ldots$, obeys an own autoregressive law, not influenced by the process Y_t, $t = 1, 2, \ldots$, of the response variables. We will be interested in the l-step predictor of the conditional probabilities $p_{t,j}$, i.e.

$$\hat{p}_{T,j}(l) = \mathbb{E}(p_{T+l,j} \mid \mathcal{F}_T), \quad \mathcal{F}_T = (\mathcal{H}_T, Y_T).$$

We will present two forecasting methods. The first is based on a multiple integral formula for $\hat{p}_{T,j}(l)$ and on its calculation by means of a recursive Monte Carlo algorithm. The second method is based on an approximation of the form

$$\mathbb{E}(h(\eta_{T+l}) \mid \mathcal{F}_T) \approx h(\mathbb{E}(\eta_{T+l} \mid \mathcal{F}_T))$$

and on an recursion formula for $\mathbb{E}(\eta_{T+l} \mid \mathcal{F}_T)$. We will close with an application of the forecasting methods to longitudinal data on forest damages. The responses Y_t are the levels of tree damages at time t, and the covariates Z_t refer to the trees, the site and the soil.

2   Modelling

Let the components of $p_t = (p_{t,1}, \ldots, p_{t,m-1})^T$ be positive, with a sum less than 1, let the vector response variable

$$W_t = (Y_{t,1}, \ldots, Y_{t,m-1})^T$$

be multinomially distributed with parameters 1 and p_t and put

$$Y_{t,m} = 1 - (Y_{t,1} + \ldots + Y_{t,m-1}), \quad p_{t,m} = 1 - (p_{t,1} + \ldots + p_{t,m-1}).$$

A regression model for categorical time series is defined by

$$p_t = h(\eta_t), \quad h : \mathbb{R}^{m-1} \to \mathbb{R}^{m-1} \tag{1}$$

where the regression term $\eta_t = (\eta_{t,1}, \ldots \eta_{t,m-1})^T$ is of the form

$$\eta_{t,j} = \alpha_j + \pi \cdot p_{t-1,j} + \lambda^T \cdot \Lambda(W_{t-1})_j - \beta^T \cdot Z_t. \tag{2}$$

It comprises as regressors the preceding probability vector p_{t-1}, a $q-$ dimensional function Λ of the preceding response W_{t-1} and the covariates Z_t. Unknown are the parameters

$$\alpha \in \mathbb{R}^{m-1}, \quad \pi \in \mathbb{R}, \quad \lambda \in \mathbb{R}^q, \quad \beta \in \mathbb{R}^r.$$

For such transition models see Fahrmeier and Kaufmann (1987) , Zeger and Qaqish (1988) . In the case of an ordinal response it is useful to introduce cumulative probabilities

$$p_{t,(j)} = p_{t,1} + \ldots + p_{t,j}$$

as well as the cumulative quantities $\Lambda(\cdot)_{(j)}$ and $\eta_{t,(j)}$. Putting

$$h_j(\eta_t) = F(\eta_{t,(j)}) - F(\eta_{t,(j-1)})$$

model (1),(2) has the form of a cumulative regression model, see McCullagh (1980) , namely

$$p_{t,(j)} = F(\eta_{t,(j)}), \quad \eta_{t,(j)} = \alpha_{(j)} + \pi \cdot p_{t-1,(j)} + \lambda^T \cdot \Lambda(W_{t-1})_{(j)} - \beta^T \cdot Z_t, \quad (3)$$

where the $\alpha_{(j)}$ stand in an increasing order, with $\alpha_{(m)} = \infty$, where F is a cumulative distribution function and where the parameters are restricted by $\pi \cdot p + \lambda^T \cdot \Lambda(w)_j > 0$ for all $w, j, p = 0, 1$. The asymptotic theory of model (3) was given in some detail in Pruscha (1993) . Two important special cases concerning Λ are

1. $q = m - 1$, $\Lambda(W_t) = W_t$,   (lagged dummy variables)

2. $q = 1$, $\Lambda(W_t) = Y_t = \sum_{j=1}^{m} j \cdot Y_{t,j}$ (lagged ordinal variable).

3   Forecasting. The General Set-Up

We adopt the following set-up from time series analysis, see Brockwell and Davies (1987, sec.5.1-5.5) . If $X_t, t = 1, 2, \ldots$, is a time series and $\mathcal{F}_T$ comprises the information on $X_1, \ldots, X_T$, we define the $l-$step predictor, the prediction error and the prediction m.s.e, respectively, by

$$\hat{X}_T(l) = \mathbb{E}(X_{T+l} \mid \mathcal{F}_T), \ \Delta_T(l) = X_{T+l} - \hat{X}_T(l), \ V_T(l) = \mathbb{E}(\Delta_T^2(l) \mid \mathcal{F}_T).$$

For $MA(\infty)-$ processes $\Delta_T(l)$ and $\mathcal{F}_T$ are independent, such that $V_T(l) = \mathbb{E}(\Delta_T^2(l))$ too. With the short-hand notation

$$\mathbb{E}_T(\cdot) = \mathbb{E}(\cdot \mid \mathcal{F}_T), \quad \mathrm{Var}_T(\cdot) = \mathbb{E}_T(\cdot - \mathbb{E}_T(\cdot))^2 \tag{4}$$

we can write

$$V_T(l) = \mathrm{Var}_T(X_{T+l}). \tag{5}$$

For regression models (1)-(3) we are firstly interested in forecasting $p_{t,j}$ and then in forecasting the derived quantity

$$\mu_t = \sum_{j=1}^{m} j \cdot p_{t,j} = \sum_{j=0}^{m-1} (1 - p_{t,(j)}).$$

For this reason we put $\mathcal{F}_T = (Y_1, \ldots, Y_T, Z_1, \ldots, Z_T)$ and -with the definition of $\mathbb{E}_T$ and Var_T as in (4)- we introduce the $l-$step predictors

$$\hat{p}_{T,j}(l) = \mathbb{E}_T(p_{T+l,j}), \quad \hat{\mu}_T(l) = \mathbb{E}_T(\mu_{T+l}) = \sum_{j=1}^{m} j \cdot \hat{p}_{T,j}(l)$$

for $p_{T+l,j}$ and μ_{T+l}, respectively. From the equation

$$p_{T+l,j} = \mathbb{P}(Y_{T+l} = j \mid \mathcal{F}_{T+l-1}, Z_{T+l})$$

we immediately obtain

$$\hat{p}_{T,j}(l) = \mathbb{P}(Y_{T+l} = j \mid \mathcal{F}_T). \tag{6}$$

Due to (5) the prediction m.s.e. $V_{T,j}(l)$ and $V_{T,\mu}(l)$ related to $\hat{p}_{T,j}(l)$ and $\hat{\mu}_T(l)$, respectively, are

$$V_{T,j}(l) = \mathrm{Var}_T(p_{T+l,j}), \quad V_{T,\mu}(l) = \mathrm{Var}_T(\mu_{T+l}).$$

For the rest of the paper we assume that the centered covariate process Z_t, $t = 1, 2, \ldots$, follows an $AR(p)$-equation of the familiar form

$$Z_t = R_1 \cdot Z_{t-1} + \ldots + R_p \cdot Z_{t-p} + e_t, \quad t = 1, 2, \ldots \tag{7}$$

where the $r \times r$ - matrices R_i fulfil the causality criterion, see Brockwell and Davies (1987, sec.11.3) , and the e_t are independently and $N(0, \Sigma_e)$-distributed.

4   Monte Carlo Simulation

In a first attempt to solve the forecast problem we will write down an precise expression for the $l-$ step predictor $\hat{p}_{T,j}(l)$ by using a multiple integral and will calculate the integral by means of Monte Carlo simulation . Note that this is not a forecast procedure in the classic sense since many forthcoming paths of the process are generated.
Let us write $x_t = (z_t, y_t)$, $X_t = (Z_t, Y_t)$ and let us denote by

$$f_T(x_{T+1}, \ldots, x_{T+l-1}, z_{T+l})$$

the conditional density of the regular conditional probability

$$\mathbb{P}(X_{T+1} \in B_1 \times \{y_1\}, \ldots, X_{T+l-1} \in B_{l-1} \times \{y_{l-1}\}, Z_{T+l} \in B_l \mid \mathcal{F}_T)$$

w.r.t. the measure $\nu = (\lambda \times \zeta)^{l-1} \times \lambda$, where λ is (only here) the Lebesgue-measure on $\mathbb{R}^r$ and ζ the counting measure on J. Then

$$\hat{p}_{T,j}(l) = \int \ldots \int f_T(x_{T+1}, \ldots, x_{T+l-1}, z_{T+l}) \cdot p_{T+l,j} \cdot$$
$$\cdot d\nu(x_{T+1}, \ldots, x_{T+l-1}, z_{T+l}), \tag{8}$$

where the integration/summation is over $(\mathbb{R}^r \times J)^{l-1} \times \mathbb{R}^r$. The right hand side of (8) can now approximately be calculated by using Monte Carlo methods to generate repeatedly a sequence $(X_{T+1}, \ldots, X_{T+l-1}, Z_{T+l})$. To achieve this the following recursive algorithm is employed:

1. From $\mathcal{F}_T$ calculate Z_{T+1} according to (7) by drawing an $N(0, \hat{\Sigma}_e)$ random vector

2. From $(\mathcal{F}_T, Z_{T+1})$ calculate

$$\eta_{T+1} = \alpha + \pi \cdot p_T + \lambda^T \cdot \Lambda(Y_T) - \beta^T \cdot Z_{T+1}$$

and then $p_{T+1} = h(\eta_{T+1})$

3. Draw Y_{T+1} according to the probability vector p_{T+1}.

Continue with 1.-3. after increasing T to $T+1$. After l steps we arrive at the vectors $\eta_{T+l}^{(1)}$ and $p_{T+l}^{(1)}$. K repetitions of this algorithm lead to vectors

$$\eta_{T+l}^{(k)} \quad \text{and} \quad p_{T+l}^{(k)}, \quad k = 1, \ldots, K,$$

then to the averaged vectors $\bar{\eta}_{T+l}$ and $\bar{p}_{T+l}$, where we have set $\bar{a} = \sum_{k=1}^{K} a^{(k)}/K$, and to $\bar{\mu}_{T+l} = \sum_{j=1}^{m} j \cdot \bar{p}_{T+l,j}$. Now $\bar{\eta}_{T+l}, \bar{p}_{T+l}$ and $\bar{\mu}_{T+l}$ are the Monte Carlo solutions for

$$\hat{\eta}_T(l) = \mathbb{E}_T(\eta_{T+l}), \quad \hat{p}_T(l) \quad \text{and} \quad \hat{\mu}_T(l), \quad \text{resp.}$$

In the application below we will further make use of the prediction m.s.e of $\hat{\mu}_T(l)$ estimated by

$$\hat{V}_{T,\mu}(l) = \sum_{k=1}^{K} (\mu_{T+l}^{(k)} - \bar{\mu}_{T+l})^2 / (K-1). \tag{9}$$

5 Approximation Procedure

In a second approach we want to gain a predictor for p_{T+l} by interchanging conditional expectation and response function h. Here we have to make use of the predictors of the covariate process Z_t. For the $AR(p)$-process Z_t an $l-$ step predictor of Z_{T+l} will be denoted by $\hat{Z}_T(l)$, see Brockwell and Davies (1987, sec.11.4). We will now calculate the $l-$ step predictor

$$\hat{p}_T(l) = \mathbb{E}_T(h(\eta_{T+l}))$$

by the approximation formula $\hat{p}_T(l) \approx \check{p}_T(l)$, where

$$\check{p}_T(l) = h(\hat{\eta}_T(l)), \quad \hat{\eta}_T(l) = \mathbb{E}_T(\eta_{T+l}). \tag{10}$$

One successively derives, by using (6) and $\hat{Z}_T(l) = \mathbb{E}_T(Z_{T+l})$,

$$\hat{\eta}_T(1) = \alpha + \pi \cdot p_T + \lambda^T \cdot \Lambda(Y_T) - \beta^T \cdot \hat{Z}_T(1)$$

$$\cdots$$

$$\hat{\eta}_T(l) = \alpha + \pi \cdot \hat{p}_T(l-1) + \sum_j \hat{p}_{T,j}(l-1)\lambda^T \Lambda(j) - \beta^T \cdot \hat{Z}_T(l), \qquad (11)$$

where $\hat{p}_T(l-1)$ is approximated by $\breve{p}_T(l-1) = h(\hat{\eta}_T(l-1))$. Thus equation (11) allows a recursive calculation of $\hat{\eta}_T(l)$ and -via (10)- of $\hat{p}_T(l) \approx \breve{p}_T(l)$. We want to give now an estimate $B_T^{(2)}(l)$ of the bias

$$B_T(l) = \hat{p}_T(l) - \breve{p}_T(l) = \mathbb{E}_T h(\eta_{T+l}) - h(\hat{\eta}_T(l))$$

produced by approximation formula (10). To this end we assume that h is twice continuously differentiable, and we start with the second-order expansions

$$h_j(\eta_{T+l}) = h_j(\eta_T) + h'_j(\eta_T) \cdot x_{T,l} + (x_{T,l})^T \cdot h''_j(\eta_T) \cdot x_{T,l}/2 + R_T(l),$$

where $x_{T,l} = \eta_{T+l} - \eta_T$, and with the same expansion for $h_j(\hat{\eta}_T(l))$, where $x_{T,l}$ is replaced by $\hat{x}_{T,l} = \hat{\eta}_T(l) - \eta_T$. This leads to

$$B_{T,j}^{(2)}(l) = \frac{1}{2}\mathbb{E}_T[(\eta_{T+l})^T \cdot h''_j(\eta_T) \cdot \eta_{T+l}] - \frac{1}{2}\hat{\eta}_T(l)^T \cdot h''_j(\eta_T) \cdot \hat{\eta}_T(l). \quad (12)$$

In the special case of the cumulative regression model (3) we can simplify formula (12). In fact, the second-order approximation for the bias $B_{T,(j)}(l) = \hat{p}_{T,(j)}(l) - \breve{p}_{T,(j)}(l)$ amounts to

$$B_{T,(j)}^{(2)}(l) = \frac{1}{2}F''(\eta_{T,(j)}) \cdot \mathrm{Var}_T(\eta_{T+l,(j)}).$$

Taking as an example the logistic distribution function $F(s) = 1/(1 + e^{-s})$, then $B_{T,(j)}^{(2)}(l)$ turns out to be positive/negative, if $\eta_{T,(j)}$ is negative/positive.

6   Application

6.1   Forest Damage Data

The cumulative logistic regression model (3) is now applied to three longitudinal data sets on damages in beech, oak and pine trees, respectively. These data were gathered by Dr.A.Göttlein, University of Bayreuth, during the last 12 years in a forest district of Spessart (Bavaria). The damage Y_t in the year t was measured on an ordinal scale consisting of $m = 8$ categories of needles/leaves lost. The longitudinal structure of the data is determined by the observation period of 12 years (1983 - 1994) and by N sites (N = 80 beech sites, N = 25 oak sites, N = 14 pine sites). For each site and each year a vector Z_t of $r = 20$ covariates were recorded concerning the trees (age, canopy, stand), the site (gradient, height, exposition), the climate and the soil (type, moisture, pH-values), see Göttlein and Pruscha (1992) and

(1995) for more details. The parameter of the model were estimated from the longitudinal data by the m.l. method for each species separately. Concerning the function Λ we made the special choice $\Lambda(W_t) = Y_t$, see special case 2 in sec.2. We further put $\pi = 0$. The covariate process Z_t is assumed to be driven by an AR(1)-equation.

6.2   Forecasting μ_t

Fixing the outcomes of the years 1983 - 1992 as known, we try to forecast the values of the mean damage category $\mu_t = \sum_{j=1}^m j p_{t,j}$ for the years 1993 - 1998, letting the years 1993 and 1994 -for which we have observations- as control. That is, we put $T = 10$ and we are interested in the $l-$ step predictors $\hat{\mu}_T(l), l = 1, 2, \ldots, 6$. The calculations of the forecasts, leading to Figure 1, are performed separately for each of the three species, beech, oak and pine trees, and for each site $i = 1, 2, \ldots, N$, followed by averaging over the N sites of the species.

First, the Monte Carlo method (MOCA) of sec.4 is applied, with $K = 200$ repetitions to calculate $\bar{\mu}_{T+l}$ as Monte Carlo solution for $\hat{\mu}_T(l)$ and the corresponding m.s.e. $\hat{V}_{T,\mu}(l)$ according to (9). A 95 per cent confidence interval for the averaged μ_{T+l} is established by the confidence limits

$$\hat{\mu}_T(l) \pm \sqrt{\hat{V}_{T,\mu}(l)} \cdot 1.960/\sqrt{N}$$

holding approximately for the years $l = 1, 2, \ldots, 6$. At the end of the forecast curves the intervals for $l = 6$ are indicating by vertical bars.

Secondly, the approximation method (APPR) of sec.5 is employed. For all three species the forecasts $\hat{\mu}_T(l)$ produced by the MOCA and by the APPR method run very similar over the 6 years 1993 to 1998, with the APPR curve below the MOCA curve. The upward trend of the pine curve at the end of the observation period is continued in a strongly damped form.

To compare the forecast solutions with the observation data of the years 1986 to 1994, we include plots for $\bar{Y}_t$ and $\bar{\mu}_t$, where Y_t is the OBServed category, $\mu_t = \sum j p_{t,j}$ is the predicted mean value at year t (as predicted on the basis of the ESTIMated cumulative regression model) and the bar means averaging over the N sites of the tree species. Note the lag-effect which is produced by the term Y_{t-1} in the regression model, especially in the oak data: a zig zag run of the $\bar{Y}_t$ values becomes apparant in the run of the $\bar{\mu}_t$ values with a lag of one year.

Acknowledgments: This work is partially supported by the DFG (SFB 386).

240   Helmut Pruscha

Plots of Observed, Predicted and Forecasted Forest Damages

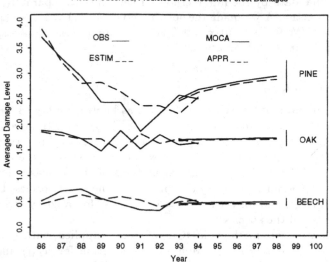

FIGURE 1. Observed and predicted forest damages

References

Brockwell,P.J. and Davis,R.A. (1987). *Time Series: Theory and Methods.* Springer, N.Y.

Fahrmeir,L. and Kaufmann,H. (1987). Regression models for non stationary categorical time series. *Journal of Time Series Analysis*, **8**, 147-160.

Göttlein,A. and Pruscha,H. (1992). Ordinal time series models with application to forest damage data. In: *Lecture Notes in Statistics*, **78**, Springer, N.Y., 113-118.

Göttlein,A. and Pruscha,H. (1995). Ergebnisse einer mehrjährigen Erfassung des Waldzustandes im Bereich Rothenbuch (submitted).

McCullagh,P. (1980). Regression models for ordinal data (with discussion). *Journal of the Royal Statistical Society*, Series B, **42**, 109-142.

Pruscha,H. (1993). Categorical time series with a recursive scheme and with covariates. *statistics*, **24**, 43-57.

Zeger,S.L. and Qaqish,B. (1988). Markov regression models for time series: a quasi likelihood approach. *Biometrika*, **44**, 1019-1031.

Standard Errors, Correlations and Model Analysis

K.L.Q. Read

ABSTRACT:
In recent years there has been a great increase in the computing power and user-friendliness of packages that may be used for statistical analysis. One effect of these developments has been to bring progressively more sophisticated models within the scope of supposedly routine analysis by people who may or may not be adequately trained in statistics. Often a major purpose of such analysis is to carry out parsimonious model selection to fit a set of data. The purpose of this paper is to illustrate the need for caution when interpreting standard errors and other diagnostics used to guide the selection process.

KEYWORDS: model selection, standard error, correlation, conditional analysis, EM algorithm.

1 Introduction

The main sources of uncertainty or inaccuracy in the calculation of standard errors are well-known to statisticians, and frequently arise in practice. Apart from sampling variation, they may be listed as follows:

(i) an invalid assumption of homoscedastic Normal errors;

(ii) unjustified use of large sample asymptotic (Normal) sampling theory;

(iii) use of analyses which are conditioned on the estimated values of some model parameters as if they were known, or which otherwise over-represent the information contained in the data;

(iv) logically identifiable models giving highly correlated estimates.

To illustrate these points we discuss analyses of three datasets that have arisen in practice, with a view to exploring the implications of the results for model selection. We warn against the uncritical use of standard errors and other diagnostics, or 'standard' procedures such as the routine inclusion of a constant term.

2 Modelling the Mains Leakage of Water

Leakage of mains water may account for 30-40% of the total supplied, and reducing this loss is of major concern to the water industry. Rates of leakage vary widely, even within the region supplied by a single water company. Many possible factors may be listed, from geology, geography and topography, pipework construction materials and measures of 'size' such as length of pipework and numbers of connections. For this study, about 20 variables were collected for a sample of 62 District Metering Areas (DMAs) in England. The underlying method of analysis is multiple regression with a view to explaining the observed night time flow (NLF) of water in terms of four explanatory variables identified in exploratory analysis. The aim is to focus on modelling procedures, rather than to advance any given model as a 'final' answer to a complex practical problem. We thus have

NLF: Night Line Flow in metres^3 (or m^3) per hour, the dependent variable in this study, being a proxy for mains leakage;

DOM: Number of Domestic Connections;

COM   Number of Commercial Connections;

AC: Length of mains made from Asbestos-Cement, in km;

AMQ   Annual metered quantity (total volume in megalitres).

The author's first contact with (a reduced set of) these data was in the form of a spreadsheet analysis assuming constant variance Normal errors and the inclusion of a constant term in the regression. Model (2.1), the fit of NLF in terms of DOM, COM, AC and AMQ, is as follows:

$$\text{NLF} = 2.11 + 0.038\text{COM} + 0.0069\text{DOM} + 0.0099\text{AMQ} + 1.71\text{AC}$$

Figure 1: Standardised Residuals for 'Spreadsheet Model' for Leakage Data

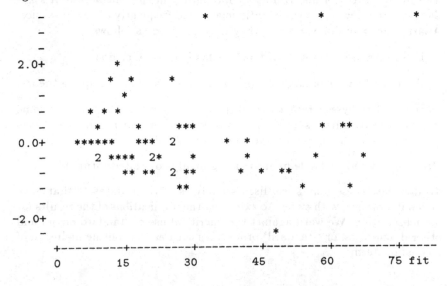

This model achieves $R^2 = 0.63$ but the constant term is not significant ($p >$ 0.5), the standardised residuals are significantly non-Normal ($p < 0.005$), three being over 3 in value, and the residual scatter increases with the fit. Inspection of the regressor variables suggests that an 'aggregation property' may apply, in the sense that if any set of neighbouring DMAs were amalgamated the total NLF should be predicted by the same regressor function evaluated at the corresponding total values of the explanatory variables. For this to be so, a linear regression through the origin must be fitted. Further, if the errors associated with leakage in different parts of the network are (as assumed) independent, then the error variance for an 'aggregated' DMA is the sum of the error variances of its component DMAs, and so, by extension, Poisson-like error variances proportional to NLF would represent a natural weighting structure. Various weightings could be based on measures of the 'size' of a DMA, but use of the Poisson weighting with no constant term yields model (2.2):

$$\text{NLF} = 0.0341\text{COM} + 0.00440\text{DOM} + 0.0294\text{AMQ} + 1.43\text{AC} \,,$$

which achieves an F for regression of 61.3. Three variables are strongly significant ($p = 0.005$ or less) whilst p for AMQ is marginal at 0.054. The standardised residuals shown in Figure 2 now appear satisfactory and can be shown to be acceptably Normal ($p \approx 0.25$).

Figure 2: Standardised Residuals for Poisson-weighted No-constant Model

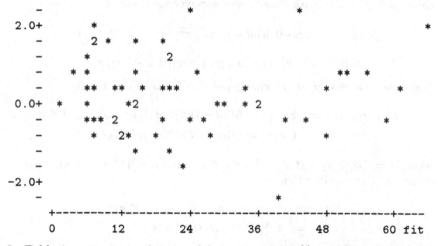

In Table 1 we compare these models in terms of 95% confidence limits for NLF for three 'small', 'medium' and 'large' DMAs that are not identified as outlying or influential. As expected, the constant-term model (2.1) is unsatisfactory for modelling leakage for a small DMA, and inferences under (2.1) are weaker than those for (2.2). The 'preferred' model (2.2) has the sharpest inferences, being well on target with the first two fits and not too far out with the third in view of an estimated prediction error of 20.2. Use of a 'standard spreadsheet' regression model on these data is misleading.

Table 1: Comparisons of Leakage Predictions for Three Regression Models

Observed NLF	Model	Predicted NLF	95% CI
1.29	(2.1)	4.03	(-2.34, 10.41)
	(2.2)	1.59	(1.14, 2.04)
12.88	(2.1)	9.56	(1.98, 17.14)
	(2.2)	10.49	(5.83, 15.15)
68.13	(2.1)	64.69	(56.22, 73.15)
	(2.2)	55.17	(47.63, 62.71)

3   Models for the Survival of Cardiac Patients

The second example concerns the analysis of survival of coronary heart disease patients after the onset of acute myocardial infarction (MI). Past work on this dataset (a subset of 921 cases from the Mather trial to compare home and hospital treatment) has established age (in years), and binary indicators of i) past history of angina or infarction and ii) clinical severity when first examined, as significant covariates z influencing survival. Patients were followed for up to about a year from the onset of MI, so that survival times are right-censored.. In the lognormal model, log(survival time) $\approx N(\beta'z, \sigma^2)$, with density and survivor function

$$f(t|\beta, \sigma) = \exp[-0.5(\ln t - \beta'z)^2/\sigma^2]/(\sigma t\sqrt{(2\pi)}), t > 0;$$

$$S(t) = 1 - \Phi[(\ln t - \beta'z)/\sigma] = \Phi[(\beta'z - \ln t)/\sigma].$$

A simple scale change at the changepoint a_1 leads to the density

$$
\begin{aligned}
f(t|\beta, \sigma_1, \sigma_2, a_1) &= k.\exp[-0.5(\ln t - \beta'z)^2/\sigma_1^2)/(\sigma_1 t\sqrt{(2\pi)}], t \le a_1; \\
&= k.\exp[-0.5(\ln t - \beta'z)^2/\sigma_2^2)/(\sigma_2 t\sqrt{(2\pi)}], t > a_1,
\end{aligned}
$$

where $k = \{\Phi[(\ln a_1 - \beta'z)/\sigma_1] + \Phi[(\beta'z - \ln a_1)/\sigma_2]\}^{-1}$ is a normalising constant. The survivor function is

$$
\begin{aligned}
S(t) &= 1 - k.\Phi[(\ln t - \beta'z)/\sigma_1], t \le a_1; \\
&= k.\Phi[(\beta'z - \ln t)/\sigma_2], t > a_1.
\end{aligned}
$$

These models, with and without a changepoint at about the time of discharge from hospital (22 days) and incorporating a linear predictor in the covariates, were fitted to the data by Read and Stasinopoulos (1994), later referred to as RS, following earlier work on changepoint models (Noura and Read, 1990; Noura, Read and Stasinopoulos (NRS), 1992). Table 2 below gives estimates and standard errors for these effects, which are broadly consistent across the models fitted and indicate increasing significance in the order A, H, S;. The simple lognormal achieves the best deviance of all the models without a changepoint fitted by RS (2687.8 against 2702.3 for the log-logistic and 2716.0 for the Weibull). The two-stage lognormal was thus worth trying, but at 2629.9 the deviance is inferior to that of a two-stage log- logistic (2608.3) fitted by RS. Corresponding estimates for the simple lognormal are effectively equivalent. The standard errors of the estimates of the β's are found for the NAG Simplex fit by inverting the matrix of the observed second derivatives of $-\ln L$ computed algebraically at the maximum, numerical differentiation being unsatisfactory; they are about twice the corresponding values given by GLIM (which thus over-states the significance of the covariates). Likely reasons for this difference are the importance of the change at $a_1 = 22$ days as seen in the disparate estimates of σ_1 and σ_2 and lower deviance for the two-stage lognormal model, along with the EM algorithm used in the single-stage GLIM fit. As almost all of the censored data (and over 75% of the total) are censored at 170 days, the censored data are on the two-stage model subject to a larger scale parameter ($\sigma_2 \approx 17$) than that of about 2.4 relating to most of the observed deaths (which are before 22 days). The EM algorithm for the one-stage model imputes expected values for the censored cases using the scale estimate $\sigma = 3.66$ which is dominated by the many deaths before 22 days, when the variability of log(survival time) from prediction is much less. Whilst the estimates of σ in the two fits of the one-stage model are close, the imputed values taken as observed deaths in GLIM overstate the information in the censored data, which is better shown by the matrix of observed second derivatives of the exact log-likelihood calculated at the NAG Simplex fit. This approach gives standard errors for σ, and σ_1 and σ_2 in the two-stage model, but not for a_1 as the likelihood is not continuous in this parameter. In each model the highest absolute correlation between two estimates is about 0.95, with no others above 0.9.

Table 2: Estimates of Parameters for Final Main Effects Models

Model	σ_1, σ_2	GM	Age	Hist	Sev	Dev
Simple Lognormal (est'd in GLIM)	$\sigma = 3.647$	14.82 (0.657)	-0.0855 (0.0114)	-1.515 (0.178)	-2.978 (0.185)	2687.8 (916 df)
Simple Lognormal (NAG simplex)	$\sigma = 3.670$ (0.211)	14.94 (1.45)	-0.0866 (0.0230)	-1.533 (0.350)	-3.006 (0.357)	2687.8 (916 df)
Two-stage Lognormal	$\sigma_1 = 2.403$ (0.169) $\sigma_2 = 17.12$ (2.52)	10.34 (1.12)	-0.0494 (0.017)	-0.9511 (0.264)	-2.077 (0.274)	2629.9 (914 df)

4   Models for Stages of Cognitive Development

Shemesh, Eckstein and Lazarowitz (1992) classified the Piagetian development of 913 adolescents aged 13-18 in Israel as concrete, transitional and formal, and the resulting data are shown in Table 3.

Table 3: 913 Israeli Adolescents Classified into 3 Stages
of Mental Development

Year Group	Stage 1		Stage 2		Stage 3		Total
	No.	(%)	No.	(%)	No.	(%)	
13	77	(63)	43	(35)	3	(3)	123
14	42	(38)	57	(52)	11	(10)	110
15	80	(30)	129	(48)	59	(22)	268
16	36	(17)	99	(46)	79	(37)	214
17	13	(12)	46	(43)	49	(46)	108
18	9	(10)	33	(37)	48	(53)	90

The data in each age group are taken as trinomial, the stage probabilities $p_i(t), i = 1, 2, 3$ being modelled as continuous functions of age t consistent with a general advance from the first stage upwards as age increases.

4.1   An Odds-Ratio Model

In the 6-parameter model of Preece and Read (1995), $p_1(t)$ and $p_2(t)$ increase with age to respective limiting proportions c_1 and c_2 of those who will never pass beyond stage 1 or beyond stage 2. The probabilities are defined by two odds-ratio conditions. According to the first,

$$(p_2(t) + p_3(t))/(p_1(t) - c_1) = \exp[k_1(t - T_1)],$$

the odds in favour of someone who will eventually progress from stage 1 being at stage 2 or above increase exponentially with age. Also, the proportion $x(t)$, of those aged t who are at stage 2 and will stay in stage 2, is a fixed proportion of those who have left stage 1. The ratio $x(t)/(1 - p_1(t))$ is thus fixed $= c_2/(1 - c_1)$, and then

$$p_3(t)/[p_2(t) - x(t)] = \exp[k_2(t - T_2)]$$

implies exponential increase with age of the odds in favour of an eventual stage 3 person being in stage 3 rather than stage 2. The goodness-of-fit χ^2 is 2.22 on 6 degrees of freedom (df) ($p > 0.8$) for the data of Table 3. All parameters are significant, the ratio of estimate to standard error is 4.8 or more, and all correlations between estimates are < 0.9 in absolute value.

4.2   Differential Equation Model

For this 4-parameter model (Eckstein and Shemesh, 1992),

$$p_1(t) = A \exp(-\alpha t), \quad p_2(t) = B \exp(-\beta t) - \alpha p_1(t)/(\alpha - \beta)$$

and $p_3(t) = 1 - p_1(t) - p_2(t)$ satisfy the differential equations

$$dp_1/dt = -\alpha p_1, \quad dp_2/dt = \alpha p_1 - \beta p_2 \text{ and } dp_3/dt = \beta p_2.$$

In contrast to the odds-ratio model, here the constants α and β imply that the rates at which individuals make transitions are independent of age. Thus for a group of 100 people in stage 1 then the expected number per year moving to stage 2 is 100α whether the age of the group is 13 or 18. All individuals appear to reach stage 3 eventually, but snags arise when age is extrapolated beyond the range of the data: thus $p_1(t)$ must be set $= 1$ for $t < ln(A)/\alpha$. The model fits well ($\chi^2 = 3.11$ on 8 df) and all parameters are significant, but the correlations between estimates of A and a, B and β, are high at 0.994 and 0.982 respectively.

4.3   Preece's Ratio IQ Model

In this 3-parameter model (Preece (1995), private communication) mental age thresholds M_1 and M_2 define passage from stage 1 to stage 2 and from stage 2 to stage 3. On a logistic approximation to the usual Normal distribution for ratio IQ ($I = M/t$ or mental age/chronological age),

$$p_1(t) = 1/\{1+\exp[-1.7(M_1/t-1)/s], \quad p_2(t) = 1/\{1+\exp[-1.7(M_2/t-1)/s],$$

and $p_3(t) = 1 - p_1(t) - p_2(t)$. We find $\chi^2 = 7.28$ on 9 df: all parameters are significant and correlations between estimates are < 0.8 in absolute value.

Here we have three contrasting psychometric models, all broadly consistent with the data. We compare these models in terms of their precision of inference on the stage probabilities at ages 14 and 17.

Table 4: Comparison of Inferences from 3 Models for Mental Development

Model	Age 14			Age 17		
	$p_1(t)$	$p_2(t)$	$p_3(t)$	$p_1(t)$	$p_2(t)$	$p_3(t)$
Odds-Ratio	0.430	0.476	0.093	0.123	0.408	0.469
(6 params)	(0.015)	(0.015)	(0.005)	(0.006)	(0.017)	(0.017)
Diff-Eq.	0.415	0.465	0.120	0.124	0.425	0.450
(4 params)	(0.012)	(0.011)	(0.004)	(0.006)	(0.012)	(0.012)
Ratio IQ	0.419	0.475	0.106	0.128	0.398	0.473
(3 params)	(0.012)	(0.012)	(0.004)	(0.005)	(0.011)	(0.013)

The probabilities are generally well identified, to within 5-10% at worst. Consistently wider standard errors are quoted for the 6-parameter Odds-Ratio model; surprisingly, however, the highly correlated estimates in the Differential Equation model do not cause great imprecision in the 18 fitted probabilities, 9 being slightly more accurate on this model than for Ratio IQ whilst for the other 9 the contrary is true. The parameters of the three models are not comparable, but in relation to standard errors those of the Ratio IQ model seem better identified, with absolute partial t-values > 30, against values of 4-5 upwards for the other models. There may thus be a hint of overfitting in the 6-parameter model, with the 4-parameter model gaining more from parsimony than it loses on correlation structure. The 3-parameter Ratio IQ model can hardly be overparameterised but incurs a small penalty in terms of lack of fit in that χ^2 is 7.28-3.11 = 4.17 worse than for the 4-parameter model. An encouraging feature of this model is the estimation of s as 18.2 percentage points of IQ, which accords with the coefficient of variation of IQ found in a large population.

5   Discussion

We recommend that model selection be based on several criteria, such as

goodness of fit and validity of stochastic assumptions;
consistency with relevant physical or other theory for the data;
ease of interpretation;
parsimony, or a preference for the 'simplest' model.

The examples considered typify many statistical modelling tasks. In the case of the leakage data misleading inferences may be obtained from standard or 'naive' models; the decision to include or exclude marginally significant explanatory variables may rest on interpretation or judgment, as in the larger survival example. Here the interplay of heavy censoring, use of the EM algorithm and a conditional analysis lead to overstatements of significance, vindicating decisions (NRS, RS) to reject parameters that were not clearly significant in conditional analyses: again, model choice should not rely solely on the operation of formal rules or standard procedures. The third dataset is well fitted by several models, but in one the ability to extrapolate over age is very limited: as George Box once said, 'All models are wrong, but some models are useful'. This example makes the point that, while a model may be discredited by a poor fit to data, a good fit does not confirm it: trials on other data may reveal that in terms of general success some models are indeed more useful than others.

References

Eckstein, S.G., Shemesh, M. (1992) Mathematical Models of Cognitive Development. *Brit.J.Math.StatPsychol*, **45**, 1-18.

Noura, A. A., Read, K. L. Q. (1990) Proportional Hazards Change- point Models in Survival Analysis. *Appl. Statist.*, **39**, 241-253.

Noura, Read, K. L. Q., Stasinopoulos, M. (1992) Parametric Proport- ional Odds Changepoint Models in Survival Analysis: a Case Study Comparison with Proportional Hazards. *In Statistical Modelling: a Selection of Papers from the Sixth International Workshop on Statistical Modelling,* Utrecht, The Netherlands, 15-19 July, 1991, van der Heijden *et al*, eds, pp 273-282.

Preece, P. F. W., Read, K. L. Q. (1995) A Stage-Theory Model of Cognitive Development. *Brit. J. Math and Statist. Psych.*, to appear.

Read, K. L. Q., Stasinopoulos, M. (1994) Parametric Changepoint Models in Survival Analysis. *Ninth International Workshop on Statistical Modelling*, Exeter, UK, 11-15 July 1994.

Mean and Dispersion Additive Models: Applications and Diagnostics

Robert A. Rigby
Mikis D. Stasinopoulos

ABSTRACT: This paper presents further applications and diagnostics of the 'Mean and Dispersion Additive Model' or 'MADAM'. This is a flexible model for the mean and variance of a dependent variable in which the variance is modelled as a product of the dispersion parameter and a known variance function of the mean, and the mean and dispersion parameters are each modelled as functions of explanatory variables using a semi-parametric Additive model. MADAM's are fitted using a successive relaxation algorithm which alternates between mean and dispersion model fits until convergence, providing diagnostics for each model. It is shown in the appendix that the algorithm maximises the penalised extended quasi-likelihood of the MADAM.

KEYWORDS: Dispersion parameter, Generalised Additive Models, Extended Quasi-likelihood, Penalised likelihood, MADAM

1 Introduction

Rigby and Stasinopoulos (1995a) presented a flexible model for the mean and variance of a dependent variable. We called it the the 'Mean and Dispersion Additive Model' or 'MADAM'. The variance is modelled as a product of a dispersion parameter and a known variance function of the mean. The dependence of each of the mean and dispersion parameter on explanatory variables is modelled using a semi-parametric Additive model (Hastie and Tibshirani, 1990).

Specifically, the model assumes that the response variable y_i is independently distributed with mean μ_i and variance $\phi_i V(\mu_i)$, where

$$g_1(\mu_i) = \eta_i = \sum_{j=1}^{p} f_j(x_{ij}), \qquad g_2(\phi_i) = \xi_i = \sum_{k=1}^{q} h_k(z_{ik}) \qquad (1)$$

where ϕ_i is the dispersion parameter and $V(\mu_i)$ is the known variance function, for $i = 1, 2, \ldots, n$. The function f_j is either a linear or a nonparametric function of explanatory variable x_j for the mean, $j = 1, 2, \ldots, p$, and the function h_k is either a linear or a non-parametric function of explanatory variable z_k for the dispersion, for $k = 1, 2, \ldots, q$. The $\mathbf{x}$'s and the $\mathbf{z}$'s are assumed to be fixed and g_1 and g_2 are known monotonic link functions, where typically g_2 is a log link.

Estimation of non-parametric smooth functions f_j and h_k can be achieved by maximising the Penalised extended Quasi-Likelihood L_{PQ}^+ (or equivalently minimising the corresponding Penalised extended Quasi-Deviance D_{PQ}^+), with penalties for lack of smoothness in the additive functions f_j and h_k in the mean and dispersion models respectively, given by

$$
\begin{aligned}
D_{PQ}^+ &= -2 \log L_{PQ}^+ \\
&= D_Q^+ + \left[\sum_{j=1}^{p} \lambda_{1j} \int_{-\infty}^{\infty} \left[f_j''(t) \right]^2 dt + \sum_{k=1}^{q} \lambda_{2k} \int_{-\infty}^{\infty} \left[h_k''(t) \right]^2 dt \right]
\end{aligned} \tag{2}
$$

where D_Q^+ is the extended Quasi-Deviance defined in Nelder and Pregibon (1987) by

$$
D_Q^+ = \sum_{i=1}^{n} \frac{d_i}{\phi_i} + \sum_{i=1}^{n} \log \left[2\pi\phi_i V(y_i) \right] \tag{3}
$$

and where $V(y_i)$ is the variance function evaluated at y_i and d_i is the deviance increment defined, for the particular variance function $V(\mu_i)$ used, by

$$
d_i = -2 \int_{y_i}^{\mu_i} \frac{y_i - u_i}{V(u_i)} du_i \tag{4}
$$

The maximising functions f_j and h_k are cubic splines. The fitting procedure is a successive relaxation algorithm, in which the mean and the dispersion models are alternately fitted until convergence. For details on the algorithm see Rigby and Stasinopoulos (1995a). It is shown in the appendix that the algorithm used minimises D_{PQ}^+. The algorithm has been implemented in GLIM4, allowing flexible and interactive modelling of both the mean and dispersion of a dependent variable. MADAM models can be used to model parametrically or non-parametrically, the variance heterogeneity in Normal regression models (Rigby and Stasinopoulos, 1994, 1995b), and the overdispersion in Poisson or Binomial logistic regression models, (Rigby and Stasinopoulos, 1995a). See these and Smyth (1988) for references to parametric submodels of MADAM. In the present paper we concentrate on diagnostics for the MADAM models. In section 2 a brief discussion of the diagnostic tools for MADAMs is given. Section 3 covers an important practical aspect of fitting MADAMs, the shifted log link function for the dispersion. Section 4 gives a practical example.

2 Diagnostics

The successive relaxation fitting algorithm provides on convergence, diagnostics for each of the mean and dispersion models and global diagnostics for the MADAM. Raw deviance residuals from the mean model are given by $r = \text{sign}(y_i - \hat{\mu}_i)\sqrt{d_i}$ where d_i is given by (4). Standardised residuals are given by $r' = r/\sqrt{\hat{\phi}_i}$. Studentised residuals are given by $r'/\sqrt{(1 - l_{1i})}$ where $\hat{\mu}_i$ and $\hat{\phi}_i$ are the fitted mean and dispersion parameters for the ith observation and where l_{1i}, the ith diagonal element of L_1, is a rough approximation to the ith leverage value for the mean model. The matrix L_1 is given by

$$L_1 = H_1 + \sum_{j=1}^{p} S_{1j} \qquad (5)$$

where H_1 is the usual Generalised Linear model Hat matrix for the linear part of the mean model (McCullagh and Nelder, 1989), with diagonal weight matrix A_1 and S_{1j} is the jth smoothing matrix, where both A_1 and S_{1j} are defined in the appendix. A justification for this type of approximation is given by Hastie and Tibshirani (1990), but the approximation requires further investigation for the case of more than one smoothing term in the mean model. Outlier observations in the response variable y may be identified by large values of the standardised or studentised residuals. High leverage observations, each of whose observed y value greatly contributes to its corresponding fitted value, may be identified by large values of the leverage l_{1i}.

The response variable used for fitting the dispersion model is d_i, the deviance increment given by (4). The dispersion model is fitted using a Gamma error distribution with scale 2 (i.e. χ_1^2), leading to standardised residuals $r'_{\phi_i} = \text{sign}(d_i - \hat{\phi}_i)\sqrt{d_{\phi_i}/2}$ where $d_{\phi_i} = (d_i/\hat{\phi}_i) - \log(d_i/\hat{\phi}_i) - 1$ is the ith deviance increment from the dispersion model fit. Studentised residuals are given by $r'_{\phi_i}/\sqrt{(1 - l_{2i})}$ where l_{2i}, the ith diagonal element of $L_2 = H_2 + \sum_{k=1}^{q} S_{2k}$, is an approximation to the ith leverage value for the dispersion model, where H_2 is the hat matrix for the linear part of the dispersion model and S_{2k} is defined in the appendix.

The contribution D_i of the ith observation to the extended Quasi-Deviance D_Q^+ is given by $D_i = d_i/\hat{\phi}_i + \log[2\pi\hat{\phi}_i V(y_i)]$. Larger values of D_i indicate observations with smaller fitted likelihoods which are therefore less likely to come from the fitted model. Hence D_i provides a global diagnostic for the MADAM.

3   Bounded dispersion link functions

One problem that can occasionally occur in certain data sets is that the residual corresponding to an extreme leverage point for the mean model is very small, leading to a very small response variable d_i for the dispersion model. If the observation is also an extreme leverage point for the dispersion model, then d_i and its fitted dispersion $\hat{\phi}_i$ may both tend to zero with D_Q^+ tending to $-\infty$, when minimising D_{PQ}^+ in the MADAM fitting algorithm. The chance of this occurring may be reduced by using a leverage adjusted response variable obtained by $d_i/(1-l_{1i})$ when fitting the dispersion model Nelder (1992). The problem may be prevented by using a link function for the dispersion model with a positive lower bound, for example the shifted log link function given by $\log(\phi-c) = \xi$. This ensures that $\phi > c$. The choice of c=1 is particularly appropriate for modelling overdispersed Binomial or Poisson data. For Normal error models, the choice of c is more difficult, however the choice of c equal to 1% of the average initial scale estimate, i.e. $c = 0.01 \sum_{i=1}^{n} \hat{\phi}_i/n$, or c chosen equal to 0.001% of the sample variance of the dependent variable, may well be reasonable in practice. [An adjustment to the derivative of the link function to $1/(\phi - c + \delta)$ with δ very small relative to c may occasionally be needed to prevent overflow for fitted ϕ values extremely close to c].

4   An application of MADAM

As an example of the use of MADAMs we reanalyse the well known Minitab tree data, previously analysed by Aitkin (1987) and McCullagh and Nelder (1989). The data consist of 31 measures of volume, diameter (at $4'6''$ above ground level) and height of black cherry trees. The interest is to try to predict the tree volume (V) from measurements of diameter (D) and height (H). If the trees were cylinder shaped we would expect to find that $\log(V) = c + \log(H) + 2\log(D)$ for constant c.
Here we assume that $\log(V) \sim N(\mu, \phi)$ where

$$\mu = \beta_0 + \beta_1 \log(H) + \beta_2 \log(D), \qquad \log(\phi_i) = h[\log(D)]$$

The model was fitted to the data for a range of smoothing parameters corresponding to a range of degrees of freedom df_ϕ, for the dispersion using the MADAM fitting algorithm (implemented in the GLIM4 macro MADAM), and the resulting fitted deviances are given in table 1. Note that the first column in table 1 with $df_\phi = 1$ corresponds to constant dispersion model. Subsequent columns correspond to models for $\log(\phi)$ which are non-parametric functions of $\log(D)$ using a total of df_ϕ degrees of freedom **including** the constant. Using a forward selection approach, (at a 5% significance level with $\chi^2_{1,0.05} = 3.8$ and $\chi^2_{2,0.05} = 6.0$), led to the choice of $df_\phi = 4$.

TABLE 1. Analysis of deviance table for the Tree data.

df_ϕ	1	3	4	5	6	7
Deviance	-70.7	-80.8	-87.2	-88.8	-90.4	-93.0

Figure 1(a) gives a partial residual plot for the fitted mean model against $\log(D)$. Figure 1(b) gives the response variable d_i and the fitted values $\hat{\phi}_i$ for the dispersion model.

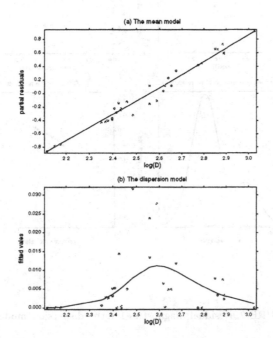

FIGURE 1. Fitted MADAM for the tree data.

Figure 2 gives plots of the standardized (deviance) residuals r'_{ϕ_i} for the dispersion model. Figure 2(a) and 2(b) show no apparent relationship of the standardised residuals with either $\log(D)$ or the fitted dispersion values $\hat{\phi}_i$ respectively. Figure 2(c) and 2(d) investigate the distribution of the standardised residuals using a Normal kernel density estimate and a Normal Q-Q plot respectively. These show a negatively skew distribution. The explanation appears to be that the deviance residuals provide an inadequate Normalising transformation for the fitted dispersion model residuals (d_i relative to $\hat{\phi}_i$) because d_i/ϕ_i is fitted using an extremely skew (J-shape) χ_i^2 error distribution. Similar plots corresponding to figure (2) of the standardised residuals $r'_i = (y_i - \hat{\mu}_i)/\sqrt{\hat{\phi}_i}$ for the mean model show no systematic departure from the mean model.

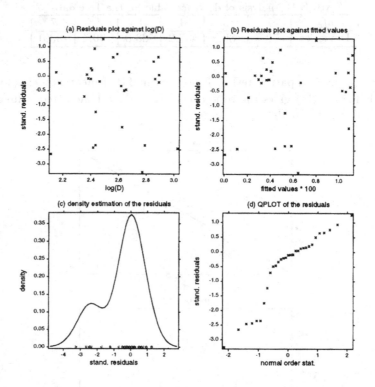

FIGURE 2. Residual diagnostics for the dispersion model.

Appendix: Maximisation of the Penalised Extended Quasi-Likelihood

Following Reinsch (1967), the minimising functions $f_j(x_j)$ and $h_k(z_k)$, for D_{PQ}^+ given by (2) are all natural cubic splines and hence the penalty integrals can be replaced by quadratic forms to give

$$\ell_{PQ} = -\tfrac{1}{2}D_i - \tfrac{1}{2}\left[\textstyle\sum_{j=1}^{p} \lambda_{1j}\mathbf{f}_j' K_{1j}\mathbf{f}_j + \sum_{k=1}^{q} \lambda_{2k}\mathbf{h}_k' K_{2k}\mathbf{h}_k\right] \qquad (6)$$

where $D_i = d_i/\phi_i + \log[2\pi\phi_i V(y_i)]$ and K_{1j} and K_{2k} are penalty matrices, obtained from the natural cubic spline basis functions, which depend only on the values of the explanatory variables $\mathbf{x}_j$ and $\mathbf{h}_k$ respectively, (Rigby and Stasinopoulos, 1995b). The first and expected second derivatives of ℓ_{PQ} are obtained to give Fishers's scoring step for maximising (6) with respect to the $\mathbf{f}_j$'s and $\mathbf{h}_k$'s.

$$\frac{\partial \ell_{PQ}}{\partial \mathbf{f}_j} = \mathbf{u}_1 - \lambda_{1j} K_{1j} \mathbf{f}_j, \qquad \frac{\partial \ell_{PQ}}{\partial \mathbf{h}_k} = \mathbf{u}_2 - \lambda_{2k} K_{2k} \mathbf{h}_k, \qquad (7)$$

where $\mathbf{u}_1 = A_1 G_1 (\mathbf{y} - \boldsymbol{\mu})$, $\mathbf{u}_2 = A_2 G_2 (\mathbf{d} - \boldsymbol{\phi})$, $A_1 = \text{diag}[\{g_1'(\mu_i)\}^2 \phi_i V(\mu_i)]^{-1}$, $A_2 = \text{diag}[2\{g_1'(\phi_i)\}^2 \phi^2)]^{-1}$, $G_1 = \text{diag}\{g_1'(\mu_i)\}$, $G_2 = \text{diag}\{g_2(\mu_i)\}$, and

$$E\left[\frac{\partial^2 \ell_{PQ}}{\partial \mathbf{f}_j \partial \mathbf{f}_i'}\right] = -A_1 - \lambda_{1j} K_{1j} if(i = j), \qquad E\left[\frac{\partial^2 \ell_{PQ}}{\partial \mathbf{h}_k \partial \mathbf{f}_j'}\right] = 0$$

$$E\left[\frac{\partial^2 \ell_{PQ}}{\partial \mathbf{h}_k \partial \mathbf{h}_i'}\right] = -A_2 - \lambda_{2k} K_{2k} if(i = k) + \left[\frac{\partial}{\partial \mathbf{h}_k'} A_2 G_2\right] E[\mathbf{d} - \boldsymbol{\phi}] if(i = k)$$

$$(8)$$

Approximating $E[\mathbf{d} - \boldsymbol{\phi}]$ to zero simplifies the Fisher Scoring step, which for illustration is given for p=q=2 by

$$\begin{bmatrix} A_1 + \lambda_{11} K_{11} & A_1 & 0 & 0 \\ A_1 & A_1 + \lambda_{12} K_{12} & 0 & 0 \\ 0 & 0 & A_2 + \lambda_{21} K_{21} & A_2 \\ 0 & 0 & A_2 & A_2 + \lambda_{22} K_{22} \end{bmatrix} \begin{bmatrix} \mathbf{f}_1^{(1)} - \mathbf{f}_1^{(0)} \\ \mathbf{f}_2^{(1)} - \mathbf{f}_2^{(0)} \\ \mathbf{h}_1^{(1)} - \mathbf{h}_1^{(0)} \\ \mathbf{h}_2^{(1)} - \mathbf{h}_2^{(0)} \end{bmatrix}$$

$$= \begin{bmatrix} \mathbf{u}_1 - \lambda_{11} K_{11} \mathbf{f}_1^{(0)} \\ \mathbf{u}_1 - \lambda_{12} K_{12} \mathbf{f}_2^{(0)} \\ \mathbf{u}_2 - \lambda_{21} K_{21} \mathbf{h}_1^{(0)} \\ \mathbf{u}_2 - \lambda_{22} K_{22} \mathbf{h}_2^{(0)} \end{bmatrix}$$

For general p and q, given the block structure of the Fisher information matrix, the Fisher Scoring step can be solved by alternately solving the upper left and lower right blocks, leading to the successive relaxation algorithm.

Using simple matrix manipulation (Hastie and Tibshirani, 1990, p 150), the solution of the upper left block leads to the backfitting equations

$$\mathbf{f}_j^{(1)} = S_{1j}(\mathbf{y}_{adj} - \sum_{i \neq j} f_i^{(1)}) \qquad (9)$$

for $j = 1, 2, \ldots, p$ where $\mathbf{y}_{adj} = \boldsymbol{\eta}^{(0)} + A_1^{-1} \mathbf{u}_1$ and $S_{1j} = (A_1 + \lambda_{1j} K_{1j})^{-1} A_1$. This is solved by fitting a GAM using the local scoring algorithm of Hastie and Tibshirani (1990) with dependent variable $\mathbf{y}$, an error distribution with variance function $V(\mu)$ and scale 1, link function $g_1(\mu)$ and prior weights $\mathbf{w}_1 = 1/\hat{\phi}$, since the resulting score function is the same as $\mathbf{u}_1 = A_1 G_1 (\mathbf{y} - \boldsymbol{\mu})$.

Similarly solving the lower right block leads to the backfitting equations

$$\mathbf{h}_k^{(1)} = S_{2k}(\mathbf{d}_{adj} - \sum_{i \neq k} h_i^{(1)}) \qquad (10)$$

for $k = 1, 2, \ldots, q$ where $\mathbf{d}_{adj} = \boldsymbol{\xi}^{(0)} + A_2^{-1}\mathbf{u}_2$ and $S_{2k} = (A_2 + \lambda_{2k}K_{2k})^{-1}A_2$. This is solved by fitting a GAM with dependent variable $\mathbf{d}$, a Gamma error with scale parameter 2, link function $g_2(\phi)$ and prior weights $\mathbf{w}_2 = \mathbf{1}$. Justification for this (in particular the Gamma error with scale parameter 2) comes from the fact that the resulting score function is the same as $\mathbf{u}_2 = A_2 G_2(\mathbf{d} - \phi)$. In conclusion maximization of the penalised extended Quasi-likelihood function is achieved by the successive relaxation algorithm, alternating between fitting GAM's for the mean and the dispersion respectively.

References

Aitkin, M. (1987) Modelling variance heterogeneity in normal regression using GLIM. *Appl. Statist.*, **36**, 332-339.

Hastie, T.J., Tibshirani, R.J. (1990) *Generalized Additive Models.* London: Chapman and Hall.

McCullagh, P., Nelder, J.A. (1989) *Generalised Linear Models* (2nd ed.), London, Chapman and Hall.

Nelder, J.A. (1992) Joint modelling of the mean and dispersion. In *Statistical Modelling*, 263-272 (P.G.M. Van der Heijden, W. Jansen, B. Francis and G.U.H. Seeber eds). North Holland: Amsterdam.

Nelder, J.A., Pregibon, D. (1987) An extended quasi- likelihood function. *Biometrika*, **74**, 211-232.

Reinsch, C. (1967) Smoothing by spline functions. *Numer. Math.*, **10**, 177-183.

Rigby, R.A. and Stasinopoulos, D.M. (1994) Robust fitting of an Additive model for variance heterogeneity. In *COMPSTAT, Proceedings in Computational Statistics*, eds R. Dutter and W. Grossmann, 261-268, Physica-Verlag.

Rigby, R.A. and Stasinopoulos, D.M. (1995a) Mean and Dispersion Additive Models. To appear in *Computational Statistics*.

Rigby, R.A. and Stasinopoulos, D.M. (1995b) A semi-parametric Additive model for variance heterogeneity. To appear in *Statistics and Computing*.

Smyth, G.K. (1989) Generalised Linear Models with Varying Dispersion. *J.R.Statist Soc.* B, **51**, 47-60.

Computational Aspects in Maximum Penalized Likelihood Estimation

Ori Rosen
Ayala Cohen

ABSTRACT: In this paper we describe the technical details for implementing maximum penalized likelihood estimation (MPLE). This includes description of software for fitting weighted cubic smoothing splines, which constitute building blocks in MPLE. An example is given for illustration.

KEYWORDS: Weighted cubic smoothing splines; Maximum penalized likelihood estimation

1 Introduction

Cubic smoothing splines are commonly used for fitting a nonparametric curve to a set of points in the plane, see e.g., Eubank (1988). They also constitute building blocks in fitting nonparametric generalized linear models and in maximum penalized likelihood estimation (MPLE) (Green and Silverman (1994)). Both of the latter areas employ weighted cubic smoothing splines (WCSS). The emphasis in this paper is on technical issues regarding use of software for fitting WCSS in the context of MPLE. In section 2 we describe various equivalent formulations of the minimization problem leading to WCSS. Section 3 deals with the relationship between MPLE and WCSS. Section 4 includes details of WCSS software and its applicability to MPLE. In section 5 we illustrate use of software for fitting WCSS. Section 6 concludes with a summary.

2 Cubic Smoothing Splines

Given pairs of data (x_i, y_i), $i = 1, \ldots, n$, where Y is a dependent variable and x is an explanatory variable, a cubic smoothing spline can be used to estimate nonparametrically the conditional expectation of Y given x, $f(x) = E(Y \mid x)$. A cubic smoothing spline is the solution to the minimization of

$$\sum_{i=1}^{n}(y_i - f(x_i))^2 + \lambda \int_a^b (f''(t))^2 dt \; , \tag{1}$$

over all functions f with continuous first and integrable second derivatives, where λ is a fixed constant and $a \le x_1 \le \ldots \le x_n \le b$ (see Hastie and Tibshirani (1990) or Green and Silverman (1994), GS in the sequel). It can be shown that the minimizer of (1) is a unique natural cubic spline with knots at the distinct values of x. Furthermore, the minimizer can be written as

$$\hat{\boldsymbol{f}} = (I + \lambda K)^{-1}\boldsymbol{y} \; , \tag{2}$$

where $\hat{\boldsymbol{f}} = (\hat{f}(x_1), \ldots, \hat{f}(x_n))^T$, $\boldsymbol{y} = (y_1, \ldots, y_n)^T$ and K is an $n \times n$ matrix, which is a function of the x_i's only (See GS, pp. 12-13). The first term in (1) measures closeness to the data, while the second term penalizes roughness of the function. The smoothing parameter λ determines the smoothing degree of the resulting curve: large values of λ produce smoother curves, while smaller values result in wigglier curves. The penalized sum of squares in (1) gives the same weight to each observation. If however, it is necessary that unequal weights be given to the observations, the following criterion should be used

$$\sum_{i=1}^{n} w_i(y_i - f(x_i))^2 + \lambda \int_a^b (f''(t))^2 dt \; . \tag{3}$$

In the case where the Y_i's are distributed with mean $f(x_i)$, but with heterogeneous variances, the w_i's should be set to be inversely proportional to the variances of the observations. As in the case of (1), the minimization of (3) yields a natural cubic spline, which can be found by the Reinsch algorithm, in $O(n)$ arithmetic operations (See Reinsch (1967) or GS). The minimizer of (3) is commonly called a weighted cubic smoothing spline (WCSS) and can be expressed by

$$\hat{\boldsymbol{f}} = (W + \lambda K)^{-1}W\boldsymbol{y} \; , \tag{4}$$

where W is a diagonal matrix with diagonal entries w_i. Another application of WCSS is to the case where there is more than one observation at each distinct point x_i. Suppose that at each distinct x_i there are m_i independent observations, Y_{ij}, $j = 1, \ldots, m_i$ with mean $\bar{Y}_i$. Then the problem of minimizing

$$\sum_i \sum_j (y_{ij} - f(x_i))^2 + \lambda \int f''(t)^2 dt$$

is equivalent to that of minimizing

$$\sum_i m_i(\bar{y}_i - f(x_i))^2 + \lambda \int (f''(t))^2 dt \tag{5}$$

(GS, p. 43 or Eubank (1988), p. 207). A third application of WCSS to MPLE is presented in the next section.

Criterion (3) is formulated by de Boor (1978) in a slightly different way

$$p \sum_{i=1}^{n} \left(\frac{y_i - f(x_i)}{\delta y_i} \right)^2 + (1-p) \int_a^b (f''(t))^2 dt , \qquad (6)$$

where $0 \le p \le 1$. The relationships between p and λ and δy_i and w_i are $\lambda = (1-p)/p$ and $1/(\delta y_i)^2 = w_i$.

The minimization of (3) is equivalent to the problem

$$\min \int_a^b (f''(t))^2 dt \text{ subject to } \sum_{i=1}^{n} w_i (y_i - f(x_i))^2 \le \Lambda . \qquad (7)$$

This is in fact the formulation given in Reinsch (1967). The parameters λ and Λ have the same meanings. Let $G(\Lambda) = \int (\hat{f}''_\Lambda(t))^2 dt$, where $\hat{f}_\Lambda$ is the solution of (7). Then $\lambda = -(G'(\Lambda))^{-1}$, where $G'(\Lambda) = dG(\Lambda)/d\Lambda$ (Härdle (1990), p. 62).

According to the bounded roughness approach (GS, p. 51), the WCSS is the solution to the problem

$$\min \sum_{i=1}^{n} w_i (y_i - f(x_i))^2 \text{ subject to } \int_a^b (f''(t))^2 dt \le C . \qquad (8)$$

Both of the constrained optimization problems (7) and (8) are solved by searching on λ in (3) until the minimizer $\hat{f}$ satisfies the constraint.

3 Maximum Penalized Likelihood Estimation

Suppose that we have n independent observations $\{y_i\}$ at corresponding covariate values $\{t_i\}$. Let l be the log-likelihood function, based on a certain probabilistic model, which depends on the y_i's and on, say, two unknown, but smooth functions $\mu(t)$ and $\sigma(t)$. An attempt to maximize l over all smooth functions $\mu(t)$ and $\sigma(t)$ is useless. It is always possible to choose $\mu(t)$ and $\sigma(t)$ such that they interpolate the data in the sense that the fitted values are identical to the observed responses. According to Cole and Green (1992) and GS, rather than maximizing l alone, $\mu(t)$ and $\sigma(t)$ are chosen to maximize the penalized log-likelihood:

$$\Pi = l - \frac{1}{2}\lambda_\mu \int (\mu''(t))^2 dt - \frac{1}{2}\lambda_\sigma \int (\sigma''(t))^2 dt .$$

The functions $\mu(t)$ and $\sigma(t)$ are in the space of functions, which are differentiable on $[a, b]$ and have absolutely continuous derivatives. The parameters λ_μ and λ_σ are smoothing parameters. See also Hastie and Tibshirani

(1990), pp. 149-151, where MPLE is dealt with in the context of generalized additive models. The MPLE approach allows one to balance fidelity to the data (high values of the log-likelihood) with smoothness of the fitted curves (low values of the roughness penalties), whose degree is governed by λ_μ and λ_σ. It is common to perform the maximization by Fisher scoring. Given μ and σ at the current iteration, updated estimates are obtained by solving the following system of equations:

$$\begin{pmatrix} W_\mu + \lambda_\mu K & W_{\mu\sigma} \\ W_{\sigma\mu} & W_\sigma + \lambda_\sigma K \end{pmatrix} \begin{pmatrix} \mu^{new} - \mu \\ \sigma^{new} - \sigma \end{pmatrix} = \begin{pmatrix} u_\mu - \lambda_\mu K \mu \\ u_\sigma - \lambda_\sigma K \sigma \end{pmatrix}.$$

(9)

The u's are the first derivatives of l. The diagonal matrices W are the expectations of the negative second derivatives of l with respect to the variables specified in the subscripts. In this case the matrix K depends solely on the t_i's (cf. (2) and (4)). The system (9) consists of $2n$ equations, therefore it may not be practical to solve directly. Instead, one can use the backfitting algorithm, an iterative method, which converges under certain conditions (See GS, pp. 67-69). The two block rows of (9) can be rewritten in the form of updating equations for μ and σ. The resulting μ and σ are obtained by applying WCSS:

$$\mu^{new} = S_1(y_1 - W_\mu^{-1} W_{\mu\sigma} \sigma^{new})$$ (10)

$$\sigma^{new} = S_2(y_2 - W_\sigma^{-1} W_{\sigma\mu} \mu^{new}),$$ (11)

where S_1 and S_2 are the WCSS operators:

$$S_1 = (W_\mu + \lambda_\mu K)^{-1} W_\mu$$ (12)

$$S_2 = (W_\sigma + \lambda_\sigma K)^{-1} W_\sigma$$ (13)

and

$$y_1 = \mu + W_\mu^{-1} u_\mu + W_\mu^{-1} W_{\mu\sigma} \sigma$$ (14)

$$y_2 = \sigma + W_\sigma^{-1} u_\sigma + W_\sigma^{-1} W_{\sigma\mu} \mu.$$ (15)

Initial estimates of $\mu(t)$ and $\sigma(t)$ must first be found by some method. The solution then involves two loops:

1. **Inner loop (backfitting)** - The vectors y_1 and y_2 are calculated at the initial estimates (equations (14) and (15)). A new vector μ^{new} is obtained by (10), applying WCSS to the pairs $(t_i, (y_1 - W_\mu^{-1} W_{\mu\sigma} \sigma^{new})_i)$ with weights $(W_\mu)_{ii}$, $i = 1, \ldots, n$. A new vector σ^{new} is found in turn by substituting μ^{new} in (11). This cycling between (10) and (11) continues until

$$\sum_{i=1}^{n} |\sigma^{new}(t_i) - \sigma(t_i)| < \epsilon_{inner}.$$

2. **Outer loop** - As the inner loop converges, y_1 and y_2 are recalculated at the values of μ and σ from the final iteration of the inner loop. The penalized log-likelihood is calculated at these estimates as well. New vectors μ and σ are obtained from the inner loop. The whole process continues until the absolute difference between the current penalized log-likelihood and the old one is less than ϵ_{outer}.

4   WCSS software

In order to perform MPLE, it is necessary that software which solves the minimization problem, formulated in (3) be used. In other words, the user must be able to specify the smoothing parameter λ appearing in (3). A statistical package meeting this requirement is, e.g., S-PLUS. Section 6.11.4 of *S-PLUS Guide to Statistical and Mathematical Analysis* version 3.2 (1993), p. 6-52 introduces smoothing splines by presenting the penalized residual sum of squares given in (1). It then says that a cubic smoothing spline is generated in S-PLUS by the function `smooth.spline`, which uses a B-spline basis. The smoothing parameter λ can be supplied using the `spar` argument of `smooth.spline`. It is also possible to determine the smoothness degree by using either ordinary or generalized cross validation, or by supplying the `df` argument. This argument specifies the degrees of freedom of the smooth (see Hastie and Tibshirani (1990)). For MPLE, we need to specify λ explicitly, that is, by using the `spar` argument. Chambers and Hastie (1993) give detailed description of the `smooth.spline` function including all its arguments. The `spar` argument is defined there as 'the usual smoothing parameter for smoothing splines, which is the coefficient of the integrated second squared derivative penalty function'. The user reading these two references (S-PLUS guide and Chambers and Hastie (1993)) might be misled to believe that $\lambda = $ `spar`, which is not the case. Chapter 8 of GS sheds some more light on this issue: the smoothing parameter is specified with respect to units in which the x_i's are rescaled to have range 1 and the weights (if any) rescaled to sum n, the number of data points. If we denote the range of the x_i's by r, $r = \max(x_i) - \min(x_i)$ and the mean of the weights by $\bar{w}$, then the relationship between λ and `spar` is given by

$$\lambda = r^3 \bar{w} \cdot \textbf{spar} . \tag{16}$$

See the appendix for the proof of (16). Thus, the `smooth.spline` function actually minimizes a criterion, which is equivalent to (3).

MPLE requires iterative fitting of WCSS. If in addition, it is desirable that MPLE be repetitively used in a simulation study or in bootstrap, heavy computations may be involved. In this case, using S-PLUS is likely to be impractical. A better alternative is to write the necessary routine in a programming language such as Fortran or C. If either of the latter is used, it will later be possible to incorporate this routine into S-PLUS. The question is which software should be used for fitting WCSS. The mathematical libraries of IMSL and NAG include the routines CSSMH and E02BEF respectively (See *IMSL User's Manual*, version 2.0, volume 2 (1991) and *NAG Fortran Library Manual*, Mark 15, volume 3 (1991)). Unfortunately, both of them perform the constrained minimization (7), rendering them inappropriate for MPLE as described in section 3. Instead, we utilize the Fortran programs appearing in de Boor (1978), pp. 240–242, also available in the pppack subdirectory of netlib. Like the subroutines of IMSL and NAG, these programs solve the constrained minimization problem. However, since their Fortran code is accessible, we can modify them to perform the unconstrained minimization. The code consists of the smooth program and its subroutines setupq and cholld. They perform the constrained minimization by minimizing criterion (6) for different values of p until the constraint $\sum_{i=1}^{n}((y_i - f(x_i))/\delta y_i)^2 \leq s$ has been met (s is the smoothing parameter selected by the user). Our modification includes the elimination of the search on p from smooth and altering the cholld subroutine, such that the program performs the minimization of

$$\sum_{i=1}^{n}\left(\frac{y_i - f(x_i)}{\delta y_i}\right)^2 + \lambda \int_a^b (f''(t))^2 dt$$

rather than (6). This is also done by Cole and Green (personal communication) in the context of the LMS method, described in Cole and Green (1992).

As in any weighted least squares software, it is important that the weights be properly supplied. For the MPLE, if one uses S-PLUS, the weights should simply be the diagonals of the W's in (9). For the same purpose, de Boor's program can be utilized letting $\delta y_i = 1/\sqrt{W_{ii}}$, $i = 1, \ldots, n$, where W_{ii} are the diagonal entries of W.

To sum up, either S-PLUS or the modified de Boor's code can be used for fitting WCSS in the context of MPLE. If intensive computations are involved, we recommend use of a programming language which may later be incorporated into S-PLUS.

5   Illustration

In this section we illustrate the use of S-PLUS and de Boor's code for fitting WCSS. For applications of MPLE, see e.g. Cole and Green (1992) or Rosen and Cohen (1995).

Suppose that it is desirable that a WCSS with a smoothing parameter $\lambda = 10$ be fitted to the data appearing in table 1. This is carried out in S-PLUS by the smooth.spline function having x,y,w and spar as arguments. The vectors x, y and w can be supplied in any order of x. Repeated observations are also allowed. In the example, spar $= 10/((10-1)^3 \cdot 5.5) = 2.494 \cdot 10^{-3}$

x	1	2	3	4	5	6	7	8	9	10
y	1	4	9	16	25	36	49	64	81	100
w	1	2	3	4	5	6	7	8	9	10

TABLE 1. Example Data

(see equation (16)). Thus, the call to the function is

$$\texttt{smooth.spline(x,y,w,spar=2.494\cdot10^{-3})}.$$

For MPLE it is also necessary to compute the value of the roughness penalty $\int (\hat{f}''(t))^2 dt$. The object returned by **smooth.spline** contains the value of **pen.crit** which is defined in Chambers and Hastie (1993) as 'penalized criterion'. It might seem to be the value of (3) evaluated at $\hat{f}$ and hence a means of calculating the roughness penalty, but in fact it gives the value of $\sum_{i=1}^{n} \frac{w_i}{\bar{w}}(y_i - \hat{f}(x_i))y_i$. Thus the roughness penalty should be computed directly. According to GS, p. 25, it can be expressed as the quadratic form $\hat{\gamma}^T R \hat{\gamma}$, where $\hat{\gamma}$ is the vector $(\hat{\gamma}_2, \ldots, \hat{\gamma}_{n-1})^T$ of the second derivatives of $\hat{f}$, evaluated at $x_2, \ldots, x_{n-1}$. The second derivatives at x_1 and x_n are both zero, therfore $\hat{\gamma}_1$ and $\hat{\gamma}_n$ are defined to be zero. The symmetric banded $(n-2) \times (n-2)$ matrix R is a function of the x_i's only. Expanding the quadratic form yields

$$\int (\hat{f}''(t))^2 dt = \frac{1}{3} \sum_{i=2}^{n-1} \{(h_{i-1} + h_i)\hat{\gamma}_i^2 + h_i \hat{\gamma}_i \hat{\gamma}_{i+1}\},$$

where $h_i = x_{i+1} - x_i$, $i = 1, \ldots, n-1$ and $\hat{\gamma}_i$ is the i'th entry of $\hat{\gamma}$. The required second derivatives are available from the **predict.smooth.spline** function. If the object name returned from **smooth.spline** is **fit**, then the call to **predict.smooth.spline** should be

$$\texttt{predict.smooth.spline(fit,x,deriv=2)}$$

The program of de Boor expects the data to be in acsending order of x. Repeated observations are not directly allowed for, but they can easily be handled exploiting equation (5). Using the modified de Boor's software, a WCSS is fitted to the example data by

$$\texttt{call smooth(x,y,dy,10,10.0,v,a)},$$

where 10 is the number of distinct data points and 10.0, the value of λ. The vector dy contains the entries $1/\sqrt{w_i}$, $i = 1, \ldots, 10$. The subroutine returns the arrays **v** and **a**: **v** is a work array needed for the other subroutines and **a** is an $n \times 4$ array whose columns contain $\hat{f}$ and its three first derivatives evaluated at the x_i's. It may be worthwhile to note that there is some discrepancy between the values of the first and third derivatives computed by S-PLUS and their counterparts from de Boor's program. Table 2 shows

$\hat{f}'$	de Boor	5.81	6.01	6.73	8.07	9.88	11.97	14.10	16.02	17.40	0.00
	S-PLUS	5.81	6.01	6.73	8.07	9.88	11.97	14.10	16.02	17.40	17.90
$\hat{f}'''$	de Boor	.40	.64	.59	.38	.16	-.07	-.35	-.75	-1.00	0.00
	S-PLUS	.40	.64	.64	.38	.16	.16	-.07	-.75	-1.00	-1.00

TABLE 2. Values of First and Third Derivatives

these values, evaluated at the x_i's. The first derivative is identical in both programs, except for $\hat{f}'(x_{10})$. The program of de Boor does not calculate the last value, but as the row dimension of a is n, it returns 0. S-PLUS gives the correct value. As for the third derivative, de Boor's code gives correct values, but there is a discrepancy between them and those of S-PLUS.

6 Summary

The contribution of this paper is in giving detailed technical account of how to carry out MPLE. This includes clarification of the ambiguities in using WCSS software due to different equivalent formulations of the associated minimization problem.

Appendix: The Connection Between λ and spar

Suppose that a WCSS is fitted to the n pairs (x_i, y_i) with weights w_i. Denote the range of the x_i's by r, $r = \max(x_i) - \min(x_i)$, and the fitted WCSS by $\hat{f}$, $\hat{f} = (W + \lambda K)^{-1} W y$.
Consider now the data (x_i^*, y_i) with weights w_i^*, such that the range of the x_i^*'s equals 1 and $\sum_{i=1}^{n} w_i^* = n$. In other words, $x_i^* = x_i/r$ and $w_i^* = w_i/\bar{w}$, where $\bar{w} = \frac{1}{n}\sum_{i=1}^{n} w_i$.

Proposition 1 *The WCSS fitted to (x_i, y_i) with weights w_i and smoothing parameter λ is identical to the WCSS fitted to (x_i^*, y_i) with weights w_i^* and smoothing parameter* spar $= \lambda/(r^3 \bar{w})$.

Proof. According to GS, pp. 12–13, $K = QR^{-1}Q^T$, where the matrices Q and R are functions of the x_i's only. It follows from their definitions that for the transformed data, $Q^* = rQ$, $R^* = \frac{1}{r}R$, hence $K^* = r^3 K$. Denote by $\hat{g}$ the WCSS fitted to (x_i^*, y_i) with weights w_i^* and smoothing parameter $\lambda/(r^3 \bar{w})$, then

$$\hat{g} = (W^* + \text{spar} \cdot K^*)^{-1} W^* y$$

$$= (\frac{1}{\tilde{w}}W + \frac{\lambda}{r^3\tilde{w}} \cdot r^3 K)^{-1}\frac{1}{\tilde{w}}W\boldsymbol{y}$$
$$= (W + \lambda K)^{-1}W\boldsymbol{y}$$
$$= \hat{\boldsymbol{f}}$$

□

References

Chambers, J.M. and Hastie, T.J. (1993). *Statistical models in S*. Chapman and Hall, New York.

Cole, T.J. and Green, P.J. (1992). Smoothing reference centile curves: the LMS method and penalized likelihood. *Statistics in Medicine*, **11**, 1305–1319.

de Boor, C. (1978). *A practical guide to splines*. Springer-Verlag, New York.

Eubank, R.L. (1988). *Spline smoothing and nonparametric regression*. Dekker, New York.

Green, P.J. and Silverman, B.W. (1994). *Nonparametric regression and generalized linear models: a roughness penalty approach*. Chapman and Hall, London.

Härdle, W. (1991). *Applied nonparametric regression*. Cambridge university press, Cambridge.

Hastie, T.J. and Tibshirani, R.J. (1990). *Generalized additive models*. Chapman and Hall, London.

Reinsch, C. (1967). Smoothing by spline functions. *Numerische Mathematik*, **10**, 177–183.

Rosen, O. and Cohen, A. (1995). Extreme percentile regression. In *COMPSTAT, proceedings in computational statistics*, eds W. Härdle and M.G. Schimek, Vienna: Physica-Verlag. To appear.

Risk Estimation Using a Surrogate Marker Measured with Error

Amrik Shah
David Schoenfeld
Victor De Gruttola

ABSTRACT: Our goal is to estimate the risk of contacting *Pneumocystis carinii* Pneumonia (PCP) in a fixed time interval based on the current observed CD4 count for an individual. The methodology used involves a linear random effects model for the trajectory of the observed CD4 counts in order to obtain predicted values at each event time. These predicted counts are imputed in the partial likelihood for estimating a regression coefficient in the Cox model. Subsequently, Monte Carlo techniques are employed to approximate the risk of PCP in a six month period. The method will be illustrated on data from AIDS patients.

KEYWORDS: Longitudinal data; Random effects model; Proportional hazards model; Monte Carlo; CD4 count

1 Introduction

The relationship of survival to CD4 counts in patients with HIV infection has been examined by Tsiatis et al (1993). Their goal was an estimate of patients instantaneous risk of an event as a function of the CD4 count in order to determine the efficacy of CD4 as a surrogate marker. We develop methods to estimate the probability that a patient will have an event in a given period of time, say six months, based on their current observed CD4 count. The estimate takes into account the varying nature of instantaneous hazard and changes in the CD4 count over the time period. When the event is an opportunistic infection for which their exists an effective prophylaxis, the estimate will help physicians make a more knowledgeable decision about when prophylaxis should begin. The estimate can also be used to develop a cost-benefit analysis of prophylaxis which could be based on the number of patient years of treatment required to prevent one infection.

We begin by introducing the necessary notation and assumptions in section 2. Section 3 is divided into three subsections: The first subsection develops a model for the trajectory of the CD4 count over time where it is assumed that there is a true CD4 count with a log-linear trajectory that is observed with error. This model is used for two purposes later on. To estimate the effect of CD4 on survival using the proportional hazards model, you need CD4 counts at each event time rather than the times when they were actually measured and to estimate a patients risk over a period of time you need to take into account the patients declining CD4 count. The second subsection describes how the imputed CD4 counts are used to estimate the hazard. The third subsection describes how the hazard estimate is used to find the risk over a specified period for a given observed CD4 count. Section 4 deals with the computing aspects of this methodology. In section 5 we apply these methods to develop risk estimates based on data from an AIDS clinical trial.

2   Methods

We begin by assuming that there is a "true" CD4 count that has a trajectory over time but is measured periodically with error, i.e., the observed CD4 count is the value of this "true" CD4 count plus random error. Let $Z(t)$ denote this true covariate value, or some transformation of it at time t for a person, and let $Z^o(t)$ be the observed covariate value measurement at time t. In what follows we suppress the index indicating patient number to improve readability. We also assume that $t = 0$ is the start of the study period for each patient. Since the Cox model relates the hazard of an event in the future with complete information about the past, we will define notation for the complete history of the Z up until time t. Let $\bar{Z}(t)$ denote history up to time t , i.e.,

$$\bar{Z}(t) = \{Z(\tau); \ \tau \leq t\}.$$

We use a random effects model to estimate the distribution of $\bar{Z}(t)$ and to estimate the predicted value of $\bar{Z}(t)$ given $\{\dots, Z^o(t_{-1}), Z^o(t_0)\}$ where $\{\dots, t_{-1}, t_0\}$ are the observation times $\leq t$. Secondly we model the risk using a proportional hazards model. We use the predicted value of $\bar{Z}(t)$ to estimate the hazard at time t given $\bar{Z}(t)$ denoted by $\lambda(t|\bar{Z}(t))$.

Finally we use the distribution of $\bar{Z}(t)$ and the estimate of $\lambda(t|\bar{Z}(t))$ to estimate the risk for a given time interval. If $\bar{Z}(t)$ were available for a patient then their risk of PCP during the interval (T_0, T_1) would be

$$P(T \in (T_0, T_1)|\bar{Z}(t)) = 1 - e^{-\int_{T_0}^{T_1} \lambda(t|\bar{Z}(t))dt}.$$

We will assume that the best information available at time t would be $Z^o(t_0)$, so our estimate will be

$$P(T \in (T_0, T_1)|Z^o(t_0)) = 1 - e^{-\int_{T_0}^{T_1} \lambda(t|Z(t))dt}|Z^o(t_0).$$

For simplicity we assume that the risk is to be predicted for the time interval that starts when the CD4 count is observed, i.e. $T_0 = t_0$. Since the above integral does not offer a closed form solution, the final step will involve the use of a Monte Carlo simulation.

2.1   Random Effects Model for CD4 counts

The first step in the modeling process is to find a linearizing transformation of the CD4 counts that also gives data with normal errors. In the data set used in the example we decided to define $Z^o(t)$ as the Log-CD4 count but other transformations might be better in different settings. The covariate data was modeled using a linear random effects model as described by Laird and Ware (1982). A model with linear effects can be represented as

$$Z^o(t) = \alpha_0 + \alpha_1 * t + \gamma_0 + \gamma_1 * t + \epsilon(t) \tag{1}$$

where α_0 = population intercept,
       α_1 = population slope,
       γ_0 = individuals intercept,
       γ_1 = individuals slope,
       $\epsilon(t)$ = measurement error at time t.

Using a linear random components model allows each subject to have individual intercept and slope parameters and also incorporates a population intercept α_0, and slope α_1. For this model, γ_0, γ_1 are considered random with

$$E(\gamma_0, \gamma_1) = (0, 0), \quad Var(\gamma_0, \gamma_1) = \mathbf{D}.$$

The measurement error ϵ is assumed to be independent of the random effects γ_0, γ_1. Also,

$$E[\epsilon(t)] = 0, \ Var[\epsilon(t)] = \sigma^2, \text{ and } Cov[\epsilon(s), \epsilon(t)] = 0, \ s \neq t.$$

After estimation of parameters, we can obtain predicted $\ln(\text{CD4})$ for a person at time t using

$$Z^p(t) = \widehat{\alpha}_0 + \widehat{\alpha}_1 * t + \widehat{\gamma}_0 + \widehat{\gamma}_1 * t \tag{2}$$

since $Z^o(t) = Z(t) + \epsilon$. In this expression $\widehat{\alpha}_0$, $\widehat{\alpha}_1$ are the same for each person and $\widehat{\gamma}_0$ and $\widehat{\gamma}_1$ vary from person to person.

2.2   Cox Proportional Hazards Model

To model the relationship of hazard to the time dependent covariates, we used the proportional hazards model introduced by Cox. This is of form

$$\lambda(t|\bar{Z}(t)) = \lambda_0(t)e^{\beta Z(t)}. \tag{3}$$

If the underlying hazard $\lambda_0(t)$ is left unspecified, then the parameter β is estimated by maximizing the partial likelihood given by Cox (1975) which only makes use of covariate values at the failure times. In order to make use of this partial likelihood, one requires, at each of the failure times, the values of $Z(t)$ for all individuals who were at risk. We replace $Z(t)$ at each failure time with its estimate $Z^p(t)$ using the model explained in the previous section. The proof that the Cox proportional hazards model yields consistent and asymptotically normal estimates depends on the fact that covariate values are *predictable* (Fleming & Harrington, 1992). This will be the case as long as CD4 measurements that occurred at times after t are not used in the estimation of $Z(t)$. Therefore we estimated the parameters of the random effects model separately at each event time. Note, the likelihood thus obtained is **not** the expected value of the partial likelihood for β but rather a first order approximation to it, since the expected marginal likelihood would be $E[Lik\{Z(t)|\bar{Z}^o(t)\}]$. The use of a second order approximation is discussed by Dafni & Tsiatis (1993) but it does not make a large difference in the estimates.

2.3   Approximating the Risk

Suppose $\beta, \alpha_0, \alpha_1$ and $\mathbf{D}$ are known or estimated, then a Cox regression model relating the hazard to the ln(CD4) counts is

$$\lambda(t|\bar{Z}(t)) = \lambda_0(t)e^{\beta Z(t)}.$$

If we knew $Z(t)$ we could obtain the risk of an event in time T_0 to T_1 by

$$P_{risk} = \quad P(T \in (T_0, T_1)|Z(t)) \quad = 1 - e^{-\int_{T_0}^{T_1} \lambda_0(t)e^{\beta Z(t)}dt}. \tag{4}$$

Our estimate of the risk based on $\bar{Z}(t)|Z^o(t_0)$ is given by

$$E[P_{risk}|Z^o(t_0)] = 1 - \int e^{-\int_{T_0}^{T} \lambda_0(t)e^{\beta(\alpha_0+\alpha_1 \cdot t+\gamma_0+\gamma_1 \cdot t)}dt} dF(\gamma_0, \gamma_1|Z^o(t_0)).$$

If $\lambda_0(t) = \lambda_0$ were to be constant, then integration by parts yields

$$E[P_{risk}|Z^o(t_0)] = 1 - \int e^{-\lambda_0 C D e^{\beta(\alpha_0+\gamma_0)}} dF(\gamma_0, \gamma_1|Z^o(t_0)) \tag{5}$$

where

$$C = [\hat{\beta}(\alpha_1 + \gamma_1)]^{-1},$$
$$D = e^{\hat{\beta}(\alpha_1 T_1+\gamma_1 T_1)} - e^{\hat{\beta}(\alpha_1 T_0+\gamma_1 T_0)}.$$

Note, we need the expectation based on the observed value $Z^o(t_0)$ since that is what is known in practice. The density $f(\gamma_0, \gamma_1|Z^o(t_0))$ is multivariate normal with the parameters estimated from the random effects model. This follows from the result that

$$\begin{pmatrix} Z^o(t_0) \\ \gamma_0 \\ \gamma_1 \end{pmatrix} \sim \text{MVN} \left[\begin{pmatrix} E[Z^o(t_0)] \\ 0 \\ 0 \end{pmatrix}, \begin{pmatrix} V(Z^o(t_0)) & \sigma_{z,\gamma_0} & \sigma_{z,\gamma_1} \\ \sigma_{z,\gamma_0} & d_{11} & d_{12} \\ \sigma_{z,\gamma_1} & d_{21} & d_{22} \end{pmatrix} \right].$$

Using standard conditional distribution principles we obtain

$$\begin{pmatrix} \gamma_0 \\ \gamma_1 \end{pmatrix} | Z^o(t_0) \sim \text{BVN} \left[\begin{pmatrix} \sigma_{z,\gamma_0} \\ \sigma_{z,\gamma_1} \end{pmatrix} [V(Z^o(t_0))]^{-1} (Z^o(t_0) - E[Z^o(t_0)]), \Sigma \right]$$

$$\text{where} \quad \Sigma = D - \begin{pmatrix} \sigma_{z,\gamma_0} \\ \sigma_{z,\gamma_1} \end{pmatrix} [V(Z^o(t_0))]^{-1} (\sigma_{z,\gamma_0}, \sigma_{z,\gamma_1}).$$

The estimates of $Cov[Z^o(t_0), \gamma_1] = \sigma_{z,\gamma_1}$, $Cov[Z^o(t_0), \gamma_0] = \sigma_{z,\gamma_0}$ and $Var[Z^o(t_0))]$ are easily obtained. It is not possible to get a closed form solution for the integral in (5) and thus Monte Carlo techniques will have to be employed.

The estimate given by (5) depends on T_0 whether or not $\lambda_0(t)$ is constant. This is a shortcoming of the analysis because ideally the estimated risk would depend only on $Z^o(t_0)$ and $T_1 - T_0$. Therefore we later examine the effect of choosing different starting times. It is not possible to test if $\lambda_0(t)$ is constant in the case of a Cox model with time dependent covariates. However, a plot of the Nelson cumulative hazard estimator $\Lambda_0(t)$ vs. *time* may provide some information in this regard.

3 Software

The estimation of the parameters for the random effects model was carried out by the fortran programs REML (Laird et al, 1987) and MIXMOD (Lindstrom and Bates, 1988). The output of these programs yields estimates for $\alpha_0, \alpha_1, \sigma, D$ and the random effects γ_0, γ_1 for each patient. Use of both programs is necessitated by the inability of one or the other to successfully compute the Bayes estimators for all risk sets. There are three basic steps in obtaining the predicted CD4 counts. We begin by constructing risk sets for each event time. The risk set for failure time t_q consists of observations of CD4 obtained up to time t_q and these are only of those patients who did not have an event prior to time t_q. Then each risk set is subjected to an estimation process yielding fixed and random effects estimates. These estimates are used to obtain predicted ln CD4 for an individual at time t. These are the empirical Bayes' estimators from the Laird-Ware model (1982). Using these in Cox's partial likelihood enabled us to get an estimate of β.

4   Application to 019 data

We had data on a group of 1362 patients, of which 874 were administered zidovudine (ZDV). Their CD4 counts were determined approximately every four weeks while on therapy. It is already known and verified that CD4 counts in the presence of ZDV exhibit an initial increase with the peak attained on or about the 8th week followed by a steady decline (Tsiatis et al, 1993). For the purpose of this analysis, the counts for the first eight weeks were discarded in order to fit a linear growth curve model. The event of interest was the first onset of PCP. The median duration of follow up was 542 days and there were 14 cases of PCP in the ZDV group and 17 cases in the Placebo arm during the course of the study. The number of patients in risk sets varied from 303 to a high of 656 for the ZDV group and from 169 to 433 for the Placebo arm. We estimated the risk of PCP during a 6 month period as a function of one observed CD4 level.

4.1   Estimation of Parameters

A random effects model was fit for each treatment group, and within treatment, for each risk set. This approach yielded estimates for $\alpha_0, \alpha_1, \sigma$, and

Group	$\widehat{\alpha}_0$	$\widehat{\alpha}_1$	$\widehat{\lambda}_0$	$\widehat{\beta}$	$s.e.(\beta)$
ZDV	5.8764	-0.000225	4.324	-2.022	0.2464
Placebo	5.7634	-0.000039	0.044	-1.047	0.3227

TABLE 1. Parameter Estimates

D for each risk set. The final estimates for α_0 and α_1 (Table 1) were chosen to be those obtained from the first six months of data for each treatment. These were not much different from those obtained by using the entire data. In our analysis, scarcity of data did not permit us to fit a model with quadratic random effects. A second order fixed effect was tried but turned out to be statistically insignificant. We also obtained estimates for the random effects γ_0, γ_1 for all patients in each risk set. Estimates such as these, calculated separately for each risk set were used in equation (2) to give predicted CD4 counts. The estimates for σ and **D**, again obtained from data for the first six months, are

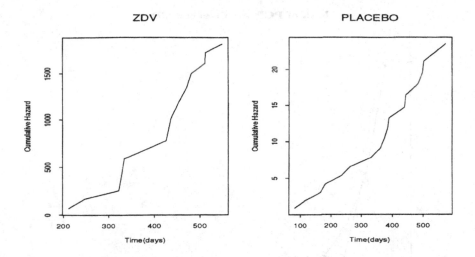

FIGURE 1. Plots of cumulative baseline hazard vs. time

ZDV	PLACEBO

$$\widehat{\sigma} = 0.084432 \qquad\qquad\qquad \widehat{\sigma} = 0.086859$$

$$\widehat{\mathbf{D}} = \begin{pmatrix} 0.20537 & -3.1 \times 10^{-4} \\ -3.1 \times 10^{-4} & 5.0 \times 10^{-6} \end{pmatrix}, \qquad \widehat{\mathbf{D}} = \begin{pmatrix} 0.10612 & 2.1 \times 10^{-4} \\ 2.1 \times 10^{-4} & 4.7 \times 10^{-6} \end{pmatrix}.$$

A SAS procedure, *Proc Phreg*, was employed to fit a proportional hazards model using the above obtained predicted counts as the time varying co-variate. This resulted in estimates for β, which are shown in Table 1. We also obtained the Nelson estimates for the baseline hazard at every event time for each treatment arm. A plot of the cumulative baseline hazard vs. time (Figure 1) supports the assumption of a constant underlying hazard. The slope obtained from each line was used as the estimate of λ_0 for that treatment.

4.2   Estimation of Risk

At this point we have estimates for $\alpha_0, \alpha_1, \beta, \lambda_0$ and $\mathbf{D}$. To approximate the value of expression (5) we generated γ_0, γ_1 using their estimated bi-variate parameters. Since the distribution of γ_0, γ_1 is dependent upon the observed CD4 value, different estimates for the parameters are obtained in the estimation of risk for different CD4 values. Figure 2 is a plot of *Risk vs CD4* for both treatment arms. The Placebo arm is seen to have a higher risk for the same observed CD4 value, but at lower values of CD4, the risks for both arms are almost the same.

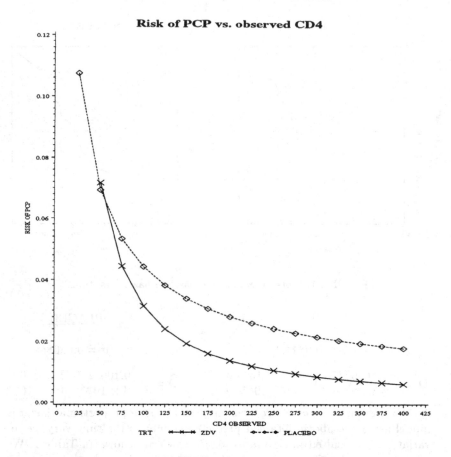

FIGURE 2. Risk vs CD4

5   Discussion

The risk of PCP in the next 180 days, using different values of T_0, was obtained for the Placebo group for an observed CD4 count of $100/mm^3$. The results (Table 2) exhibited a negligible increase in risk with increasing values for T_0. The plot for *Risk vs CD4* implies that the risk of PCP depended on whether the patient received ZDV as well as their CD4 count which was similar to the finding of Tsiatis et al in regards to mortality. The number of patients that need to be treated for every PCP prevented is the inverse of the risk. Thus for patients with CD4 counts of $100/mm^3$ we would treat 30 patients to prevent one case of PCP, whereas for a CD4 count of $200/mm^3$, 75 patients would need to be treated. The analysis can be extended to cases in which there are multiple measurements of the CD4 count over time. This methodology is applicable to other surrogate markers

Risk of PCP in 180 days for CD4 = 100					
T_0(days)	0	30	60	120	180
Risk	0.0432	0.0483	0.0535	0.0623	0.0678

TABLE 2. Effect of T_0 on risk estimate(Placebo Group)

and other diseases where the risk of mortality or morbidity is a function of a laboratory measurement.

Acknowledgments: This work was funded by NIAID AI-95030 and partial support provided by AI-30885.

References

Anderson, P.K., Gill, R. (1982). Cox's regression model for counting processes, *Annals of Statistics*, vol **10**

Cox, D.R. (1975). Partial Likelihood. *Biometrika*, vol **69**

Dafni, U.G., Tsiatis, A.A. (1993). A Bias-reduced Method for Estimating Association of Survival to Time varying Covariates Measured with Error in the Cox Model. *Ph.D. Thesis, Harvard School of Public Health*

Fleming, P., Harrington, D. (1992). Counting Processes and Survival Analysis. *Wiley*, pp 132-137

Laird, N.M., Ware, J.H. (1982). Random-effects models for longitudinal data. *Biometrics*, vol **38**

Laird, N.M., Lange, N., Stram, D. (1987). Maximum Likelihood Computations With Repeated Measures: Application of the EM Algorithm, *Journal of the American Statistical Association*, vol **82**

Lindstrom, M.J., Bates, D.M. (1988). Newton-Raphson and EM algorithms for linear mixed-effects models for repeated measures data, *Journal of the American Statistical Association*, vol **83**

Tsiatis, A.A., De Gruttola, V., Wulfsohn, M. (1993). Modeling the relationship between Survival and CD4-lymphocytes in Patients with AIDS and AIDS-related Complex. *Journal of AIDS*

Estimating Attributable Risks under an Arithmetic Mixture Model

Karen Steindorf
Jay Lubin

ABSTRACT: The concept of attributable risks can be used to estimate the number of lung cancers attributable to residential radon in a population. Recent studies indicate that smoking might modify the effect of radon on lung cancer. In this paper, three different approaches for incorporating this kind of information into attributable risk calculations, one of which is using an arithmetic mixture model, are presented and discussed. For illustration purposes, a risk assessment for the population of the former West Germany is conducted.

KEYWORDS: Arithmetic Mixture Models; Attributable Risks; Modelling of Interactions; Radon

1   Introduction

This paper is motivated by the need to estimate the number of lung cancers attributable to residential radon for populations of smokers and never-smokers. There is substantial epidemiological and experimental animal evidence that high cumulative exposure to radon (more precisely radon-222 and its short-lived decay progeny) can cause lung cancer (IARC, 1988; BEIR IV, 1988).   In recent years there has been increasing concern about the health risks due to low exposures, after it had been recognized that radon can enter homes from subsoils and accumulate, in rare instances, to levels that exceed occupational standards for mines. Several studies indicate that smoking, as the most important cause of lung cancer, might modify the effect of radon on lung cancer. In their recent analysis of 11 cohort studies of radon-exposed underground miners, including over 2,700 lung cancer deaths, nearly 1.2 million person-years of follow-up and 6 cohorts with smoking information, Lubin et al. (1994) reinforced the hypothesis that the joint relative risk model for the two exposures is most consistent with an intermediate relationship between multiplicative and additive. To incorporate this kind of information into attributable risk calculations different modelling approaches will be presented and discussed. Some of the methodological issues as well as the data used for the risk calculation in this paper are described in more detail in Lubin and Steindorf (1995) and Steindorf et al. (1995).

2   The Radon Risk Model

2.1   The model by Lubin et al. (1994)

Lung cancer risks from residential radon and its progeny are currently predicted from relative risk regression models developed from studies of underground miners. According to the results of the pooled analysis by Lubin et al. (1994), the dose-response relationship between radon and lung cancer, can be described as

$$RR\,(w,\,age,\,dur)\;=\;1+\beta_t\cdot(w_{5-14}+\theta_2 w_{15-24}+\theta_3 w_{25+})\cdot\phi_{age}\cdot\gamma_{dur} \qquad (1)$$

where $\beta_t = 0.0039$, $\theta_2 = 0.76$, $\theta_3 = 0.31$,

$$\phi_{age} = \begin{cases} 1.00 & \text{for } age < 55 \\[4pt] 0.57 & \text{for } 55 \le age < 65 \\[4pt] 0.34 & \text{for } 65 \le age < 75 \\[4pt] 0.28 & \text{for } 75 \le age \end{cases},$$

and

$$\gamma_{dur} = \begin{cases} 1.00 & \text{for duration} < 5 \text{ years} \\[4pt] 3.17 & \text{for } 5 \le \text{duration} < 15 \text{ years} \\[4pt] 5.27 & \text{for } 15 \le \text{duration} < 25 \text{ years} \\[4pt] 9.08 & \text{for } 25 \le \text{duration} < 35 \text{ years} \\[4pt] 13.6 & \text{for } 35 \le \text{duration} . \end{cases}$$

The parameters ϕ_{age} and γ_{dur} refer to the category-specific effects for attained age and duration of radon exposure in years. The cumulative radon progeny exposure w is defined in Working Level Months (WLM), a standard unit for occupational radon exposures. It is decomposed into time windows, where w_{5-14}, w_{15-24}, and w_{25+} are cumulative exposures received 5-14, 15-24, and 25 and more years prior. The lag interval, which is the minimum time from exposure to a resultant effect, is assumed to be 5 years.

This risk model is developed using data from studies of miners, where exposure is measured in units of WLM. Radon levels in homes are generally measured in units of Bq/m^3 (or pCi/l), a count of the number of atomic transformations per second per cubic meter (or liter). To apply the risk model to a general population, the indoor concentrations of radon in Bq/m^3 can be converted to WLM/year by assuming that 50 Bq/m^3 of indoor exposure corresponds to 0.2 WLM/year (IARC, 1988, p.177). Further, the age-specific exposures are multiplied by the so called K-factor which takes into account the dosimetry differences between mines and homes for the exposure-dose-relationship at the target cells in the respiratory tract. This factor was estimated by the National Research Council (1991) to be 0.8 for children under age 10 years and 0.7 for ages over 10.

Adjustment for smoking

Risk model (1) has not incorporated the effects of tobacco use, because information on smoking was not available for all 11 miners cohorts. However, analyses of smoking data from 6 cohorts did enable Lubin et al. to suggest a *post hoc* adjustment of model (1) for smoking status. Under the assumption that the influence of smoking status acts to modify only β_t, while leaving the other parameter estimates unchanged, they suggested to replace β_t by $\beta_s = \beta_t \times \tau_s$ for smokers and by $\beta_n = \beta_t \times \tau_n$ for never-smokers, where τ_s and τ_n are adjustment factors for smokers and never-smokers, respectively. To obtain estimates for these factors, Lubin et al. fitted simple linear excess relative risk models ($RR = 1 + \beta \times w$) to the data of the 6 cohorts with smoking information: for smokers, for never-smokers and for the combined group of smokers and never-smokers, ignoring smoking status. The slope parameters were estimated to be 0.0103 (95% CI: [0.002, 0.057]) for never-smokers, 0.0034 ([0.001,0.015]) for smokers and 0.0037 ([0.001,0.017]) for the miners mixed miner population. Thus, τ_s as the adjustment for smoking, was taken to be 0.9 (=0.0034/0.0037), while τ_n was taken to be 3.0 (=0.0103/0.0037). For model (1), this means that the estimate of $\beta_t = 0.0039$ was reduced to $\beta_s = 0.0035$ for smokers and increased to $\beta_n = 0.0117$ for never-smokers. A more theoretical justification for this *ad hoc* adjustment is given in the next section.

2.2   The arithmetic mixture model

The *ad hoc* approach can be developed by assuming that the joint relative risk for smoking status (s) and radon exposure (w), RR(s,w), is given by an arithmetic mixture model (Thomas, 1981), where the joint relative risk is a weighted mean of a multiplicative and additive relative risk for smoking and radon progeny exposure, namely,

$$RR(s, w) = \lambda \, \psi_s \, (1 + \beta^* \, w) + (1 - \lambda) \, (\psi_s + \beta^* \, w) \,, \tag{2}$$

where ψ_s is the relative risk for smoking status (ψ_1 is the relative risk for smokers and ψ_0 (=1) for never-smokers), and β^* is the slope parameter for the excess relative risk in never-smokers. The quantity λ specifies the mixing parameter which characterizes the joint relationship. In miner data, λ was estimated as 0.12, with a likelihood-based 95%-confidence interval of [0.1; 0.6]. Thus, the analysis suggested that the joint relative risk for radon progeny exposure is most likely intermediate between an additive and a multiplicative model. This result agreed with previous investigations (e.g. Moolgavkar et al., 1993).

Starting with this arithmetic mixture model, subgroup-specific models for smokers ($RR_s(w) = RR(1, w)/RR(1, 0)$), for never-smokers ($RR_n(w) = RR(0, w)/RR(0, 0)$), and for the mixed miners population of smokers and never-smokers ($RR_t(w) = [\pi RR(1, w) + (1 - \pi)RR(0, w)]/[\pi RR(1, 0) + (1 - \pi)RR(0, 0)]$) can be written as linear excess relative risk models with $RR_s(w) = 1 + \beta^* \mu w$, $RR_n(w) = 1 + \beta^* \mu w$, and $RR_t(w) = 1 + \beta^* \nu w$, where $\mu = [\lambda(\psi_1 - 1) + 1]/\psi_1$ and $\nu = [1 + \lambda(\psi_1 - 1)\pi]/[1 + (\psi_1 - 1)\pi]$ and π denotes the proportion of smokers. From this it follows that $\beta^* = \beta_n$ and that the adjustment factors are given by

$$\tau_n = \frac{1 + \pi(\psi_s - 1)}{1 + \lambda(\psi_s - 1)\pi} \tag{3}$$

and

$$\tau_s = \frac{\lambda(\psi_s - 1) + 1}{\psi_s} \cdot \frac{1 + \pi(\psi_s - 1)}{1 + \lambda(\psi_s - 1)\pi}. \tag{4}$$

From these calculations it can be seen that, under the mixture model, the adjustment factors actually include population-specific parameters, such as the relative risk (ψ_s) for smokers compared to never-smokers and the prevalence of smoking (π). Our interest is in applying the radon risk model for attributable risk estimation to a non-miner target population, which might have different smoking characteristics than the miners population. The ad hoc adjustment by Lubin et al. is unable to incorporate population-specific smoking characteristics, while the adjustment factors derived under the mixture model allow for the use of additional information about population smoking characteristics. In particular, this procedure allows for gender-specific attributable risk calculations which account for differences in smoking behaviour in males and females.

3   Calculation of attributable risks

The attributable risk (AR) for lung cancer is defined as the reduction in the lung cancer mortality rate in a population after elimination of exposure as a fraction of the total lung cancer-specific rate (Levin, 1953). If I is the total lung cancer mortality rate in a population and I_0 is the rate in the same population with exposure eliminated, then

$$AR \ = \ \frac{I - I_0}{I} \ = \ \frac{\int \{RR(X) - 1\}f(X)dX}{\int RR(X)f(X)dX} \ ,$$

where f is the probability density function for the residential radon concentration and $RR(X)$ is the relative lifetime risk in the presence of competing risks for a lifetime exposure at X Bq/m^3. The relative lifetime risk is the lifetime probability of dying of lung cancer for a given radon exposure profile divided by the corresponding lifetime probability of lung cancer for a non-exposed individual. The calculation of lifetime risks incorporates lifetable methods together with age-specific relative risks for radon progeny exposure and is e.g. described in the BEIR IV Report (1988) or in Thomas et al. (1992).

The radon distribution f can be approximated by using percentiles. For each interval with probability 0.01, the conditional mean value X_i ($i = 1, .., 100$) of radon exposure in Bq/m^3, is calculated and the first exposure interval X_1 is taken as the baseline category for the relative risk calculation. From this, the attributable risk is computed as

$$AR \ = \ \frac{\sum_{i=1}^{100}\{RR(X_i) - 1\}}{\sum_{i=1}^{100} RR(X_i)} \ . \tag{5}$$

The corresponding confidence interval can be calculated by applying the multivariate delta method to logit AR (Leung and Kupper, 1981) and assuming that the variability arises primarily from the parameters in the age-specific relative risk model.

As smoking was identified as an interacting factor for the relative risk function for radon, the attributable risk calculations under the adjusted radon model have to be performed separately for smokers and never-smokers. Thus, the age-specific lung cancer mortality rates, h, have to be adjusted to reflect rates for never-smokers, h_n, and for smokers, h_s. This can be done by solving the approximation $h = \tilde{\pi} \, h_s + (1-\tilde{\pi}) \, h_n = \tilde{\pi} \, \tilde{\psi}_s \, h_n + (1-\tilde{\pi}) \, h_n$ for h_n, and then by calculating $h_s = h_n \, \tilde{\psi}_s$, where the tilde-notation denotes that the values refer to the target population of the risk assessment.

4   Example: Indoor Radon Exposure in West Germany

These methods will now be applied in an attributable risk estimation for the the German population. The radon distribution f will be approximated by a log-normal distribution with geometric mean 40 Bq/m^3 and geometric

Table 1: Attributable risk calculations under different adjustment models for smoking - by gender

	Multiplicative Model (a)	Model with ad hoc adjustment (b)	Arithmetic Mixture Model (c)
Females			
Never-smokers	0.073	0.188	0.135
Smokers	0.072	0.066	0.040
Males			
Never-smokers	0.073	0.188	0.223
Smokers	0.068	0.063	0.055

standard deviation 1.8, as it was estimated by the 1984 national survey of residential concentrations of radon in homes of the former West Germany (Schmier, 1984). The calculations of the attributable risks will be based on the age-, gender- and cause-specific mortality rates for 1985-89 for the former West Germany, for which the estimated lifetime probability of lung cancer death is 0.012 for females and 0.065 for males. According to national survey data from 1978, it can be assumed that the proportion of smokers are 0.26 for females and 0.58 for males. The year 1978 was chosen to allow for latency effects. As relative risks for smoking the quantities 6.0 and 10.0 were used for females and males, respectively. These values represent a crude average of the results from major epidemiologic studies in Germany. More details about the underlying data are given in Steindorf et al. (1995).

Ignoring the smoking status, the attributable risks can be calculated from (5) under model (1), The overall estimated attributable risk for females is 0.073 (95% CI: [0.01; 0.34]) and 0.070 ([0.02; 0.27]) for males. In 1991, there were 28,700 lung cancer deaths in the western part of the Federal Republic of Germany, 6,120 in females and 22,580 in males. Thus, the estimated attributable risk correspond to a total of about 2,000 radon attributable lung cancer deaths (95% CI [500; 8,200]), about 400 in females and 1,600 in males.

In order to apply the arithmetic mixture model with the specific smoking information to the German population, the adjustment factors τ_n and τ_s have to be calculated for males and females. The calculation according to (3) and (4) result in $\tau_s = 0.53$ and $\tau_n = 1.99$ for females, and $\tau_s = 0.79$ and $\tau_n = 3.79$ for males. Table 1 gives the results for the discussed adjustment procedures for smoking. Compared are the results under the arithmetic mixture model incorporating the population-specific smoking information, with two other models: (a) the multiplicative model, which is the standard approach taken in most risk assessments, where the same model is applied to smokers and never-smokers with the intrinsic assumption of no interactions for smoking and radon progeny exposure, and (b) the ad hoc modification according to Lubin et al. (1994).

It can be seen, that approaches (b) and (c) substantially affect the attributable risks for never-smokers and smokers, compared with the multiplicative assumption of approach (a). Adjusting for the intermediate relationship for smoking and radon, the attributable risk is estimated to be about 4-7% for smokers and 14-22% for never-smokers, in comparison to 7% for both smokers and never-smokers under adjustment (a). Furthermore, the results for approach (c) tend to be more extreme than those for procedure (b) in this example. The standard error of these estimates (not shown) is about a factor of two, so that 95 % confidence limits are obtained by multiplying and dividing by a factor of four.

5   Discussion

It has been demonstrated how the interacting factor smoking can be incorporated in attributable risk estimation. The ad hoc approach by Lubin et al. was based on an empirical adjustment. It is applicable to the general population if smoking behaviour is approximately the same in miners and in the general population. If this assumption is not fulfilled, with this approach it is not clear how τ_s and τ_n should be modified. In contrast, the approach under the arithmetic mixture model allows for population differences in smoking, but is based on the assumption that the joint relative risk for smoking and radon progeny exposure is specified by an arithmetic mixture model with a fixed value for λ (here:0.12). For the empirical adjustment by Lubin et al., no assumptions on the joint relative risk are required. For both approaches, the comparison with the multiplicative model as it was used as standard model in former risk assessment showed how important it is to incorporate information on interacting factors in the attributable risk calculation. However, further work is needed to develop adjustment procedures which allow for population-specific information but which do not rely on stringent model assumptions.

The results for radon are subject to a variety of uncertainties which could not be discussed in this paper. The reader is urged to use caution when interpreting the estimates of attributable risks, mindful that confidence intervals reflect only statistical uncertainty in model parameters, while ignoring other sources of uncertainty, such as misspecification of exposures, an incorrect risk model, estimation of τ_n and τ_s and dosimetry differences in radon exposure in mines and homes. For more details, the interested reader is referred to Lubin et al. (1994).

References

BEIR IV Committee (1988). *Health effects of radon and other internally deposited alpha emitters.* National Research Council. Washington DC, National Academy Press

International Agency for Research on Cancer (IARC) (1988). *Evaluation of carcinogenic risk to humans: Man-made mineral fibers and radon.* Lyon: IARC Monographs 43

Leung, H.M., Kupper, L.L. (1981). Comparisons of confidence intervals for attributable risk. *Biometrics*, **37**, 293-302

Levin,M. L. (1953). The occurence of lung cancer in man. *Acta Unio International Contra Cancrum*, **243**, 531-41

Lubin, J.H., Boice, J.D., Edling, C. et al. (1994). *Lung cancer and radon: a joint analysis of 11 underground miners studies.* U.S. National Institutes of Health , Publication No. 94-3644

Lubin, J.H., Steindorf, K. (1995). Cigarette use and the estimation of radon- attributable lung cancer in the U.S. *Radiation Research*, **141**, 79-85

Moolgavkar, S.H., Luebeck, E.G., Krewski, D., Zielinski, J.M. (1993). Cigarette smoke and lung cancer: a re-analysis of the Colorado Plateau uranium miners' data. *Epidemiology*, **4**, 417-30

National Research Council (1991). *Comparative dosimetry of radon in mines and homes.* Washington DC, National Academy Press

Schmier, H. (1984). *Die Strahlenexposition in Wohnungen durch die Folgeprodukte des Radon und Thoron.* In: Bericht zur Informationstagung des Bundesministers des Innern über die Aufgaben und Ergebnisse der Strahlenschutzforschung, p.37-54

Steindorf, K., Lubin, J.H., Wichmann, H.-E., Becher, H. (1995). Lung cancer deaths attributable to indoor radon exposure in West Germany. *International Journal of Epidemiology*, **24**, in press

Thomas, D.C. (1981). General relative risk models for survival time and matched case-control analysis. *Biometrics*, **37**, 673-86

Thomas, D., Darby, S., Fagnani, F. et al. (1992). Definition and estimation of lifetime detriment from radiation exposures: principles and methods. *Health Physics*, **63**, 259-72

A Nonparametric Method for Detecting Neural Connectivity

Klaus J. Utikal

ABSTRACT: We apply graphically a method (Mip-method) of inference on dependence among the firing times of a biological neural network. This we compare to the traditional method based on the cross correlogram (CCH).
In the analysis of firing times from two artifical networks it is shown that the CCH can be a very poor device for short as well as arbitrary long series of observations while the Mip-methods recovers the dependence clearly. We further analyse an empirically obtained series and thus illustrate that the proposed method can be a useful alternative to cross correlation analysis.

KEYWORDS: Neural Firings; Connectivity; Spike Trains; Counting Processes; Intensities; Censoring; Survival Analysis.

1 Introduction and Summary

The statistical analysis of firing activities of an ensemble of neurons (i.e. nerve cells of an organism) is an important problem in neurology. The observed firing times, recorded in form of a multivariate series of spike trains, are naturally described by multivariate counting processes, see e.g. the survey of Brillinger (1992).

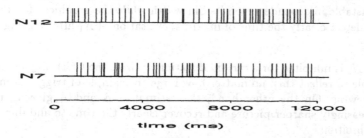

Figure 1: Recorded firing times of a pair of neurons see Example 3 of Section 2.

Neurons are said to be connected if their firing activities are not independent of each other. If the firings of a (trigger) neuron leads to an increased firing intensity of a target neuron this connection is called excitatory. It is called inhibitory if the trigger firings lead to a decreased firing activity of the target. This way neurons are "wired" to form a network that is studied in order to understand the behavior of the underlying organism.

The statistical problem is to detect dependence among those components of a counting process whose jumps lead to a decreased or increased firing probability of others. For an account on different methods in use see Eggermont (1990).

A new method of analyzing statistical aspects of the firing activities of a neural ensemble was introduced in Utikal(1994b) and called Mip-method. The Mip-method has shown promising results in detecting connectivity among pairs of neurons, see Utikal(1994c).

The method is summarized as follows. For a pair of neurons N_1, N_2, at the time of a firing of N_2 we consider the distribution F until the next firing of N_1. We denote the hazard rate of F by α and the cumulative hazard rate by A. The hazard rate α is a measure of the firing intensity of the neuron. It characterizes the distribution of waiting time for a firing. When graphed as a function of time, it is even more informative than the distribution itself because it reveals the times of extreme hazard, which are of main interest to us. For these parameters we introduce nonparametric functional estimators $\hat{\alpha}$ and $\hat{A}$. Inhibition of N_1 by N_2 is recognized as total flatness of $\hat{A}$ over a certain region, or equivalently as a pronounced local minimum of $\hat{\alpha}$. Excitation of N_1 by N_2 is manifested as a sudden pronounced jump of $\hat{A}$, respectively a prominent local maximum of $\hat{\alpha}$. A formal definition of the estimators is given in Section 3.

In the present article we compare the Mip-method to the traditional cross-correlation approach using the cross-correlogram (CCH) described in Perkel et al(1967). In two examples, simulating the firing activity of pairs of neurons (N_1, N_2) of known circuitry, we show

- In small samples the CCH (cross correlation histogram) can be very unstable in dependence on its bin width while the graphics of $\hat{A}$ and $\hat{\alpha}$ detect clearly the time of maximal excitation of N_1 after a firing of N_2.

- If N_2 is inhibited for a certain duration of time, the CCH may be unable to reflect this, no matter how large the sample of trigger events growths. On the other hand, the estimators $\hat{\alpha}$ and $\hat{A}$ give an increasingly sharper picture and recover clearly the time to and during inhibition.

In a third example, analyzing an empirically obtained bivariate spike train series of neural firing times, it can be seen that

- Even if the CCH depicts the effect, this can be seen clearer using the Mip-method; moreover, we can estimate the time after a trigger firing to its maximal effect on the target as well as the strength and duration of the effect.

The estimator $\hat{A}$ used is common in survival analysis, see Andersen et al(1993). The estimator $\hat{\alpha}$ is obtained by kernel function smoothing of $\hat{A}$ and may give a sharper picture for the graphical assessment of connectivity relations.

From a statistical point of view the proposed method is novel and different in the following respect. For estimating the intensity of the target, given the time since the last trigger, we consider times between the first trigger after a target firing and the following target firing. If the trigger fires again before its previous firing could show an effect on the target, we consider this effect of the first trigger firing as no longer observable from the time of the second firing on. However, that the event has not taken place during the observed time of two successive trigger firings needs to be taken into account in the estimation since it is vital information. This approach, familiar from problems of estimating life distributions from censored observations, leads to superior estimates. Even in most recent studies on firing time distributions, neural spike train analysts do not seem to have recognized this situation.

In summary we recommend the use of nonparametric hazard estimators as a new tool in the analysis of neuronal spike train data and, as the examples illustrate, an alternative to traditional cross-correlation analysis for the detection of neuronal interactions.

A study of the mathematical foundations of this method has been carried out by the author, see Utikal (1994b), under the assumption that the counting processes are a Markovian interval process. Extensions to include other covariate information as well as generalizations to more complex networks are currently investigated using semiparametric methods, see Utikal (1994a), (1995).

2   Results

We start by presenting the results of analyzing simulated neuronal firing activities using the CCH and using nonparametric hazard estimators. In the first two examples we consider the following pair of neurons N_1, N_2. N_2 fires independently of N_1 as homogeneous Poisson process of rate 1. The intensity (rate) of N_1 changes accordingly to the time passed since the last firing of N_2 before the next firing of N_1.

In Example 1 this rate remains constant 1.0 for 0.5 units after a firing of N_2, then jumps to a higher level for a duration of 0.1 units of time then it falls back to its previous level thereafter.

In Example 2 the rate of N_1 remains constant 1.0 for 1.0 unit after a firing of N_2, then drops to zero for a duration of 0.4 units of time and thereafter jumps back to its previous level. After a firing of N_1 in both examples the process is standard Poisson until the next firing of either N_1 or N_2. For an algorithmic description of the processes we refer to the appendix.

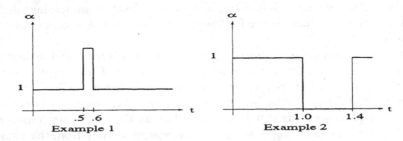

Figure 2.1: Intensity of N_1 after a firing of N_2 before the next firing of either N_1 or N_2.

Example 1: A small number of 40 trigger events was generated from the pair (N_1, N_2) where N_2 excites N_1 after a time of 0.5 units. The CCH does not lead to any conclusive evidence as to the nature of this interaction or to the exact time when it takes place. A plot of the cumulative hazard estimate of the time to the next firing of N_1 after a firing of N_2 before another firing of N_2 was generated next, using the LIFETEST procedure of SAS. A clear jump is recognizable at the correct time 0.5. From kernel function smoothing this estimate we obtain an even clearer picture of the nature and time of this interaction. Using a triangular kernel of bin width 0.15 we obtain the plot of a curve that depicts a maximum at the correct time of 0.5.

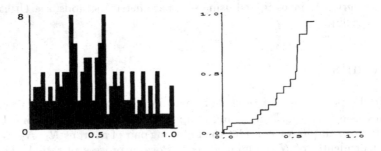

Figure 2.2: CCH [left] and cumulative intensity estimate $\hat{A}$ [right].

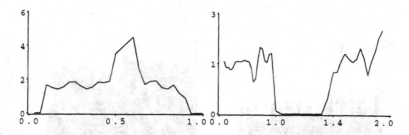

Figure 2.3: hazard rate estimate $\hat{\alpha}$ for Example 1 [left] and Example 2 [right].

Example 2: A sample of 470 triggers was generated from the pair (N_1, N_2) where N_2 inhibits N_1 after a time of 1.0 and for a duration of 0.4 units. The CCH does not show any conclusive evidence of this interaction or its duration. Using the same procedure as in the previous example the cumulative hazard estimate shows very clearly a flatness over the region from 1.0 to 1.4 units. An even clearer picture of this interaction is obtained from the plot of the hazard rate estimate. Over the range of time $(1.0, 1.4)$ there is zero firing activity observed. In this estimate a bin width of .10 was used.

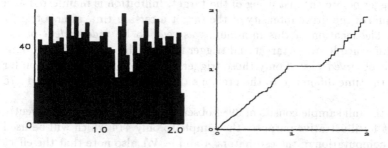

Figure 2.4: CCH [left] and cumulative intensity estimate $\hat{A}$ [right].

In the following we show how increasing the sample might or might not improve the analysis based on the CCH. After generating ten times as many observations from the networks described in the previous two examples we compute the CCH from the enlarged samples. As can be seen from Figure 2.6 [left] the CCH clearly reveals a spike at $\approx .50$, from which excitation fo N_1 by N_2 is inferred. On the other hand, increasing the sample tenfold from the second network we see that the CCH is unable to reveal very clearly the inhibition effects of N_2 on N_1 which sets in at 1.0 and extends to 1.4. This can be explained by the presence of higher order correlation terms which spill into the trough and cover it almost completely.

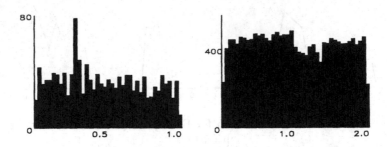

Figure 2.5: CCH after tenfold increase of sample size, Example 1 [left] and Example 2 [right].

Example 3: We apply the method illustrated in the previous examples on a set of data obtained by Lindsey et al. (1992) who recorded firing times of a pair of neurons located in the midline brain stem of a cat.

Different aspects of this data have previously been analyzed using more elaborate Mip-methods, see Utikal (1994a,c). One of the main findings was that the firing intensity of the target depends not only on the time since the last firing of the trigger but also on the time since the last firing of the trigger before the last firing of the target. Inhibition is manifested as a minimum of the firing intensity of the target a certain time after a trigger firing. The location of this minimum was seen to be *independent* of the time difference between target and trigger firing within certain limits. This effect is observed best if only those triggers are used in the estimation for which the time difference to the previous target firing does not exceed 175 ms.

While the full sample consits of 3984 observed trigger events, we presently restrict for illustrative purposes the sample to only 414 which will be used in the computation of the estimators $\hat{A}$ and $\hat{\alpha}$. We also note that the effect is observed to change its location depending on where this sample is taken; this suggest transience of the underlying network. From that sample of 414 trigger events we selected those whose time difference to the previous firing did not exceed 175 ms. Enlarging this range, it can be seen that long times between firings of the target unduly increase the probability of firing and distort the minimum caused by the inhibiting trigger firing which can be observed for moderate times, as is demonstrated in Utikal(1994c). The number of trigger events is therefore reduced to 197. In the estimation of $\hat{\alpha}$ a triangular kernel of 40 ms bin width was used. As in the previous examples we observe a sharper estimate of the effect when using the Mip-method as compared to the CCH, see figures 2.6, 2.7.

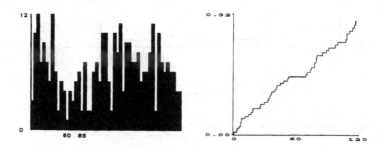

Figure 2.6: CCH [left] and cumulative intensity estimate $\hat{A}$ [right].

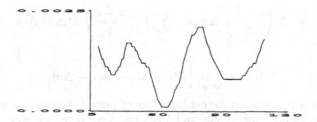

Figure 2.7: hazard rate estimate $\hat{\alpha}$.

3   Estimators

We briefly present the relevant formulas for the hazard estimator $\hat{\alpha}$ and the cumulative hazard estimator $\hat{A}$ computed and graphed in the previous section. For a study of the statistical properties of the estimators, i.e. consistency etc. we refer to Utikal(1994b).

Denote the jump times of N_1 by $\tau_1^{(1)}$, $\tau_2^{(1)}$, ... Setting the clock to zero at $\tau_i^{(1)}$ we observe the time T_i until the next jump of either N_1 or N_2. If this jump comes from N_1 we set $\delta_i = 1$ and $\delta_i = 0$ otherwise. In other words we consider T_i to be censored or uncensored according to if N_2 or N_1 fires next. It is therefore natural to estimate the cumulative hazard A of T using methods from survival analysis, see for example Andersen et al (1993).

Relabeling the observations (T_i, δ_i) for $i = 1, 2, \ldots, \nu$ by $T_1, T_2, \ldots$ and ordering them into $T_{(1)} < T_{(2)} < \ldots < T_{(\nu)}$ we compute the Nelson-Aalen estimator

$$\hat{A}(s) = \sum_{T_{(i)} \leq s} \frac{\delta_{(i)}}{\nu - i + 1}. \tag{1}$$

All throughout we use the convention $0/0 = 0$. We obtain the hazard rate estimator $\hat{\alpha}$ from this using the method of kernel function smoothing.

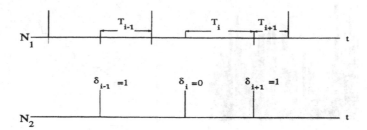

Figure 3.1: data for computing the estimators $\hat{A}$ and $\hat{\alpha}$.

$$\hat{\alpha}(s) = \sum_v K\left(\frac{s-v}{b}\right) \Delta\hat{A}(v) \Big/ \sum_v K\left(\frac{s-v}{b}\right) I\{\Delta\hat{A}(v) \neq 0\} \qquad (2)$$

where

$$\Delta\hat{A}(v) = \lim_{\Delta v \to 0} \left[\hat{A}(v + \Delta v) - \hat{A}(v - \Delta v)\right]$$

where K is a nonnegative kernel functions of bounded variation with compact support and integral one, and b is a band width parameter.

For completeness we briefly define next the CCH as used in the computations of Section 2. For a partition of the time axis into intervals of equal size with midpoints t_i, define $\mathrm{CCH}(N_1, N_2) = \sum_i \Delta N_1(t_i)\Delta N_2(t_{i+j})$, where $j = \ldots -2, -1, 0, 1, 2\ldots$ A maximum (minimum) of the CCH at $t_{i+j} - t_i$ indicates increased (decreased) firing probability and therefore suggests excitation (inhibition) of N_2 by N_1 for $j > 0$ and of N_1 by N_2 for $j < 0$.

Acknowledgement: This work would not have been started without Bruce Lindseys unconditional readiness to provide data and biological expertise.

4   Appendix: Simulations

We include a description of the networks constructed for generating the bivariate spike train series used in the analysis in Example 1 and Example 2. The programs are written in SAS and are available from the author. A series of firing times τ is generated togther with indicator *ind* showing the procedence of the firings.

Algorithm for simulating modulated Poisson process

```
tau(0)=0
next: generate x ~ exponential(rate = lambda1)
      generate y ~ exponential(rate = lambda2)
```

```
if (min(x,y) < delay) tau(i+1)=tau(i)+min(x,y)
                      if(x<y) ind(i+1)=target
                      if(x>y) ind(i+1)=trigger
                      i=i+1
                      goto next
if (min(x,y) > delay) tau(i+1)=tau(i)+delay
      generate x ~ exponential(rate=lambda1+hi)
      generate y ~ exponential(rate=lambda2)
      if (min(x,y) < duration) tau(i+1)=tau(i)+min(x,y)
         if(x<y) ind(i+1)=target
         if(x>y) ind(i+1)=trigger
         i=i+1
         goto next
      if (min(x,y) > duration) tau(i+1)=tau(i)+duration
         goto next
```

References

ANDERSEN, P.K., BORGAN, O., GILL, R.D., KEIDING, N. (1992). *Statistical Models Based on Counting Processes.* Springer, New York.

BRILLINGER, D.R. (1992). Nerve cell spike train data analysis: a progression of technique. *Journal Amer. Statist. Assoc.* **87**, 260-271.

EGGERMONT, J. (1990). *The Correlative Brain.* Springer-Verlag, Berlin.

LINDSEY, B.G., HERNANDEZ, Y.M., MORRIS, D.F., SHANNON, R. (1992). Functional connectivity between brain stem midline neurons with respiratory-modulated firing rates. *Journal Neurophysiology.* **67**, 890-904.

PERKEL, D.H., GERSTEIN, G.L., MOORE, G.P. (1967). Neuronal spike trains and stochastic point process. II. Simultaneous spike trains. *Biophys. J.*, **7**, 419-440.

UTIKAL, K.J. (1994a). Semiparametric inference on neural connectivity. Tech.Report #350, Dept. Statist. Univ. Kentucky.

UTIKAL, K.J. (1994b). Markovian interval processes I: nonparametric inference. Tech.Report #347 Dept. Statist. Univ. Kentucky.

UTIKAL, K.J. (1994c). Markovian interval processes II: applications to the analysis of neural spike trains. Tech.Report #348, Dept. Statist. Univ. Kentucky.

UTIKAL, K.J. (1995). A semiparametric model for biological neural networks. *to appear.*

On The Design Of Accelerated Life Testing Experiments

A. J. Watkins

ABSTRACT: This paper considers the design of constant stress accelerated life testing experiments, using a recently outlined framework based on
various classes of stress-parameter models. For brevity, we focus on the two
parameter Weibull distribution, and the use of likelihood-based estimation
techniques. We present formulae useful in various numerical methods, and
discuss the role of the Fisher information in the design of experiments.
We also present an illustrative example of experimental design, and briefly
discuss the validation of proposed designs by stochastic simulation.

KEYWORDS: Accelerated Life Test, Experimental Design; Fisher Information; Maximum Likelihood Estimation; Weibull distribution

1 Introduction

When items operate satisfactorily for long periods of time, the relatively
small number of failures occurring in any reasonable test period causes
difficulties in quantifying the long-term reliability. The idea behind accelerated life testing is to subject items to extreme stresses, so that items
under test have much shorter lives than usual; the results obtained from
experimental data are then extrapolated to provide information on potential performance at design stresses. For instance, electro-mechanical items
are usually affected to some extent by ambient operating temperature, so
that increases in temperature produce considerable reductions in the lives
of items. Thus, items tested in a controlled high-temperature environment
usually fail relatively quickly, and data on these times to failure may be
used to establish a link between temperature and life; this, in turn, provides
information about item performance at design temperatures.
Here, we consider the design of accelerated life testing experiments for a
recently outlined framework based on various classes of stress-parameter
modelling assumptions; see Nelson (1990) for some overall context for this
discussion. For brevity, we focus on the case of a single constant stress, and
the assumption of the two parameter Weibull distribution. Throughout,
we assume the use of likelihood-based estimation techniques, and present

formulae useful in various numerical methods. We consider the role of the Fisher information in the design of accelerated life testing experiments, present an illustrative example of experimental design, and outline the validation of proposed designs by stochastic simulation.

2   Data and Underlying Assumptions

We let X denote a single stress factor, and assume that each of N items is tested with X set at one of k distinct levels $x_1, ..., x_k$. The data for analysis comprises the information on the N_i items tested at $X = x_i$, for $i = 1, ..., k$, where n_i items fail (with observed times to failure $f_{ij}, j = 1, ..., n_i$) and $m_i = N_i - n_i$ are censored (with observed times in service $c_{ij}, j = 1, ..., m_i$). Although $k = N$ is theoretically possible, in practice $k \ll N$ is more usual; for instance, all examples in Hirose (1993) and Watkins (1994) have $2 \leq k \leq 5$ and $24 \leq N \leq 44$. In certain cases, the experiment will be concerned with the performance of items at a single *design* stress x_s; if x_s is also an experimental stress, then the test is said to be *partially* accelerated; see Bai and Chung (1992). In other cases, the experimental data may be used to consider a range of design stresses. Nelson (1990) gives examples of published data in the above format, with temperature the usual stress factor.

3   A Framework for Accelerated Life Testing Experiments

Watkins (1995) notes that the following assumptions allow a likelihood-based statistical analysis of the data generated in a constant stress accelerated life testing experiment:

- at any level of the stress factor, the distribution of lifetimes follows the same *type* of distribution,

- as stress changes, the parameters of the distribution change,

- the relationship between stress and parameters is smooth, possibly monotonic,

and that the Weibull distribution, with shape parameter β and scale parameter θ, is widely regarded as a sensible point for the analysis of reliability data; the second assumption is then $\beta \equiv \beta(X)$ and $\theta \equiv \theta(X)$. Various models for data may then be based on the following assumptions for β and θ:

- there is no connection between X and the parameter,

- there is a simple relationship between X and the parameter,

- there is a complex relationship between X and the parameter.

The third assumption leads to k parameters - one for each x_i - while the second assumption is based on relationships with known or identifiable functional form, characterised by $\leq k$ quantities and generally leading to an overall reduction in the number of model parameters; the first assumption leads to a single parameter. For β, it seems usual to make either the first or third assumption; for θ, the first assumption is of little interest, unless the experiment has been poorly designed. The assumption of a scale-stress relationship facilitates extrapolation to operating stresses, while analyses based on the third assumption may be used to assess the efficacy of proposed $\theta - X$ relationships. Various examples of this relationship in Nelson (1990) have the form

$$\theta = \theta_0 g(X; \phi_1, ..., \phi_m) = \exp(\alpha) g(X; \phi_1, ..., \phi_m), \qquad (1)$$

in which g is a function of growth parameters ϕ_i but is independent of the overall scale θ_0; in practice, it is sometimes more convenient to use the reparameterisation $\theta_0 = \exp(\alpha)$. Some relationships - such as the Arrhenius and Eyring - are based on empirical physical evidence, while others are selected for more pragmatic reasons, such as general plausibility and mathematical tractability. Watkins (1995) remarks that graphical methods for identifying g are typically based on rather sparse plots, and classifies constant stress accelerated life testing models as follows:

1. distinct shape and scale parameters for each experimental stress level; $2k$ parameters.

2. distinct shape parameter for each experimental stress level; functional relationship between scale and stress; $k + m + 1$ parameters.

3. constant shape over experimental stress levels; distinct scale parameter for each experimental stress level; $k + 1$ parameters.

4. constant shape over experimental stress levels; functional relationship between scale and stress; $m + 2$ parameters.

Models in the first class require k separate sub-analyses, and provide a basis for gauging the worth of models with fewer parameters. Models in the final class are generally the most parsimonious; if such models prove to be too inflexible, then models in the second or third class may be considered. Watkins (1995) presents the various stages in a typical analysis of data based on the above classification, gives formulae for log-likelihoods l_i ($i = 1, ..., 4$) for these models, and notes two natural hierarchies of analyses which arise in practice.

4 Maximum Likelihood Estimation

In general, the log-likelihoods l_i $(i = 1, ..., 4)$ - and, where appropriate, corresponding profile log-likelihoods l_i^* - for the above models must be maximised numerically. An important class of methods, including classical and quasi-Newton procedures, requires derivatives of the objective function; for (1), we now assume that g is differentiable with respect to the ϕ_i, and write

$$\frac{\partial g}{\partial \phi_r} = g_r, \frac{\partial(\log g)}{\partial \phi_r} = \frac{g_r}{g} = h_r;$$

we also need to assume the existence of second derivatives of g, but these do not appear explicitly in our discussion. Reference to the forms for l_i, l_i^* (Watkins, 1995) prompts the definitions

$$\Sigma_c^i(\beta_i) = \sum_{j=1}^{n_i} f_{ij}^{\beta_i}(\log f_{ij})^c + \sum_{j=1}^{m_i} c_{ij}^{\beta_i}(\log c_{ij})^c, \qquad (2)$$

and

$$S_{abc}^i(\beta_i) = \{g(x_i)\}^{-\beta_i - a}[-\log\{g(x_i)\}]^b \Sigma_c^i(\beta_i) \qquad (3)$$

for $i = 1, ..., k$; (2) is used in all four log-likelihoods, while (3) is used for models in classes 2 and 4. It is now possible to write all log-likelihoods and their derivatives in relatively compact forms; central formulae are

$$\frac{\partial \Sigma_c^i}{\partial \beta_i} = \Sigma_{c+1}^i, \qquad\qquad \frac{\partial S_{abc}^i}{\partial \beta_i} = S_{a,b,c+1}^i + S_{a,b+1,c}^i$$

and

$$\frac{\partial S_{abc}^i}{\partial \phi_r} = -g_r(x_i)\left(bS_{a+1,b-1,c}^i + (\beta_i + a)S_{a+1,b,c}^i\right).$$

These formulae show that, while the derivatives of likelihoods are not particularly elegant, they only involve summations, and hence are readily computed in $O(N)$ operations using subroutines or procedures in programming languages. The notations Σ_c^i, S_{abc}^i are useful in such implementations, extend those previously introduced in Watkins and Leech (1989) and Watkins (1991), and can be considerably simplified for certain choices of g.

5 Fisher Information Matrices

In addition to enabling various numerical procedures to be implemented, the Σ_c^i, S_{abc}^i and their expectations also appear in the Fisher information matrices corresponding to various log-likelihoods; they thus contribute to the assessment of precision in any set of maximum likelihood estimates, and the subsequent design of experiments based on that assessment. For each class of models above, with log-likelihood l_i, we may define a differential operator Δ_i, and then write the observed Fisher information matrix as $-\Delta_i'\Delta_i l_i$; the expected Fisher information is then $-E[\Delta_i'\Delta_i l_i]$. It is now sufficient to note the large sample arguments in Cox and Hinkley(1974) which imply that the observed information converges to its expectation, and that the inverse of this information - calculated at the true, unknown parameter values - is the variance-covariance matrix of the maximum likelihood estimators; this, in turn, yields an approximation to the variance of any function of these estimators. We also note that, although the observed information may only be calculated from experimental data, the expected information for a given test design may be calculated in advance.

6   Fisher Information Matrices For Class 4 Models

For brevity, we restrict attention here to the expected Fisher information matrices for the fourth class above, under the simplifying assumption that all items are observed to failure ($N_i = n_i$), for a general form for g in (1). For the final class of models, we have, using the second parameterisation in (1),

$$\Delta_4 = \left[\frac{\partial}{\partial\beta}, \frac{\partial}{\partial\alpha}, \frac{\partial}{\partial\phi_1}, ..., \frac{\partial}{\partial\phi_m}\right]$$

so that the expected Fisher information is, for $m = 1$,

$$\left[\begin{array}{cccc} N\beta^{-2}\left\{\frac{\pi^2}{6} + (\gamma - 1)^2\right\} & & & symmetry \\ N(\gamma - 1) & N\beta^2 & & \\ (\gamma - 1)\sum_{i=1}^k N_i h_1(x_i) & \beta^2\sum_{i=1}^k N_i h_1(x_i) & \beta^2\sum_{i=1}^k N_i h_1^2(x_i) \end{array}\right], \quad (4)$$

where $\gamma = 0.57721...$ is Euler's constant. The three entries in the last row of (4) generalise to

$$-E\left[\frac{\partial^2 l_4}{\partial\beta\partial\phi_j}\right] = (\gamma - 1)\sum_{i=1}^k N_i h_j(x_i),$$

$$-E\left[\frac{\partial^2 l_4}{\partial\alpha\partial\phi_j}\right] = \beta^2\sum_{i=1}^k N_i h_j(x_i)$$

and

$$-E\left[\frac{\partial^2 l_4}{\partial\phi_j\partial\phi_k}\right] = \beta^2\sum_{i=1}^k N_i h_j(x_i)h_k(x_i);$$

these formulae enable the general form for the expected Fisher information matrix to be written down, and its inverse may then be considered. Although some general analytical remarks on entries in the inverse are now possible - for instance, the inverse of (4) will *always* have 0 in the final place of its first row and column, which implies that the maximum likelihood estimators of the shape parameter β and growth parameter ϕ_1 are asymptotically independent - in practice, we seek first to exploit any structure in the selected form for g. For instance, the log-linear relationship - as used in Watkins (1991,1995) - is $g(X;\phi_1) = \exp(\phi_1 X)$, so that $h_1(X;\phi_1) = X$, and the last row of (4) is then

$$\left[(\gamma - 1) \sum_{i=1}^{k} N_i x_i \quad \beta^2 \sum_{i=1}^{k} N_i x_i \quad \beta^2 \sum_{i=1}^{k} N_i x_i^2 \right].$$

The interested reader may note that the derivation of the corresponding matrices for other cases is based on similar calculations, although considerably more manipulations are required to deal with censoring.

7   Design of Accelerated Life Testing Experiments

In practice, we may wish to design experiments which yield accurate information on particular parameters, such as the ϕ_i in (1). In general, we note that strategic choices of experimental design are often constrained by laboratory logistics, but may include

- the number of stress levels k, and the choice of stress levels $x_1, ..., x_k$,

- the number of items N_i allocated for testing at x_i,

- the time to (simple or overall) Type I censoring,

- the times to (complex or staggered) Type I censoring, and

- proportions of failures under Type II censoring.

We focus here only on the first choice, and suppose that a total of $N = 60$ items are available for test, with a stress factor X of temperature ($^\circ C$). We further assume that the log-linear stress-scale relationship is appropriate; from the inverse of (4), the theoretical variance of the maximum likelihood estimator of β is $6N^{-1}\pi^{-2}\beta^2$, which is independent of k, x_i, but the corresponding quantities for the other parameters are

- overall scale α
$$\frac{6(1 - \gamma)^2}{N\pi^2\beta^2} + \frac{s_2}{\beta^2 (Ns_2 - s_1^2)},$$

  where $s_j = \sum_{i=1}^{k} N_i x_i^j$; the first term is independent of k, x_i, but the form of the second term implies that a large *coefficient of variation* in the x_i will lead to more accurate estimation of α, and

- growth ϕ_1

$$\frac{N}{\beta^2 \left(N s_2 - s_1^2\right)};$$

this implies that large *variation* in the x_i will lead to more accurate estimation of ϕ_1.

This example is somewhat unusual, since some analytical progress is possible; a more typical use of the above information is exemplified in the following calculations, in which we first investigate the effect of increasing k, with equal N_i and equi-spaced x_i over the test range $X \in [200, 260]$, for fixed model parameters $\beta = 2, \alpha = 15, \phi_1 = -0.05$. This investigation is summarised in Table 1, where we give the theoretical standard deviations for scale and growth parameters; these calculations may be validated by stochastic simulation, and we also give the observed standard deviations of the estimators based on 10000 replications of each experiment. Overall agreement is good, and we see that it is not necessarily better - from a statistical viewpoint, at least - to have a large number of experimental levels. Relaxing the constraints in the above illustrative example, it is straightfor-

k		2	3	4	5	6
α	theory	0.49953	0.60990	0.66742	0.70316	0.72761
	practice	0.50562	0.61190	0.67921	0.71486	0.73590
ϕ_1	theory	0.00215	0.00264	0.00289	0.00304	0.00315
	practice	0.00218	0.00264	0.00295	0.00309	0.00318

TABLE 1. Theoretical and observed standard deviations for α and ϕ_1 for various k with $x_1 = 200, x_k = 260$ and the other x_i uniformly spread over $[200, 260]$.

ward to obtain the corresponding theoretical results for other cases; these results, which will be reported in detail elsewhere, show, for instance, that it is not necessarily optimal - again, purely from a statistical viewpoint - to have equally spaced x_i.

8   Discussion

We first remark that, although differences in theoretical standard deviations are often small, accurate estimation of parameters - particularly *growth* parameters, such as ϕ_1 above - is important, since the precise role of ϕ_i in (1) often means that small changes in these quantities lead to significant changes in θ, and hence in subsequent calculations. In practice, the statistical contribution to the design of constant stress accelerated life testing experiments should comprise an initial examination of all variable factors under experimental control, and an interactive - for example, *graphical* - study of the effect of varying these factors. The efficacy and robustness of the proposed experimental design may then be verified by stochastic simulation. This approach is possible for all common experimental designs, although reasonably sophisticated numerical tools become increasingly important for non-trivial designs.

The above framework accommodates much of the recent research on the design of constant stress accelerated life testing experiments, and the subsequent analysis of data, in a unified manner. Furthermore, it is sufficiently flexible to accommodate further refinements to outlined approach; for example, the relationships (1) between stress factor X and the scale parameter θ can be generalised - with two stress factors X, Y, we may write $\theta = \exp(\alpha)g(X, Y; \phi_1, ..., \phi_m)$ - and the above analyses extended as appropriate.

References

Bai, D.S., Chung, S.W. (1992). Optimal Design of Partially Accelerated Life Tests for the Exponential Distribution under Type-I Censoring. *IEEE Transactions on Reliability*, **41**, 400-406.

Cox, D.R., Hinkley, D.V. (1974). *Theoretical Statistics*. Chapman & Hall, New York.

Hirose, H. (1993). Estimation of Threshold Stress in Accelerated Life Testing. *IEEE Transactions on Reliability*, **42**, 650-657.

Nelson, W. (1990). *Accelerated Testing: Statistical Models, Data Analysis and Test Plans*. John Wiley & Sons, New York.

Watkins, A.J. (1991). On the Analysis of Accelerated Life-testing Experiments. *IEEE Transactions on Reliability*, **40**, 98-101.

Watkins, A.J. (1994). Likelihood Method For Fitting Weibull Log-Linear Models To Accelerated Life-Test Data. *IEEE Transactions on Reliability*, **43**, 361-365.

Watkins, A.J. (1995). Frameworks For Accelerated Life Testing Experiments. In: Pham, H. (ed). *Proceedings of the Second ISSAT Conference on Reliability and Quality in Design*, Orlando, Florida.

Watkins, A.J., Leech, D.J. (1989). Towards Automatic Assessment of Reliability for Data from a Weibull Distribution. *Reliability Engineering and System Safety*, **24**, 343-350.

Splitting Criteria in Survival Trees

Heping Zhang

ABSTRACT: A new splitting criterion is explored for the tree-based method of analysis of censored data. This criterion is an extension of those used in Classification and Regression Trees (CART). It can also be applied to analyze data with multiple responses of mixed types that arise frequently in practice. We use a simple simulation experiment to compare various splitting criteria and propose a performance score to measure the capability of the splitting criteria for discovering the data structure.

KEYWORDS: Tree-Based Methods; Survival Function; Log-Rank Test; Impurity; Kaplan-Meier Curves

1 Introduction

Regression analyses of censored data offer important tools for numerous real world problems. The proportional hazard model (Cox, 1972) has been widely used for such purposes. However, it has been recognized that the proportional hazard model does not always provide adequate analyses of practical problems because of its restrictive assumptions. In recent years, there have been considerable research activities to eliminate some of the restrictive assumptions (Gordon and Olshen, 1985; Segal, 1988; Davis and Anderson, 1989; LeBlanc and Crowley, 1993).

The present paper focuses on the regression tree methodology to the analysis of censored data. This methodology is based on a recursive partitioning technique introduced by Morgan and Songuist (1963). Also see Breiman et al. (1984) for general information on classification and regression trees (CART). Gordon and Olshen (1985) first extended the tree-based method to analyze censored data. Different proposals have been made since then, e.g., Segal (1988) and Davis and Anderson (1989).

Large sample theory and simulations are available to support those existing proposals, but their performance has not been explored when compared with each other. The main purpose of this paper is to shed light on this important issue with the use of a simple experiment and a simple performance score. In addition, we introduce a new splitting criterion that shows overall better performance than the existing criteria based on our simulation. Our new criterion is a straightforward extension from the ones used by Breiman *et al.* (1984) who deal with a response without censoring. The simplicity of the criterion allows computational efficiency. Moreover, this criterion may be further extended to analyze data with multiple responses that may contain both qualitative and quantitative variables.

2   Tree-Based Methods in Censored Data

The tree-based methods involve two major steps: growing a large tree and pruning to a final tree. The first step is done by the recursive partitioning technique and the second step is done with sample reuse methods or *ad hoc* testing procedures. The present study concentrates on the first step only and investigates whether a splitting criterion is capable to reveal a latent structure.

Suppose that a response y is observed together with a battery of explanatory variables $x_1, \cdots, x_p$ for N subjects. The problem of interest is to infer the relationship between a set of explanatory variables and the response variable. The tree-based method is useful to explore such a relationship, complementary to other more traditional statistical methods.

The idea of recursively partitioning is to divide the study population based on $x_1, \cdots, x_p$ and to form a number of homogeneous subpopulations. The homogeneity is measured by the distribution of the response in each subpopulation, which is also called a node. Following Breiman *et al.* (1984), we use the impurity measure, $i(t)$, of a node t, as opposed to the homogeneity. Unfortunately, with censored data the underlying survival time is not completely observed. The observed response is a pair of observed time, y, and censoring indicator, δ. $\delta = 1$ or 0 depending on whether y is censored or not. The impurity used by Breiman *et al.* is not directly applicable. One solution is to define an analogue using Kaplan-Meier curves. Other approaches are also proposed in the literature. Instead of minimizing the impurity of the daughter nodes, they tend to separate the Kaplan-Meier curves of the two daughter nodes.

In this article, we also explore a new impurity that does not make the use of Kaplan-Meier curves. It is based on the observation that an ideally homogeneous node should consist of subjects whose observed times are close and who are mostly censored or mostly uncensored. Therefore, the impurity of a node should reflect the homogeneity in both observed time and the proportion of censoring. More precisely, we use the following simple impurity criterion

$$i(t) = w_1 i_y(t) + w_2 i_\delta(t), \tag{1}$$

where w_1 and w_2 are prespecified weights, and $i_y(t)$ and $i_\delta(t)$ are the impurities of node t for observed time and censoring, respectively. Here, we use

$$i_y(t) = \sum_{\text{node } t} (y_i - \bar{y}(t))^2 / \sum y_i^2, \tag{2}$$

where $\bar{y}(t)$ is the average of y in node t, and the sum in the numerator is over the subjects in node t, whereas the sum in the denominator is either over the subjects in node t or over the entire study population. The denominator is introduced to normalize the scaling. When the sum is over node t, it is called adaptive normalization. When the sum is over the entire sample, it is called global normalization. The impurity $i_\delta(t)$ is defined by

$$i_\delta(t) = -p_t \log(p_t) - (1 - p_t) \log(1 - p_t), \tag{3}$$

where p_t is the proportion of censoring in node t.

3   Simulation

In this section, we examine the performance of the splitting rules proposed by Davis and Anderson (1989), Gordon and Olshen (1985), and Segal (1988). Davis and Anderson use the likelihood of an exponential distribution that assumes a constant hazard rate in each daughter node. This is a convenient assumption, but in logic it has to be violated in the partitioning process. The resulting rule is refered to as LL (likelihood) in Table 3. Gordon and Olshen suggest to split a node based on minimized distances between Kaplan-Meier curves and point masses within each daughter node, while Segal uses the log-rank statistic as a measure of separation between the two daughter nodes. See the references for complete descriptions of those rules. The rules of using Kaplan-Meier distance and log-rank test are called KMD and LR rules in Table 3. In addition, we also investigate the effect of weights and scaling in the definition of impurity (1) on the structure of trees.

3.1   The Experiment Design

The design of our experiment follows that of Davis and Anderson (1989). The latent tree structure and the censoring procedure are the same as those of Davis and Anderson. They considered only the exponential survival distribution. We enhance the experiment by allowing Weibull and mixtures of exponential survival distributions.

Figure 1 displays the latent tree structure that is determined by four categorical variables: x_1 to x_4. x_1 takes a random value from 1 to 5, x_2, x_3, and x_4 are dichotomous variables with equal probability of being 0 and 1. To challenge the methods, we introduce four noise variables: x_1' to x_4'. Their values are generated in parallel to those of x_1 to x_4.

We consider three types of survival distributions: exponential, Weibull, and

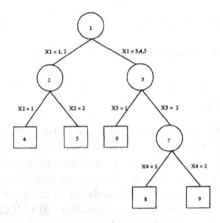

FIGURE 1. The latent Tree Structure for Simulations .

TABLE 1. Distributional Parameters For the Simulations

Terminal Node	Exponential	Weibull		Mixture Exponential	
Number	Hazard λ	α	λ	λ_1	λ_2
4	0.35	0.5	1.0	1.0	10.0
5	0.60	1.0	1.0	0.3	0.3
6	0.80	2.0	0.5	2.0	5.0
8	1.00	3.0	0.3	1.0	1.0
9	1.75	3.0	1.0	1.0	4.0

mixture exponential. Table 1 specifies the distributional parameters for the five terminal nodes in Figure 1.
Two types of censoring distributions (uniform in (2,4) and (0.25, 1.25)),

TABLE 2. Group Numbers of Simulation Data. The First Number in Parentheses Is the Sample Size and the Second the Censoring Percentage

Censoring Distributions	Survival Distributions		
	Exponential	Weibull	Mixture Exponential
Unif(2,4)	I(500, 15%)	V(500, 16%)	IX(500, 10%)
	II(250, 15%)	VI(250, 16%)	X(250, 10%)
Unif(0.25, 1.25)	III(500, 57%)	VII(500, 68%)	XI(500, 37%)
	IV(250, 57%)	VIII(250, 68%)	XII(250, 37%)

the same as those of Davis and Anderson (1989), are incorporated into our simulations. In addition, we used two kinds of sample sizes (500 and 250). Putting these choices together gives a total of 12 combinations of survival, censoring, and sample size. Each combination is called a data group in Table 2. Simulations are repeated 250 times in every data group.

Let us use group I to illustrate the experiment in detail. The steps are similar for other groups. This group has a sample size of 500, its survival distribution is exponential, and its censoring distribution is uniform in (2, 4). The hazard rates of survival for the five terminal nodes in Figure 1 are given in the second column of Table 1.

3.2 Data Generation

Each of the 250 data sets for group I is generated by the following 4 steps: (a) Generate 500 subjects. For each subject, we produce the values of the 4 covariates (x_1 to x_4) and the 4 noise covariates (x'_1 to x'_4) as described before. (b) Assign each of the 500 subjects into one of the 5 terminal nodes based on the generated values of x_1 to x_4. For example, a subject with $x_1 = 1$ and $x_2 = 1$ will be in terminal node 4. (c) After a subject is assigned to one terminal node, a random survival time is generated from the assumed survival distribution. (d) Randomly censor the survival time obtained in the previous step using uniform distribution in (2, 4) in the same way as Davis and Anderson (1989) did.

3.3 Performance Score of Splitting Rules

Every splitting rule, s, can now be applied to the 250 data sets in the group, resulting in 250 recovered trees, T_i, ($i = 1, \cdots, 250$). The performance of a splitting rule is measured by the similarity between the 250 recovered trees and the latent tree. First, let us describe how to measure the similarity between a tree T_i and the latent tree. Match tree T_i with the latent tree from the top to the bottom. We call a node of T_i matched if all of its ancestor nodes are split exactly as in the latent tree. Then, the performance score, denoted by $g(s, T_i)$, is the number of the matched nodes. The average, $\bar{g}(s)$, of $g(s, T_i)$ for $i = 1, \cdots, 250$ will be used as a measure of similarity.

3.4   Simulation Results

The impurity (1) depends on a pair of prespecified weights and the choice of normalization. After considering three pairs of weights 1:2, 1:1, and 2:1 for each of the global and adaptive normalizations, we find that the adaptive normalization appears to be better than the global one, and the performance of adaptive normalization with weight 1:2 is reliable overall, although it does not have advantages over the others for Weibull survivals. This splitting rule will be referred to as an AN (adaptive normalization) rule.

The global normalization is slightly easier to implement than the adaptive normalization and retains the subadditivity (Breiman *et al.*, Proposition 2.14, 1984). So, the global normalization is still worth considering. The choice of equal weights 1:1 is generally better than the other two. The resulting rule will be called GN (global normaliztion) rule.

Experiment 1. Exponential survival function.

We consider the exponential survival function

$$\exp(-\lambda t),$$

in which λ is the hazard rate. Note from Table 1 that the hazard rates in terminal nodes 4, 5, 6, 8, and 9 are in an increasing order. It is not surprising that all rules work very well for these data, except that the KMD rule falls a little behind others; see Groups I to IV in Table 3. Although motivated from the exponential survival distribution, the LL rule does not have an advantage over LR, AN, and GN rules. Moreover, we also examined the complete tree structures produced by LR and AN rules and found that most of them are practically the same, especially for the top 20 nodes.

Experiment 2. Weibull distribution.

Weibull survival function

$$\exp\{-(\lambda t)^{\alpha}\},$$

is chosen as the next simplest survival distribution. Four out of 5 pairs of α and λ given in Table 1 are from Miller (1981, page 13). All rules have some difficulty in recovering the complete latent structure, perhaps due to the non-monotonicity of the hazards among the nodes; see Groups V to VIII in Table 3. For this experiment, LL and LR rules are less satisfactory than KMD, AN, and GN rules.

Experiment 3. Mixtures of exponential survivals.

TABLE 3. Performance Scores of Splitting Rules. Inside Parentheses Are S.E.s of the Scores. LL=Likelihood; KMD=Kaplan-Meier Distance; LR=Log-Rank; AN=Adpative Normalization; GN=Global Normalization

			Splitting Rules		
Group	LL	KMD	LR	AN	GN
I	8.07(.10)	7.41(.14)	8.02(.10)	8.07(.10)	7.76(.10)
II	6.62(.13)	5.42(.16)	6.54(.14)	6.55(.14)	6.22(.13)
III	6.58(.14)	5.69(.17)	6.55(.15)	6.42(.14)	6.40(.14)
IV	4.60(.16)	3.84(.16)	4.39(.16)	4.54(.16)	4.38(.17)
V	3.00(.09)	4.09(.08)	2.62(.08)	3.54(.10)	4.05(.08)
VI	2.57(.08)	3.13(.09)	2.32(.07)	3.01(.10)	3.30(.09)
VII	3.66(.06)	3.85(.07)	3.71(.06)	3.74(.06)	3.72(.06)
VIII	3.38(.05)	3.44(.07)	3.37(.06)	3.46(.06)	3.42(.05)
IX	7.08(.19)	6.88(.19)	5.78(.23)	7.51(.16)	7.06(.15)
X	5.87(.20)	5.22(.21)	4.94(.22)	6.37(.18)	5.95(.18)
XI	4.42(.24)	4.42(.24)	3.96(.24)	4.61(.25)	4.19(.24)
XII	3.87(.22)	3.52(.21)	3.57(.22)	4.00(.22)	3.75(.22)

The mixtures of exponential distributions are useful, but difficult to handle in many applications. For example, the assumption for the proportional hazard model would be inappropriate for such distributions. After a conversation with Professor Richard Olshen, we consider the following mixtures

$$0.5\exp(-\lambda_1 t) + 0.5\exp(-\lambda_2 t),$$

for which 5 pairs of λ_1 and λ_2 are given in Table 1. Because we deliberately mixed up the order of the hazard rates among the 5 terminal nodes of the latent tree, it caused some problems for all rules to find the first correct split. If the first split is correct, it is relatively easy for all rules to pick up more correct splits. The AN rule recovered more nodes than anyone else. The LL and GN rules did reasonably well, the LR rule did a fine job, but the KMD rule was not so impressive.

These experiments suggest that AN rule offers the most satisfactory results in recovering the latent tree structures. The order of performance for the other rules are GN, LL, KMD, and LR rules. It is interesting to point out that LL, AN, and GN rules are much more convenient to compute than KMD and LR rules. Our experiments do not indicate that the performance score is proportional to the comptutational complexity. Although we have simulated a large number of data sets for a typical design of clinical trials, and over a common range of survivals, they are still restricted and preliminary. Further theoretical and simulation studies are still warranted.

4   Discussion

Many practical problems warrant analyses of data with multiple responses of mixed (continuous and categorical) types in connection with a number of covariates, while the existing statistical methods generally apply to data with either multivariate normal dependent variables or multiple binary outcomes. The flexibility of the tree-based method allows us to construct homogeneous groups even though both the covariates and the responses involve mixed types. To deal with this general problem, a key step is to define an impurity for a node. Although it may not be ideal, (1) seems to be still a reasonable choice based on the empirical results for analyzing censored data, although $i_y(t)$ and $i_\delta(t)$ are now the measures of impurity attributed by the continuous and categorical outcomes, respectively. The procedure will be described in detail elsewhere.

In this paper, we have emphasized the splitting rule that affects the growing of a large tree. We have not discussed the pruning procedure. Because the tree growing and pruning are in fact two independent processes, all existing pruning procedures are obviously applicable to any large tree that is grown by any of the splitting rules discussed in this paper. In addition to the existing splitting rules, we introduced a new set of splitting rules. They are simple and promising based on our simulation and some examples. These methods have been applied to data from the Western Collaborative Group Study and the results are reported elsewhere.

Acknowledgments: The work is supported in part by grant HD30712 from NIH. The author wishes to thank Drs. Richard Olshen, Dorit Carmelli, and Mark Segal for their comments.

References

Breiman, L., Friedman, J.H., Olshen, R.A. and Stone, C.J. (1984). *Classification and Regression Trees.* Wadsworth, California

Cox, D.R. (1972). Regression models and life tables (with discussion). *Journal of Royal Statistical Society B,* **34**, 187-220

Davis, R. and Anderson, J. (1989). Exponential Survival Trees. *Statistics in Medicine,* **8**, 947-962

Gordon, L. and Olshen, R.A. (1985). Tree-Structured Survival Analysis. *Cancer Treatment Reports,* **69**, 1065-1069

LeBlanc M. and Crowley J. (1993). Survival Trees by Goodness of Split. *Journal of the American Statistical Association,* **88**, 457-467

Miller, R. G. (1981). *Survival Analysis.* Wiley, New York.

Morgan, J. and Sonquist, J. (1963). Problems in the Analysis of Survey Data and a Proposal. *Journal of the American Statistical Association,* **58**, 415-434

Segal, M.R. (1988). Regression Trees for Censored Data. *Biometrics,* **44**, 35-48

The Different Parameterizations of the GEE1 and the GEE2

Andreas Ziegler

ABSTRACT: The purpose of this paper is to give a systematic presentation of the various Generalized Estimating Equation (GEE) approaches. They can be derived by using the Pseudo Maximum Likelihood (PML) approach which has been extensively discussed by Gourieroux and Monfort (1993). Furthermore, it is shown, that the Generalized Method of Moments (Hansen, 1982) can be applied to obtain estimators which are asymptotically equivalent to the GEE estimators.

KEYWORDS: Generalized Estimating Equations, Pseudo Maximum Likelihood Estimation, Generalized Method of Moments

1 Introduction

The independence of sample units is a key assumption of the classical generalized linear model. This assumption cannot be maintained if a survey is replicated at different time points or if clusters are sampled. For that reason Liang and Zeger (1986) proposed an approach for the regression analysis of correlated data that has been discussed intensely in recent years. It is based on generalized linear models or quasi–likelihood models and has primarily been used in biometrics.

The approach of Liang and Zeger is termed Generalized Estimating Equations (GEE) and has been extended in several ways. The Estimating Equations can be derived by using Pseudo Maximum Likelihood estimation (PML; Gourieroux and Monfort, 1993). Estimation of mean structures is based on the linear exponential family (PML1), whereas estimation of mean– and association parameters is based on the quadratic exponential family (PML2). For the connection of the GEE and the PML1 approach the reader is referenced to Ziegler (1994). In this paper only PML2 is considered.

2   The Quadratic Exponential Family

In this section two different formulations of the quadratic exponential family are presented.

Following Gourieroux and Monfort (1993) the density of the T dimensional quadratic exponential family with mean μ and variance Σ is given by

$$f(y, \mu, \Sigma) = \exp\left\{ c(\mu, \Sigma)' y + a(\mu, \Sigma) + b(y) + y' D(\mu, \Sigma) y \right\} \qquad (1)$$

where y and $\mu \in \mathbb{R}^T$, Σ is a positive definit $T \times T$ matrix, a and b are scalar functions, $c \sim T \times 1$ and $D \sim T \times T$. Note, that the exponent is quadratic in y with coefficient matrix $D(\mu, \Sigma)$. All moments greater than two are omitted.

Using a reparameterization several properties of the linear exponential family can be transferred to the quadratic exponential family:

$$y' D(\mu, \Sigma) y = \sum_{i=1}^{T} y_i^2 [D(\mu, \Sigma)]_{ii} + 2 \sum_{i>j} y_i y_j [D(\mu, \Sigma)]_{ij} = \lambda' w \qquad (2)$$

where $\lambda = ([D]_{11}, 2[D]_{12}, \ldots, [D]_{TT})'$ and $w = (y_1^2, y_1 y_2, \ldots, y_T^2)'$. For simplicity, let $\nu = (\nu_{11}, \nu_{12}, \ldots, \nu_{TT})'$ with $\nu_{ij} = \sigma_{ij} + \mu_i \mu_j$, $\sigma_{ij} = \mathbb{C}ov(y_i, y_j)$, $c(\mu, \Sigma) = \vartheta$ and $d(\mu, \nu) = \lambda$. Then, equation (1) can be rewritten as

$$f(y, \mu, \nu) = \exp\left\{ c(\mu, \nu)' y + a(\mu, \nu) + b(y) + d(\mu, \nu)' w \right\} \qquad (3)$$

If the parameter vectors ϑ and λ are collected to a single vector, the properties of the linear exponential family can be applied to the quadratic exponential family with vectorized variance structure. All moments exist, integration and differentiation w.r.t. $(\vartheta', \lambda')'$ can be exchanged. Hence, mean and variance can be derived by using differentiation and are given by:

$$\mathbb{E}([y', w']') = \begin{pmatrix} \mu \\ \nu \end{pmatrix}, \quad \mathbb{V}([y', w']') = \begin{pmatrix} \mathbb{V}(y) & \mathbb{C}ov(y, w) \\ \mathbb{C}ov(w, y) & \mathbb{V}(w) \end{pmatrix} \qquad (4)$$

The classical example for the quadratic exponential family is the T dimensional normal distribution. An often used discrete example is the joint distribution of dichotomous data.

3   Pseudo Maximum Likelihood Estimation Based on the Quadratic Exponential Family

In the following an independently and identically distributed (i.i.d.) sample (y_i, X_i) of $T \times 1$ stochastic vectors y_i of dependent variables and $T \times p$ matrices $X_i, i = 1, \ldots, n$ of deterministic and/or stochastic regressors is considered.

The observation y_i is decomposed into a systematic and an error term: $y_i = \mu(X_i, \beta_0) + \epsilon_i$ with $\mathbb{E}(\epsilon_i|X_i) = 0$. The conditional expectation of y_i, given X_i, is given by $\mathbb{E}_{f^*}(y_i|X_i) = \mu(X_i, \beta_0)$. The expectation is taken over the true, but unknown density $f^*(y_i|X_i)$ of y_i, given X_i. The mean structure is a function of a $p \times 1$ parameter vector β_0. In the following, only a linear connection between X_i and β_0 is considered: $\mu(X_i, \beta_0) = \mu(X_i\beta_0)$. The variance matrix of y_i, given X_i, is given by $\mathbb{V}_{f^*}(y_i|X_i) = \Omega_0(X_i, \beta_0, \alpha_0)$. It is a function of β_0 and an additional $q \times 1$ parameter vector α_0 and need to be correctly specified for simultaneous consistent estimation of β_0 and α_0. For simplicity, let $(\beta', \alpha') = \gamma'$. If X_i is stochastic, its density $m(X_i)$ need not be parameterized in γ_0. All required regularity conditions can be found in White (1982) and Gourieroux et al. (1984).

The idea of PML2 estimation is that an assumed density f from the multivariate quadratic exponential family is used instead of the true but unknown density f^*. To yield a consistent estimate of the parameter vector γ_0 the mean and the variance structures of the conditional assumed density $f(y_i|X_i)$ and the conditional true density $f^*(y_i|X_i)$ must be equal: $\mathbb{E}_f(y_i|X_i) = \mathbb{E}_{f^*}(y_i|X_i) = \mu(X_i\beta_0)$ and $\mathbb{V}_f(y_i|X_i) = \mathbb{V}_{f^*}(y_i|X_i) = \Omega(X_i, \alpha_0, \beta_0)$.

The estimating equations for γ_0 are derived by using the assumed, but possibly incorrect specified density f. The Maximum Likelihood (ML) technique is applied to the "pseudo density" f. Because of the i.i.d. assumption for (y_i, X_i) and the independence of $m(X_i)$ from γ one has to maximize

$$l(\gamma) = \frac{1}{n} \sum_{i=1}^{n} \left\{ c(\mu_i, \nu_i)'y + a(\mu_i, \nu_i) + d(\mu_i, \nu_i)'w_i \right\}, \qquad (5)$$

which is the kernel of the normed pseudo loglikelihood function. The maximum is found by derivating the normed pseudo loglikelihood function w.r.t γ. This leads to normal equations which incorporate the third and fourth moments of the assumed density f. Hence, let $\mathbb{V}_f([y_i', w_i']') = V_i$, which is called "working variance matrix" (Prentice and Zhao, 1991), since it is the variance matrix of the assumed or working density f. Computing the first derivatives of $l(\gamma)$ gives the normal equations (cf. Ziegler, 1994)

$$0 = \frac{1}{n} \sum_{i=1}^{n} M_i' V_i^{-1} \epsilon_i =: \frac{1}{n} M V^{-1} \epsilon, \qquad (6)$$

where M_i is the matrix of partial derivatives and ϵ_i is the vector of individual residuals:

$$M_i = \begin{pmatrix} \dfrac{\partial \mu_i}{\partial \beta'} & 0 \\ \dfrac{\partial \nu_i}{\partial \beta'} & \dfrac{\partial \nu_i}{\partial \alpha'} \end{pmatrix} \sim \frac{T(T+3)}{2} \times p + q, \qquad \epsilon_i = \begin{pmatrix} y_i - \mu_i \\ w_i - \nu_i \end{pmatrix} \qquad (7)$$

Correspondingly, $M = (M_1', \ldots, M_n')' \sim n\frac{T(T+3)}{2} \times p + q$, V is the block diagonal matrix of the working variance matrices V_i and $\epsilon = (\epsilon_1', \ldots, \epsilon_n')'$. Note, that the upper right block of M_i equals 0, since μ_i is independent of α.

A value $\hat{\gamma} = \hat{\gamma}_{PML} = (\hat{\beta}', \hat{\alpha}')'$, that maximizes the kernel of the normed pseudo loglikelihood function or equivalently solves the normal equations is called PML2 estimator for $\gamma_0 = (\beta_0', \alpha_0')'$.

Under suitable regularity conditions a unique, strongly consistent PML estimator $\hat{\gamma}$ exists, that is asymptotically normally distributed with mean γ and variance matrix that is given by

$$\left(M_0' V_0^{-1} M_0\right)^{-1} \left(M_0' V_0^{-1} \Omega_0 V_0^{-1} M_0\right) \left(M_0' V_0^{-1} M_0\right)^{-1}, \qquad (8)$$

where $M_0 = \partial(\mu_0', \nu_0')'/\partial\gamma'$, $V_0 = V(X, \gamma_0)$ and $\Omega_0 = \Omega(X, \gamma_0)$. A consistent estimator for the asymptotical variance is given by

$$(\hat{M}'\hat{V}^{-1}\hat{M})^{-1} \left(\frac{1}{n}\sum_{i=1}^{n} \hat{M}_i'\hat{V}_i^{-1} \hat{\epsilon}_i \hat{\epsilon}_i' \hat{V}_i^{-1} \hat{M}_i\right) (\hat{M}'\hat{V}^{-1}\hat{M})^{-1}. \qquad (9)$$

The estimating equations (6) are formulated in the second ordinary moments and were used by Liang et al. (1992). A second set of estimating equations that uses the second central moments instead of the second ordinary moments can be derived by applying an orthogonal transformation to equation (6). Let $s_i = (s_{i11}, s_{i12}, \ldots, s_{iTT})'$ and $\sigma_i = (\sigma_{i11}, \sigma_{i12}, \ldots, \sigma_{iTT})'$, where $\mathbb{C}ov(y_{it}, y_{it'}) = \sigma_{itt'}$ and $s_{itt'} = (y_{it} - \mu_{it})(y_{it'} - \mu_{it'})$. Analogous to the formulation using the second ordinary moments set $\tilde{M}_i = \partial(\mu_i, \sigma_i)/\partial\gamma'$, $\tilde{\epsilon}_i = ([y_i - \mu_i]', [s_i - \sigma_i]')'$ and $\tilde{V}_i$ correspondingly. Then the equation system (6) is equivalent to the set of normal equations given by

$$0 = \frac{1}{n}\sum_{i=1}^{n} \tilde{M}_i'\tilde{V}_i^{-1}\tilde{\epsilon}_i =: \frac{1}{n}\tilde{M}\tilde{V}^{-1}\tilde{\epsilon}. \qquad (10)$$

The equivalence of (6) and (10) is shown e.g. in Prentice and Zhao (1991) or Ziegler (1994).

It is impossible by using the PML approach to derive estimating equations that are equivalent to (6) or (10), which use the second standardized moments (Ziegler, 1994).

4   The Generalized Method of Moments

The Generalized Method of Moments (GMM; Hansen, 1982; Hall, 1993) is a different approach for the deduction of the most important properties of the estimator that is yielded by using the Generalized Estimating Equations. In this section the GMM is briefly described.

For the definition of the class of GMM estimators assume that moment conditions can be established and written as

$$\mathbb{E}[\psi(\boldsymbol{y}_i, \boldsymbol{X}_i, \boldsymbol{\beta}, \boldsymbol{\alpha})] = \mathbb{E}[\psi(\boldsymbol{y}_i, \boldsymbol{X}_i, \boldsymbol{\gamma})] = 0, \quad i = 1, \ldots, n, \qquad (11)$$

where $\psi \sim r \times 1$. Then, the GMM estimator is defined by solving the minimization problem

$$\hat{\boldsymbol{\gamma}} = \underset{\boldsymbol{\gamma}}{argmin} \left[\frac{1}{n} \sum_{i=1}^{n} \psi(\boldsymbol{y}_i, \boldsymbol{X}_i, \boldsymbol{\gamma}) \right]' \boldsymbol{W}_n \left[\frac{1}{n} \sum_{i=1}^{n} \psi(\boldsymbol{y}_i, \boldsymbol{X}_i, \boldsymbol{\gamma}) \right] \qquad (12)$$

with a weight matrix $\boldsymbol{W}_n > 0$ that converges a.s. against a deterministic regular matrix $\boldsymbol{W}$. The problem is to find the appropriate moments which define the function $\psi(\boldsymbol{y}_i, \boldsymbol{X}_i, \boldsymbol{\gamma})$. In this paper conditional moment restrictions are considered $\mathbb{E}(\epsilon_i | \boldsymbol{X}_i) = 0$, $i = 1, \ldots, n$, where ϵ_i is the ith residual vector. These conditional moment restrictions can be re-expressed as a set of unconditional moment restrictions $\mathbb{E}(\boldsymbol{C}(\boldsymbol{X}_i)\epsilon_i) = 0$, where $\boldsymbol{C}(\boldsymbol{X}_i) \sim r \times T$ (cf. Newey, 1993).

Under suitable regularity conditions (Gourieroux and Monfort, 1989, p. 339) the GMM estimator $\hat{\boldsymbol{\gamma}}$ is consistent and asymptotically normally distributed with mean $\boldsymbol{\gamma}_0$ and variance matrix which is given by

$$\left(\boldsymbol{\Psi}' \boldsymbol{W} \boldsymbol{\Psi} \right)^{-1} \left(\boldsymbol{\Psi}' \boldsymbol{W} \boldsymbol{Z} \boldsymbol{W} \boldsymbol{\Psi} \right)^{-1} \left(\boldsymbol{\Psi}' \boldsymbol{W} \boldsymbol{\Psi} \right)^{-1} \qquad (13)$$

where $\boldsymbol{\Psi} = \mathbb{E}(\boldsymbol{C}(\boldsymbol{X}) \cdot \partial \epsilon / \partial \boldsymbol{\gamma}')$, $\boldsymbol{Z} = \mathbb{E}(\mathbb{E}[\boldsymbol{C}(\boldsymbol{X})\boldsymbol{\Omega}(\boldsymbol{X})\boldsymbol{C}'(\boldsymbol{X})|\boldsymbol{X}])$ and $\boldsymbol{\Omega}(\boldsymbol{X}_i)$ $= \mathbb{E}(\epsilon_i \epsilon_i'|\boldsymbol{X}_i)$ (cf. Newey, 1993, p. 422). The asymptotic variance depends on the weight matrix $\boldsymbol{W}$ as well as on $\boldsymbol{C}(\boldsymbol{X})$. The optimal choice of $\boldsymbol{W}$ is $\boldsymbol{Z}^{-1}$, which minimizes the asymptotic variance (cf. Hansen, 1982). Hence, $\boldsymbol{W}$ should be chosen as "close" as possible to $\boldsymbol{Z}$. However, consistence and asymptotic normality do not depend on the specific choice of $\boldsymbol{W}_n$ (cf. Newey, 1993, p. 422).

In general, it will not be possible to specify $\boldsymbol{\Omega}$ correctly. Hence, $\boldsymbol{V}$ is used instead of $\boldsymbol{\Omega}$. In this case, $\boldsymbol{C} = \boldsymbol{M}'\boldsymbol{V}^{-1}$. Finally, if $\boldsymbol{W} = (\boldsymbol{M}'\boldsymbol{V}^{-1}\boldsymbol{M})^{-1}$, then the asymptotical variance matrix is equivalent to eq. (8), since $\mathbb{E}(\partial \epsilon / \partial \boldsymbol{\gamma}_0') = \boldsymbol{M}_0$ and hence $\boldsymbol{\Psi}$ and $\boldsymbol{W}^{-1}$ coincide. In practical applications, $\boldsymbol{\Psi}$ and $\boldsymbol{W}^{-1}$ have to be replaced by their estimates.

Summing up, it follows, that the GMM estimator $\hat{\boldsymbol{\gamma}}_{GMM}$ with instruments (cf. Newey, 1993) $\boldsymbol{M}_i'\boldsymbol{V}_i^{-1}$ and weight $(\boldsymbol{M}_i'\boldsymbol{V}_i\boldsymbol{M}_i)^{-1}$ is asymptotically equivalent to the PML estimator, which solves equation (6). Furthermore, the GMM estimator using instruments $\tilde{\boldsymbol{M}}_i'\tilde{\boldsymbol{V}}_i^{-1}$ and weight $(\tilde{\boldsymbol{M}}_i'\tilde{\boldsymbol{V}}_i\tilde{\boldsymbol{M}}_i)^{-1}$ is asymptotically equivalent to the PML estimator, that is obtained by solving (10).

An advantage of the class of GMM estimators might be the ability of formulating estimating equations, which use the second standardized moments. Let $\tilde{\boldsymbol{M}}_i = \partial(\boldsymbol{\mu}_i, \boldsymbol{\xi}_i) / \partial \boldsymbol{\gamma}'$, $\tilde{\boldsymbol{\epsilon}}_i = ([\boldsymbol{y}_i - \boldsymbol{\mu}_i]', [\boldsymbol{z}_i - \boldsymbol{\xi}_i]')'$ and $\tilde{\boldsymbol{V}}_i$ correspondingly, where $\boldsymbol{z}_i = (z_{i11}, z_{i12}, \ldots, z_{iTT})'$ with $z_{ijk} = [(y_{ij} - \mu_{ij})(y_{ik} -$

$\mu_{ik})/(\sigma_{ij}\sigma_{ik})]$, $\boldsymbol{\xi}_i$ is the corresponding vector consisting of $\xi_{ijk} = \mathbb{E}(z_{ijk})$, $j \leq k$. Then, a GMM estimator $\boldsymbol{\gamma}_{GMM}$ can be obtained from the instruments $\tilde{\boldsymbol{M}}_i' \tilde{\boldsymbol{V}}_i^{-1}$. A consistent estimate for the variance matrix is obtained by replacing $\tilde{\boldsymbol{M}}_0$ and $\tilde{\boldsymbol{V}}_0$ by their estimates and $\boldsymbol{\Omega}_0$ by using $\tilde{\boldsymbol{\epsilon}}_i \tilde{\boldsymbol{\epsilon}}_i'$.

5   Formulation and Parameterization of the Generalized Estimating Equations

In this section the stochastic model and different estimating equations for the GEE2 are formulated. Furthermore, the connection between the parameters of the quadratic exponential family and the regression and association parameters which have to be estimated are discussed in detail.

All estimating equations for the GEE2 are based on the following stochastic model: $\boldsymbol{y}_i = \boldsymbol{\mu}(\boldsymbol{X}_i\boldsymbol{\beta}_0) + \boldsymbol{\epsilon}_i$, with $\mu(\boldsymbol{x}_{it}'\boldsymbol{\beta}_0) = g(\eta_{it})$, $\eta_{it} = \boldsymbol{x}_{it}\boldsymbol{\beta}_0$ and $\boldsymbol{\Omega}(\boldsymbol{X}_i, \boldsymbol{\beta}_0, \boldsymbol{\alpha}_0)$ is correctly specified.

The simplest formulation of the GEE2 uses the multivariate normal distribution. Assume, $\boldsymbol{y}_i|\boldsymbol{X}_i \sim N(\boldsymbol{\mu}(\boldsymbol{X}_i\boldsymbol{\beta}), \boldsymbol{\Omega}(\boldsymbol{X}_i, \boldsymbol{\beta}, \boldsymbol{\alpha}))$. Then, the estimating equations of the GEE2 using the second central moments are given by equation (10). The advantage of this formulation is, that the third and fourth moments are determined by the second moments (cf. Anderson, 1984, p. 49). The estimation will be almost efficient, if the true distribution is similar to the assumed normal distribution. Furthermore, an implementation on a computer is easy, because the third moments are 0 and the fourth moments are determined by the second moments.

If dichotomous data are considered the estimation using a normal distribution will not be very efficient. Even though the GEE2 with assumed normal distribution can be applied to binary data. Note, that the moments σ_{itt} are included in the formulation of the GEE2 based on the normal distribution. If binary data are considered, the connection used in generalized linear models between mean and variance of y_{it} can be used. Hence, σ_{itt} need not be estimated separately and can be omitted in the formulation of the estimating equations (Ziegler, 1994).

For a complete formulation of the GEE2 the first four moments have to be specified. In the case of binary data the moments σ_{itt} are omitted. The mean is typically specified by using the logit or the probit link. The second central moments are usually specified by using Fisher's z, i.e. inverse hyperbolic (cf. Lipsitz et al. 1991)

$$\mathbb{C}orr(y_{it}, y_{it'}) = \frac{\exp[k(\boldsymbol{x}_{it}, \boldsymbol{x}_{it'})^T \boldsymbol{\alpha})] - 1}{\exp[k(\boldsymbol{x}_{it}, \boldsymbol{x}_{it'})^T \boldsymbol{\alpha})] + 1}, \tag{14}$$

and the property, that $\mathbb{C}ov(y_{it}, y_{it'}) = \sigma_{it}\sigma_{it'}\mathbb{C}orr(y_{it}, y_{it'})$, where $k(\boldsymbol{x}_{it}, \boldsymbol{x}_{it'})$ is any function, that correctly specifies the dependence between the correlation and $\boldsymbol{\alpha}$.

Since the third and fourth moments need not be correctly specified, Prentice and Zhao (1991) consider arbitrarily chosen quadratic exponential families and propose different working covariance matrices for the third and fourth moments. The easiest specification is to use $\mathbb{C}ov(\boldsymbol{y}_i, \boldsymbol{s}_i) = 0$ and $\mathbb{V}(\boldsymbol{s}_i) = \boldsymbol{I}$, which is called specification for practical applications.

A disadvantage of the central moments is, that the correlations of binary data are restricted, even if $\boldsymbol{y}_i$ is only bivariate. These restrictions can be very severe and can lead to inadmissible estimated correlation coefficients. Hence, Liang et al. (1992) proposed to choose the log Odds Ratio to specify the association. Their approach is based on the second ordinary moments, since it is possible to connect them to the Odds Ratio (cf. Bishop et al. 1975). Usually, the log Odds Ratio is modelled as a linear function of $\boldsymbol{\alpha}$, that is independent of $\boldsymbol{\beta}$. For this approach the set of normal equations given by (6) is used. Here, the interpretation of $\boldsymbol{\alpha}$ is straight forward. But, like the correlation coefficient, the Odds Ratio is bounded if $\boldsymbol{y}_i$ consists of more than two elements. An example is given in Liang et al. (1992). These restrictions lead to the same problems as the GEE2 approach based on the second central moments. Hence, Fitzmaurice and Laird (1993) propose a full likelihood approach in order to avoid these problems.

The asymptotic properties result only, if both the mean and the covariance structure are correctly specified. If $\boldsymbol{M}_i$ and $\boldsymbol{V}_i$ are block diagonal, $\boldsymbol{\beta}$ can be estimated consistently, even if $\boldsymbol{\alpha}$ has not been correctly specified. Prentice (1988) uses this property and formulates the estimating equations by using the second standardized moments. Liang and Zegers (1986) estimating equations are special cases of the Prentice estimating equations by using a working matrix with a lower right block that is equal to the identity matrix. Prentice' estimating equations can not be derived by using PML2 estimation. But it is possible to use the PML1 approach to derive the estimating equations Prentice proposed. The disadvantage of PML1 is, that the joint asymptotical normality of $(\hat{\boldsymbol{\beta}}', \hat{\boldsymbol{\alpha}}')'$ can not be established. Only $\hat{\boldsymbol{\beta}}$ is asymptotically normally distributed (cf. Ziegler, 1994). Since this approach provides consistent estimates of $\hat{\boldsymbol{\beta}}$, regardless of whether Ω has been correctly specified, these estimating equations are attractive, if emphasis is set on estimating $\boldsymbol{\beta}$.

6   Illustration

The GEE parameter estimation methods are illustrated by analyzing panel data from the GSOEP. The employment status of a cohort of 1246 men from 1985 to 1988 is considered as dependent variable over four waves. The codification is: 0 = employed, 1 = unemployed. A simple model for the disposition to become unemployed is treated.

The following variables are used as exogenous variables for waves 1 to

4: ALD = duration of unemployment between 1974 and 1984 in months, ALDSQ = ALD squared and divided by 100, ALH = frequency of unemployment between 1974 and 1984, ALT = age in years, ALTSQ = ALT squared and divided by 100, GZ = 1 if a person is severely handicapped and 0 otherwise, BB1 = 1 if a person has finished some professional education and 0 otherwise, BB2 = 1 if a person has a university degree and 0 otherwise, BST2 = 1 if a person is a white collar employee and 0 otherwise, FST (Family status) = 1 if a person is married and 0 otherwise. Effects like economic situation are taken into account by including a constant for every single wave. For illustration time independent parameters and time varying regression constants are considered. As link the logit is used. The first model is the GEE2 of Prentice and Zhao (correlations), where k is chosen to be independent of X_i. For this model the estimated correlation for y_{it} and $y_{it'}$ can be obtained by calculating: $(e^{\alpha_{tt'}} - 1)/(e^{\alpha_{tt'}} + 1)$. The second model is the GEE2 in the log Odds Ratio parameterization, where like in the first model the connection is established without using X_i. The third model is the GEE1 of Liang and Zeger with an unspecified working correlation matrix. There, no standard errors for the association parameters are calculated.

Variable	GEE2 Prentice/Zhao		GEE2 Liang/Zeger/Qaqish		GEE1 Liang/Zeger	
CONST (wave 1)	6.740	[2.415]	6.183	[2.402]	7.272	[2.286]
CONST (wave 2)	7.091	[2.423]	6.518	[2.409]	7.631	[2.289]
CONST (wave 3)	6.946	[2.429]	6.376	[2.414]	7.480	[2.295]
CONST (wave 4)	7.026	[2.425]	6.491	[2.405]	7.563	[2.293]
ALD	0.161	[0.021]	0.166	[0.021]	0.169	[0.021]
ALDSQ	-0.138	[0.040]	-0.148	[0.038]	-0.147	[0.040]
ALH	0.032	[0.068]	0.012	[0.067]	0.015	[0.066]
ALT	-0.518	[0.122]	-0.494	[0.121]	-0.550	[0.116]
ALTSQ	0.668	[0.152]	0.637	[0.151]	0.708	[0.145]
GZ	1.333	[0.316]	1.391	[0.318]	1.306	[0.321]
BB1	-0.358	[0.210]	-0.317	[0.211]	-0.373	[0.212]
BB2	-0.427	[0.491]	-0.376	[0.486]	-0.346	[0.485]
BST2	-0.766	[0.289]	-0.712	[0.291]	-0.772	[0.290]
FST	-0.391	[0.246]	-0.299	[0.252]	-0.351	[0.241]
α_{12}	1.298	[0.118]	3.701	[0.600]	0.328	
α_{13}	0.665	[0.100]	1.111	[0.530]	0.223	
α_{14}	0.612	[0.092]	1.570	[0.516]	0.150	
α_{23}	1.315	[0.118]	2.833	[0.460]	0.316	
α_{24}	1.071	[0.103]	2.509	[0.443]	0.224	
α_{34}	1.972	[0.175]	3.536	[0.493]	0.521	

Table 1. Probitmodel for unemployment status
(robust standard errors in brackets)

The variables duration of unemployment, its squared, age, age squared, being severely handicapped and being a white collar employee are significant at the 5% test level for all three models. The signs of the parameter coefficients are all in the expected direction. However, the regression parameter estimates differ only slightly. Note, the estimates of the association parameters have to be interpreted as values from Fisher's z, as values from the log Odds Ratio parameterization and as correlations.

Acknowledgments: The author thanks Christian Kastner for preparing the example.

References

Anderson, T.W. (1984). *An Introduction to Multivariate Statistical Analysis*, 2nd ed. New York: Wiley.

Bishop, Y.M.M., Fienberg, S.E. & Holland, P.W. (1975). *Discrete Multivariate Analysis: Theory and Practice.* Cambridge (Mass.): MIT Press.

Breitung, J. & Lechner, M. (1994). GMM–Estimation of Nonlinear Models on Panel Data. *Beiträge zur angewandten Wirtschaftsforschung, Discussion Paper 500-94.*

Fitzmaurice, G. & Laird, N. (1993). A Likelihood–Based Method for Analysing Longitudinal Binary Responses. *Biometrika*, 80, pp. 141–151.

Gourieroux, C. & Monfort, A. (1989). *Statistique et Modeles Econmetriques.* Paris: Economica.

Gourieroux, C. & Monfort, A. (1993). Pseudo–likelihood Methods. In Maddala, G., Rao, C. & Vinod, H., editors, *Handbook of Statistics, Vol. 11*, pp. 335–362. Amsterdam, Elvesier, 1993.

Gourieroux, C., Monfort, A. & Trognon, A. (1984). Pseudo Maximum Likelihood Methods: Theory. *Econometrica*, 52, pp. 682–700.

Hall, A. (1993). Some Aspects of Generalized Method of Moment Estimation. In Maddala, G., Rao, C. & Vinod, H., editors, *Handbook of Statistics, Vol. 11*, pp. 393–417. Amsterdam, Elvesier.

Hansen, L. (1982). Large Sample Properties of Generalized Methods of Moments Estimators. *Econometrica*, 50, pp. 1029–1055.

Liang, K.Y. & Zeger, S. (1986). Longitudinal Data Analysis Using Generalized Linear Models. *Biometrika*, 73, pp. 13–22.

Liang, K.Y., Zeger, S. & Qaqish, B. (1992). Multivariate Regression Analysis for Categorical Data. *Journal of the Royal Statistical Society, Series B*, 54, pp. 3–40.

Lipsitz, S., Laird, N. & Harrington, D. (1991). Generalized Estimating Equations for Correlated Binary Data: Using the Odds Ratio as a Measure of Association. *Biometrika*, 78, pp. 153–160.

Newey, W. (1993). Efficient Estimation of Models with Conditional Moment Restrictions. In Maddala, G., Rao, C. & Vinod, H., editors, *Handbook of Statistics, Vol. 11*, pp. 419–454. Amsterdam, Elvesier.

Prentice, R. (1988). Correlated Binary Regression with Covariates Specific to Each Binary Observation. *Biometrics*, 44, pp. 1033–1048.

Prentice, R. & Zhao, L. (1991). Estimating Equations for Parameters in Means and Covariances of Multivariate Discrete and Continuous Responses. *Biometrics*, 47, pp. 825–839.

White, H. (1982). Maximum Likelihood Estimation of Misspecified Models, *Econometrica*, 50, pp. 1–25.

Zhao, L. & Prentice, R. (1990). Correlated Binary Regression Using a Quadratic Exponential Model. *Biometrika*, 77, pp. 642–648.

Ziegler, A. (1994). *Verallgemeinerte Schätzgleichungen zur Analyse korrelierter Daten*. Dissertation, Universität Dortmund, Fachbereich Statistik.

List of Authors

AERTS, M.: Department of Biostatistics, Limburgs Universitair Centrum, Diepenbeek, Belgium

AITKIN, M.: Department of Mathematics, University of Western Australia, Nedlands, Australia

ATKINSON, A.C.: Department of Statistics, London School of Economics and Political Science, UK

BERCHTOLD, A.: Département d'Econométrie, Université de Genève, Switzerland

BIGGERI, A.: Department of Statistics, University of Florence, Italy

BINI, M.: Department of Statistics, University of Florence, Italy

BLUNDELL, R.: Department of Economics, University College London, UK

BÖHNING, D.: Department of Epidemiology, Free University of Berlin, Germany

BOOTH, J.: Department of Statistics, University of Florida, Gainesville, USA

CHENG, R.C.H.: Institute of Mathematics and Statistics, University of Kent at Canterbury, UK

CLARKE, G.P.Y.: Department of Statistics and Biometry, University of Natal, Pietermaritzburg, South Africa

COHEN, A.: Faculty of Industrial Engineering and Management, Technion, Haifa, Israel

DAVIES, R.: Centre for Applied Statistics, Lancaster University, UK

DE GRUTTOLA, V.: Department of Biostatistics, Harvard University, School of Public Health, USA

DECLERCK, L.: Department of Biostatistics, Limburgs Universitair Centrum, Diepenbeek, Belgium

DIETZ, E.: Department of Epidemiology, Free University of Berlin, Germany

DITTRICH, R.: Department of Statistics, Vienna University of Economics, Austria

EILERS, P.H.C.: DCMR Milieudienst Rijnmond, Schiedam, The Netherlands

ENGEL, B.: Agricultural Mathematics Group DLO, Wageningen, The Netherlands

ENGEL, J.: Pädagogische Hochschule Ludwigsburg and Universität Bonn, Germany

FARRINGTON, C.P.: Statistics Unit, Public Health Laboratory Service, London, UK

GILCHRIST, R.: School of Mathematical Sciences, University of North London, UK

GRIFFITH, R.: Institute for Fiscal Studies, London, UK

GRÖMPING, U.: Fachbereich Statistik, Universität Dortmund, Germany

HAALAND, P.: Becton Dickinson Research Center, RTP, North Carolina, USA

HAINES, L.M.: Department of Statistics and Biometry, University of Natal, Pietermaritzburg, South Africa

HARDY, S.: Becton Dickinson Research Center, RTP, North Carolina, USA

HATZINGER, R.: Department of Statistics, Vienna University of Economics, Austria

HELFENSTEIN, U.: Institut für Sozial und Präventiv Medizin, Universität Zürich, Switzerland

HEYSE, J.F.: Merck Research Laboratories, West Point, USA

KATZENBEISSER, W.: Department of Statistics, Vienna University of Economics, Austria

KEEN, A.: Agricultural Mathematics Group DLO, Wageningen, The Netherlands

KRISHNAKUMAR, J.: Department of Econometrics, University of Geneva, Switzerland

LIU, W.B.: Institute of Mathematics and Statistics, University of Kent at Canterbury, UK

LUBIN, J.: Epidemiologic Methods Section, US National Cancer Institute, USA

MADERBACHER, M.: Department of Statistics, Vienna University of Economics, Austria

MARX, B.D.: Department of Experimental Statistics, Louisiana State University, USA

MICHIELS, B.: Department of Biostatistics, Limburgs Universitair Centrum, Diepenbeek, Belgium

MINDER, CH.E.: Institut für Sozial und Präventiv Medizin, Universität Bern, Switzerland

MINKIN, S.: Division of Epidemiology and Statistics, Ontario Cancer Institute, Canada

MOLENBERGHS, G.: Department of Biostatistics, Limburgs Universitair Centrum, Diepenbeek, Belgium

MÜLLER, W.: Department of Statistics, Vienna University of Economics, Austria

NYCHKA, D.: Becton Dickinson Research Center, RTP, North Carolina, USA

O'BRIEN, T.E.: Department of Statistics, Washington State University, USA

O'CONNELL, M.: Becton Dickinson Research Center, RTP, North Carolina, USA

OSKROCHI, GH.R.: Centre for Applied Statistics, Lancaster University, UK

PIGEON, J.G.: Department of Mathematical Sciences, Villanova University, USA

PORTIDES, G.: School of Mathematical Sciences, University of North London, UK

PRUSCHA, H.: Mathematisches Institut, Universität München, Germany

READ, K.L.Q.: Department of Mathematical Statistics and Operational Research, University of Exeter, UK

RIGBY, R.A.: School of Mathematical Sciences, University of North London, UK

ROSEN, O.: Faculty of Industrial Engineering and Management, Technion, Haifa, Israel

SCHOENFELD, D.: Department of Biostatistics, Harvard School of Public Health, USA

SHAH, A.: Department of Biostatistics, Cleveland Clinic Foundation, USA

STASINOPOULOS, M.D.: School of Mathematical Sciences, University of North London, USA

STEINDORF, K.: German Cancer Research Center, Heidelberg, Germany

UTIKAL, K.J.: Department of OR, University of Bonn, Germany

WATKINS, A.J.: Statistics and Operational Research Group, University of Wales, Swansea, UK

WINDMEIJER, F.: Department of Economics, University College London, UK

ZHANG, H.: Department of Epidemiology and Public Health, Yale University School of Medicine, USA

ZIEGLER, A.: Institute of Medical Sociology, Phillipps-University of Marburg, Germany

Lecture Notes in Statistics

For information about Volumes 1 to 16
please contact Springer-Verlag